Advances in Neurobiology

Volume 14

Series Editor
Arne Schousboe

More information about this series at http://www.springer.com/series/8787

Hardy J. Rideout
Editor

Leucine-Rich Repeat Kinase 2 (LRRK2)

Springer

Editor
Hardy J. Rideout
Biomedical Research Foundation
of the Academy of Athens (BRFAA)
Athens, Greece

ISSN 2190-5215 ISSN 2190-5223 (electronic)
Advances in Neurobiology
ISBN 978-3-319-49967-3 ISBN 978-3-319-49969-7 (eBook)
DOI 10.1007/978-3-319-49969-7

Library of Congress Control Number: 2017932445

Printed on acid-free paper

This Springer imprint is published by Springer Nature
The registered company is Springer International Publishing AG
The registered company address is: Gewerbestrasse 11, 6330 Cham, Switzerland

Foreword

This volume represents a comprehensive compilation of up-to-date knowledge on LRRK2 biology and its link to Parkinson's disease (PD), written by authorities in the field.

The connection between LRRK2 and PD, of course, starts from genetics. Monfrini and di Fonzo go into exhaustive detail in systematically analyzing the multitude of published studies in various ethnic backgrounds. They categorize the nucleotide changes into definitely pathogenic, possibly pathogenic, and risk factor variants. They make the important points of incomplete penetrance, the link to sporadic disease, and the variable neuropathological picture, in which the only unifying feature is the nigral neurodegeneration. They conclude with recommendations for genetic testing that have to be highly individualized and adjusted to the specific ethnic background.

Kestenbaum and Alcalay pick up the torch and delve more deeply into the clinical aspects of LRRK2-associated PD. There are inconsistencies in the published literature, but LRRK2-associated PD is largely indistinguishable clinically from the idiopathic form; some subtle differences may exist, however, for example, with less olfactory dysfunction and possibly cognitive deficits reported in LRRK2 carriers. It is interesting that there is substantial variability in the clinical phenotype, even among carriers of the same mutation. The important point is made that such variability may depend in part on the variable involvement of extranigral regions, which, in turn, may relate to the neuropathological correlate of synucleinopathy. In any case, the fact that clinically LRRK2-associated PD is so similar to the sporadic condition underscores its relevance to the study of PD at large.

Several key functions of LRRK2 are regulated by its phosphorylation status, including its dimerization, interactions with other proteins such as 14-3-3, and its turnover. Nichols introduces a comprehensive review of this process, focusing in particular on the potential enzymes that regulate this phosphorylation and the interplay between phosphorylation and ubiquitination; this interplay provides a potential link between phosphorylation and dephosphorylation of LRRK2 to its degradation. The complicated relationship of particular PD-associated mutants with kinase activity and phosphorylation at specific sites is analyzed in detail, providing links to the disease which may form the basis of targeted therapies.

Nguyen and Moore provide a comprehensive review of the importance of LRRK2 GTPase activity. They review the evidence that this activity, mediated by the corresponding Roc-COR tandem domain, is closely linked to both physiologic and pathological effects of LRRK2, in part through modulation of LRRK2 kinase activity. Given the accumulating evidence that the relevant domain is important for mutant LRRK2 pathogenic effects, there is an opportunity of targeting GTPase activity as a therapeutic strategy in PD.

Manzoni and Lewis discuss the evidence that LRRK2 modulates autophagy pathways. Indeed, multiple studies have shown an association of LRRK2 overexpression or downregulation with an alteration in macroautophagy or chaperone-mediated autophagy indices. However, the direction of the effect is quite variable, depending on the study, and the molecular pathway involved has not been identified. Consequently, a unifying picture regarding the reciprocal relationship of LRRK2 and autophagy has not yet emerged. Identifying this link and its mechanistic underpinnings may be very important in tying LRRK2 with other PD-associated proteins, which may also relate to autophagy.

In addition to the regulation of LRRK2 dimerization by its phosphorylation, the GTPase activity of LRRK2 is also closely linked to this process, which is the subject of the chapter by Civiero, Russo, Bubacco, and Greggio. The ROC domain of LRRK2, as in other Roco proteins, is essential for dimerization, and dimerization influences GTPase activity. Dimerization appears to occur in a cellular context in conjunction with altered localization close to membranes and is associated with enhanced LRRK2 activity. Such data suggest the possibility that inhibiting LRRK2 dimerization may have therapeutic potential.

The physiologic and pathological relationship of LRRK2 to the immune system is reviewed by Dzamko. Cells of the adaptive and innate immune system, in particular under proinflammatory conditions, express LRRK2, which appears to have a physiologic role in inflammatory responses. This may be quite relevant not only to Parkinson's disease but also to other disorders of the immune system. Indeed, genetic evidence exists linking LLRK2 polymorphisms to inflammatory bowel disease and to susceptibility to leprosy. The mechanisms that may underlie such links are succinctly reviewed, and aspects of immune responses and LRRK2 intracellular effects are touched upon, which may be context-dependent. This aspect of LRRK2 biology may be very important not only as a therapeutic target but also when considering LRRK2-targeted therapies, as one has to bear in mind potential untoward side effects in the immune system.

Taymans reviews the complicated issue of the multiple phosphorylation sites present on LRRK2 and their regulation by phosphatases, highlighting the role of at least one, PPA1, in heterologous phosphorylation in the ANK-LRR interdomain region. Other phosphatases are likely to regulate other heterologous but also autophosphorylation events, and this deserves further study. Such knowledge is essential, as PD-associated LRRK2 mutants display in general altered phosphorylation status at the various sites compared to the WT protein, while the employed kinase inhibitors lead to general hypophosporylation of LRRK2. In this context, LRRK2 phosphatases may be important as therapeutic targets and biomarker indices.

In the final section of this volume, the chapters focus on neurotoxic mechanisms of mutant LRRK2, modeling this as well as basic LRRK2 functions in different in vivo settings, and potential therapeutic strategies to reverse LRRK2-mediated neurodegeneration. First, Xiong, Dawson, and Dawson go on to review the available LRRK2 animal models. They describe in great detail the models developed in various organisms, mainly *Drosophila*, *C. elegans*, mice, and rats. The overarching conclusion is that LRRK2 knockout leads to no discernible effects on the dopaminergic system, although it may have some impact on nonneuronal tissues. In contrast, overexpression of mutant forms of LRRK2 in some models leads to dopaminergic neurodegeneration, supporting a gain of function toxic effect that may depend on both cell-autonomous and non-cell-autonomous mechanisms.

Rideout and Re go on to dissect the relationship between mutant LRRK2 and cell death pathways. Such a relationship may underlie the neurotoxic effects seen upon mutant LRRK2 overexpression in various cell culture systems. Rideout and Re review studies that have linked LRRK2 to both the extrinsic and, more recently, the intrinsic apoptotic cell death pathways and suggest a link to necroptosis. This latter link derives from the potential functional homology of LRRK2 to RIP kinase proteins and, in particular RIP1 and RIP3, components of the necroptosis pathway. Although the idea that mutant LRRK2 may trigger cell death through necroptosis is just a hypothesis, it is an intriguing one. Non-cell-autonomous processes mediating various forms of neuronal death through microglial cells may also come into play and complicate the picture even further. Assuming direct links between pathological LRRK2 and components of the cell death machinery, such links may be therapeutically tractable.

The aforementioned link of LRRK2 to autophagy may play a particular role in its relationship with α-synuclein, the other major protein involved in autosomal dominant PD and in Lewy body formation. Daher goes over this and other possible mechanisms through which this interrelationship may occur, placing major emphasis on inflammation and vesicular trafficking, as well as on the particular brain region studied, in order to disentangle the discrepancies that exist in the literature. In part through Daher's own work in the lab of Andy West, evidence suggests that LRRK2 proinflammatory effects mediate dopaminergic neurotoxicity induced by α-synuclein. Deciphering the pathway that connects these two molecules will be a very important goal for future studies.

In his chapter discussing the toxic effects of LRRK2, Cookson argues that the unifying feature that ties all alterations in LRRK2 to PD, be they point mutations, risk factors, or GWAS hits, is the fact that they lead to increased signaling of the molecule and, therefore, to a toxic gain of activity of the WT protein. Some lingering inconsistencies in this all-encompassing view are, however, pointed out, suggesting that excessively low LRRK2 levels may also be detrimental. Cookson suggests that the underlying effects are mediated in both neuronal and nonneuronal, and in particular microglial, cells, likely consisting of modulation of vesicular dynamics or the cytoskeleton. This conceptual framework ties together neatly various aspects of LRRK2 pathobiology.

The many issues that have hampered the development of inhibitors of LLRK2 kinase activity, despite intense ongoing activity, are reviewed by Hatcher, Choi, Alessi, and Gray from a chemical perspective. Chemical structures and molecular models of the compounds are presented. Selectivity of the developed compounds in relation to other kinases; stoichiometry; ability to target LRRK2 mutants; on-target undesired effects, mainly in the lung; and brain penetration form only a part of the considerations that have to be addressed in each case. Furthermore, there has been a difficulty in assessing biological effects, due to the, up till recently, paucity of identification of biologically relevant LRRK2 substrates. Despite this, selective compounds with prospects of clinical development are beginning to emerge.

Overall, the chapters in this volume provide a rich source of information on most current aspects of LRRK2 biology, presented in a critical fashion by leaders in the field. Every chapter ends with future perspectives, such that the reader will get a sense not only of current knowledge but also of the major questions that lie ahead and will occupy the field for years to come. It is hoped that such insights will soon enable LRRK2-based therapeutics to be applied to the clinical setting.

Leonidas Stefanis
Professor of Neurology and Neurobiology
University of Athens Medical School
Director, 2nd Department of Neurology
Hospital Attikon, Athens, Greece

Preface

The *PARK8* locus on chromosome 12, identified from linkage analyses of a large Japanese family with individuals from multiple generations diagnosed with the progressive neurodegenerative disorder Parkinson's disease (PD), was first reported in 2002 by Funayama and colleagues. It was not until two years later that independent groups led by A. Singleton and T. Gasser cloned the responsible gene within this locus known as leucine-rich repeat kinase 2, or *LRRK2*. Typically, PD develops sporadically with the greatest risk factor being age. However, for approximately 10% of cases, a classical Mendelian inheritance with both autosomal dominant and recessive transmissions is present. The six clearly pathogenic single amino acid substitutions that are causative for PD collectively comprise the most frequent genetic causes of PD, with the most common mutation being Gly2019Ser.

The protein encoded by the LRRK2 gene is a large 2527 amino acid multi-domain protein comprised of several well-defined protein interaction domains as well as a Ser/Thr kinase domain and a small GTPase-like domain (ROC; Ras of complex proteins). The kinase and GTPase domains are interrupted by a COR (C-terminal of ROC) domain characteristic of the ROCO protein family. Likewise, the domain structure and kinase domain in particular of LRRK2, and the related LRRK1, bear significant similarity to the receptor-interacting protein kinase (RIPK) family. Clearly, its complex domain structure and widespread expression, including, in addition to the brain, high levels in the kidney and lung as well as circulating immune cells, predicts a wide range of cellular functions and activities. In the short period of time since its identification, our understanding of LRRK2 biology, as well as our tools to investigate its function, is rapidly evolving. Historically, only the G2019S mutant form of LRRK2, when overexpressed in cell lines, would exhibit a significant alteration in kinase activity, displaying approximately two- to fivefold increases in autophosphorylation or, later on, phosphorylation of model peptide substrates. Multiple studies of the remaining pathogenic LRRK2 mutations reported mixed effects on kinase function. Recently, however, multiple members of the large Rab GTPase family of proteins were identified as true physiological phospho-substrates of LRRK2; and strikingly, virtually all of the pathogenic LRRK2 mutants, as well as several risk factor variants, show elevated phosphorylation of these substrates. While this represents an important step in our understanding of LRRK2 function, what is still lacking is a better understanding of how mutations in LRRK2 alter its function in such a way, in what cell types, and in response to what external stimuli that results in the progressive loss of dopamine neurons of the substantia nigra pars compacta—the neurodegenerative hallmark underlying the motor symptoms of PD. With a prominent focus on its role in PD, but always in the larger context of a broader range of activities of LRRK2, this book was envisioned to provide a window into the current state of understanding of this complex protein by some of the leaders in the field of LRRK2.

In Part I, we begin with discussions of the genetic and clinical considerations of LRRK2-associated PD, with contributions by Di Fonzo and colleagues and Alcalay and colleagues, respectively. In a gene of this size, there are, expectedly, dozens of sequence variants that show varying degrees of association with developing PD. However, relatively few such mutations, located primarily within the central ROC-COR-kinase signaling core of the protein, demonstrate clear pathogenicity. Even among carriers of those specific mutations, there is considerable variability in the penetrance across different ethnic backgrounds. Despite this, the clinical features of LRRK2-associated PD are remarkably similar to the much more common idiopathic manifestation of the disease; however, some important clinical and pathological differences have been reported among the distinct mutations, including altered progression of motor symptoms as well as the absence of classical Lewy body-like inclusions in some cases.

LRRK2 possesses an extraordinarily broad range of cellular functions, dictated not only by its expression in specific cell types but also by a complicated and coordinated regulation of its activity. This aspect of LRRK2 biology is covered in detail in Part II of the book. One regulatory mechanism occurs via the phosphorylation of LRRK2, by itself and other kinases, at multiple domains throughout the protein (discussed in the chapter by Nichols). This is in turn kept in check via the action of specific cellular phosphatases (Taymans). While the kinase activity of LRRK2 has received considerable attention in terms of substrate profiling, its requirement for neurodegeneration, and the obvious opportunity for targeted therapeutic strategies, our understanding of the GTPase function of LRRK2, in terms of its reciprocal regulation of kinase activity as well as its activity underlying the pathological effects of mutant LRRK2, is rapidly increasing. This is discussed in detail in the chapter by Moore and colleagues. One of the earliest systems shown to be affected by LRRK2 function, or dysfunction, is autophagic/lysosomal protein degradation. Interestingly, as is the case for many proteins, including another dominantly inherited gene linked to PD, α-synuclein, in addition to being degraded in part through by the autophagic machinery, LRRK2 (particularly mutant forms of the protein) can also modulate the activity of multiple forms of autophagy (Lewis and colleagues).

An important regulatory protein interacting with LRRK2 is 14-3-3. It binds to a cluster of phosphorylated residues located in the N-terminal region of the protein. While the phosphorylation of these residues is dependent upon LRRK2 kinase activity, pharmacological inhibition leads to dephosphorylation at these sites; they are not true autophosphorylation sites within LRRK2. Multiple other kinases, downstream of LRRK2 activity, have been identified that can phosphorylate these residues. The phospho-dependent binding of 14-3-3 at these sites appears to play a critical role in the subcellular localization of LRRK2. Upon dissociation, LRRK2 redistributes into discrete filamentous structures of unknown function, similar to those seen upon overexpression of certain mutant forms of LRRK2. Although it remains unclear what the composition of these structures is, there is strong evidence supporting the existence of LRRK2 in a dimeric state. Further, it is believed that, at least in terms of kinase activation, the LRRK2 dimer is the active conformation. A comprehensive review of the evidence, and implications, of LRRK2 dimer formation is provided in the chapter by Greggio and colleagues. Beyond its role in neurodegeneration, LRRK2 plays an important role in many other pathways as well as a result of its prominent activity in the immune system (Dzamko).

The final section of the book focuses on modeling LRRK2 neurodegeneration, the potential links to other PD-related proteins, the mechanisms of neurotoxicity and cellular implications of LRRK2 dysregulation, and efforts to develop therapeutically viable inhibitors of LRRK2 activity. Despite the generation of many in vivo models of mutant LRRK2 overexpression, including traditional transgenic models, BAC transgenic lines, and knockin lines expressing mutant LRRK2, as well as viral models, the plurality of in vivo models fails to show evidence of a progressive loss of dopaminergic neurons (Dawson and colleagues). That is not to say that gains have not been made from these efforts; far from it. Using specific promoters, or viral vector approaches, the progressive degeneration of dopaminergic neurons can be elicited by overexpression of mutant forms of LRRK2; and other neuronal pathologies have been reported in lines even in the absence of neuronal loss. Conversely, clues to the function of LRRK2 have been discerned from knockout models in non-neuronal tissues; and its important interaction in the neurodegenerative phenotype triggered by α-synuclein overexpression or inflammatory insults (Daher) has been discovered in LRRK2-deficient rats. This is highlighted by the critical discussion of non-cell-autonomous effects of LRRK2, as well as its role in vesicular trafficking (Cookson). While the nature of cell death observed in isolated neuronal cell models of LRRK2 neurodegeneration appears to be apoptotic, whether the same is true at the systems level remains to be seen. The similarity of LRRK2 to the RIP kinase protein family and its interaction with, and activation of, extrinsic death pathway components raise the possibility that other modes of cell death may contribute to the loss of neurons in PD (Rideout). Finally, the book closes with a discussion of the efforts to develop small molecule inhibitors of LRRK2 kinase activity that could potentially be utilized in the clinic (Gray and colleagues). Multiple cellular and in vivo models indicate that mutant LRRK2-induced neuronal death is dependent upon its kinase activity, although the substrates of this activity linked to cell death remain unknown. These efforts, both academic- and industry-wide, have resulted in the discovery of ever more potent and selective inhibitors of LRRK2 kinase activity that are currently being evaluated in safety studies.

This book would not be possible without the time and effort of the contributing authors. I am personally grateful for their generosity, support, and enthusiasm for this project. I would also like to acknowledge the publisher Springer-Nature and especially would like to thank Simina Calin and Jeffery Taub for their support and guidance. It is my hope and belief that this book will serve as an in-depth introduction and snapshot of the current state of the art in LRRK2 biology and, as highlighted in each of chapters' discussion of future directions, will stand as a foundation for the next steps taken in this exciting field.

Athens, Greece Hardy J. Rideout

Contents

Contributors

Roy N. Alcalay Department of Neurology, Columbia University Medical Center, New York, NY, USA

Dario R. Alessi MRC Protein Phosphorylation and Ubiquitylation Unit, University of Dundee, Sir James Black Centre, Dundee, DD1 5EH, UK

Luigi Bubacco Department of Biology, University of Padova, Padova, Italy

Hwan Geun Choi New Drug Development Center, Daegu Gyeongbuk Medical Innovation Foundation, Daegu, South Korea

Laura Civiero Department of Biology, University of Padova, Padova, Italy

Mark R. Cookson Laboratory of Neurogenetics, National Institute on Aging, National Institutes of Health, Bethesda, MD, USA

João Paulo Lima Daher Faculty of Medicine, School of Medical Sciences, University of New South Wales, Sydney, NSW, Australia

Neuroscience Research Australia, Randwick, NSW, Australia

Ted M. Dawson Department of Pharmacology and Molecular Sciences, Johns Hopkins University School of Medicine, Baltimore, MD, USA

Neuroregeneration and Stem Cell Programs, Institute for Cell Engineering, Johns Hopkins University School of Medicine, Baltimore, MD, USA

Department of Neurology, Johns Hopkins University School of Medicine, Baltimore, MD, USA

Solomon H. Snyder Department of Neuroscience, Johns Hopkins University School of Medicine, Baltimore, MD, USA

Adrienne Helis Malvin Medical Research Foundation, New Orleans, LA, USA

Valina L. Dawson Solomon H. Snyder Department of Neuroscience, Johns Hopkins University School of Medicine, Baltimore, MD, USA

Adrienne Helis Malvin Medical Research Foundation, New Orleans, LA, USA

Department of Physiology, Johns Hopkins University School of Medicine, Baltimore, MD, USA

Neuroregeneration and Stem Cell Programs, Institute for Cell Engineering, Johns Hopkins University School of Medicine, Baltimore, MD, USA

Department of Neurology, Johns Hopkins University School of Medicine, Baltimore, MD, USA

Nicolas L. Dzamko School of Medical Sciences, University of NSW, Kensington, NSW, Australia

Neuroscience Research Australia, Randwick, NSW, Australia

Alessio Di Fonzo IRCCS Foundation Ca' Granda Ospedale Maggiore Policlinico, Dino Ferrari Center, Neuroscience Section, Department of Pathophysiology and Transplantation, University of Milan, Milan, Italy

Nathanael S. Gray Department of Cancer Biology, Dana Farber Cancer Institute, Boston, MA, USA

Elisa Greggio Department of Biology, University of Padova, Padova, Italy

John M. Hatcher Department of Cancer Biology, Dana Farber Cancer Institute, Boston, MA, USA

Meir Kestenbaum Department of Neurology, Columbia University Medical Center, New York, NY, USA

Patrick A. Lewis School of Pharmacy, University of Reading, Reading, UK

Department of Molecular Neuroscience, UCL Institute of Neurology, London, UK

Claudia Manzoni School of Pharmacy, University of Reading, Reading, UK

Department of Molecular Neuroscience, UCL Institute of Neurology, London, UK

Edoardo Monfrini IRCCS Foundation Ca' Granda Ospedale Maggiore Policlinico, Dino Ferrari Center, Neuroscience Section, Department of Pathophysiology and Transplantation, University of Milan, Milan, Italy

Darren J. Moore Center for Neurodegenerative Science, Van Andel Research Institute, Grand Rapids, MI, USA

An Phu Tran Nguyen Center for Neurodegenerative Science, Van Andel Research Institute, Grand Rapids, MI, USA

R. Jeremy Nichols The Parkinson's Institute, Sunnyvale, CA, USA

Diane B. Re EHS Department and Motor Neuron Center, Columbia University, New York, NY, USA

Hardy J. Rideout Division of Basic Neurosciences, Biomedical Research Foundation of the Academy of Athens, Athens, Greece

Isabella Russo Department of Biology, University of Padova, Padova, Italy

Jean-Marc Taymans Université de Lille, Inserm, CHU Lille, UMR-S 1172—JPARC Jean-Pierre Aubert Research Center, Neurosciences and Cancer, Lille, France

Yulan Xiong Neuroregeneration and Stem Cell Programs, Institute for Cell Engineering, Johns Hopkins University School of Medicine, Baltimore, MD, USA

Department of Neurology, Johns Hopkins University School of Medicine, Baltimore, MD, USA

Department of Anatomy and Physiology, Kansas State University, College of Veterinary Medicine, Manhattan, KS, USA

Part I
Clinical and Genetic Considerations of LRRK2 Associated Parkinson's Disease

Chapter 1
Leucine-Rich Repeat Kinase (*LRRK2*) Genetics and Parkinson's Disease

Edoardo Monfrini and Alessio Di Fonzo

Abstract The discovery of *LRRK2* mutations as a cause of Parkinson's disease (PD), including the sporadic late-onset form, established the decisive role of genetics in the field of PD research. Among *LRRK2* mutations, the G2019S, mostly lying in a haplotype originating from a common Middle Eastern ancestor, has been identified in different populations worldwide. The G2385R and R1628P variants represent validated risk factors for PD in Asian populations. Here, we describe in detail the origin, the present worldwide epidemiology, and the penetrance of *LRRK2* mutations. Furthermore, this chapter aims to characterize other definitely/probably pathogenic mutations and risk variants of *LRRK2*. Finally, we provide some general guidelines for a *LRRK2* genetic testing and counseling. In summary, *LRRK2* discovery revolutionized the understanding of PD etiology and laid the foundation for a promising future of genetics in PD research.

Keywords Leucine-rich repeat kinase 2 • *LRRK2* • Dardarin • Parkinson's disease • PARK8 • Parkinson's disease genetics • Familial Parkinson's disease • *LRRK2* mutations

Until the discovery of leucine-rich repeat kinase 2 (*LRRK2*) mutations as a genetic cause of Parkinson's disease (PD), the hereditary influences on PD were limited to observation of rare autosomal dominant familial cases harboring highly penetrant *SNCA* (alpha-synuclein) mutations and juvenile or young onset autosomal recessive forms carrying *PRKN*, *PINK1*, and *DJ-1* mutations. This scenario was more suggestive of a minor role played by genetic factors in PD, especially considering the common sporadic late-onset form. The innovative finding of *LRRK2* low penetrant mutations in common forms of PD revolutionized this outdated view.

E. Monfrini • A. Di Fonzo (✉)
IRCCS Foundation Ca' Granda Ospedale Maggiore Policlinico, Dino Ferrari Center, Neuroscience Section, Department of Pathophysiology and Transplantation, University of Milan, Milan, Italy
e-mail: alessio.difonzo@policlinico.mi.it

H.J. Rideout (ed.), *Leucine-Rich Repeat Kinase 2 (LRRK2)*,
Advances in Neurobiology 14, DOI 10.1007/978-3-319-49969-7_1

Genetic Contribution in Etiology of PD

Epidemiological studies reveal that 10–15% of PD have a positive familial history for the disease, while the majority of cases are sporadic. Through linkage analysis and positional cloning approaches, five genes have been definitely implicated in the etiology of PD. Mutations in the *SNCA* [1, 2], *LRRK2* [3, 4], and *VPS35* [5, 6] genes cause autosomal dominant forms, whereas mutations in the *PRKN* [7], *DJ-1* [8], and *PINK1* [9] genes cause autosomal recessive forms of PD. Furthermore, mutations in the *ATP13A2* [10], *PLA2G6* [11], *FBXO7* [12, 13], *DNAJC6* [14], and *SYNJ1* [15] have been reported as rare causes of early-onset parkinsonism with atypical clinical features which might be mechanistically distinct from classical PD. Finally, mutations in *UCH-L1* [16], *Omi/HtrA2* [17], *GIGYF2* [18], *EIF4G1* [19], and *DNAJC13* [20] genes have also been described in PD cases, but their role in the disease remains uncertain. Another three PD loci have also been mapped (PARK3, PARK10, PARK12, PARK16) [21–23], but the defective genes remain unknown (Table 1.1).

Table 1.1 List of loci, genes, patterns of inheritance, and clinical presentations of genetic forms of Parkinson's disease

Locus	Gene	Chromosome	Inheritance	Clinical presentation
PARK1-4	α-Synuclein	4q21	AD	PD, Lewy body dementia
PARK2	*PRKN*	6q25.2-27	AR	Early-onset PD
PARK3	Unknown	2p13	AD	Typical PD
PARK5	*UCH-L1*[a]	4p14	AD	Typical PD
PARK6	*PINK1*	1p36	AR	Early-onset PD
PARK7	*DJ-1*	1p36	AR	Early-onset PD
PARK8	*LRRK2*	12q12	AD	Typical PD
PARK9	*ATP13A2*	1p36	AR	Early-onset parkinsonian–pyramidal syndrome with dementia
PARK10	Unknown	1p32	Unknown	Typical PD
PARK11	*GIGYF2*[a]	2q36-37	AD	Typical PD
PARK12	Unknown	Xq21-q25	Unknown	Typical PD
PARK13	*Omi/HTRA2*[a]	2p12	Unknown	Typical PD
PARK14	*PLA2G6*	22q	AR	Adult-onset dystonia-parkinsonism
PARK15	*FBXO7*	22q	AR	Early-onset parkinsonian-pyramidal syndrome
PARK16	Unknown	1q32	Unknown	Typical PD
PARK17	*VPS35*	16q11.1	AD	Typical PD
PARK18	*EIF4G1*[a]	3q27.1	AD	Typical PD
PARK19	*DNAJC6*	1p31.3	AR	Early-onset PD atypical
PARK20	*SYNJ1*	21q22	AR	Early-onset PD atypical
PARK21	*DNAJC13*[b]	3q21.3-q22.2	AD	Typical PD
PARK22	*CHCHD2*[a]	7p11.2	AD	Typical PD
PARK23	*VPS13C*	15q22.2	AR	Early-onset PD atypical

[a]Not confirmed by other studies
[b]Needs confirmation by independent studies

In addition to the Mendelian forms of PD, genetic risk factors for the disease have been investigated in several candidate genes and, more recently, in genome-wide association studies [24]. With the exceptions of *SNCA*, microtubule-associated protein tau (*MAPT*), and HLA region [25–32], none of the loci reported have so far been convincingly replicated in independent studies.

Another exception is represented by the glucocerebrosidase gene (*GBA*) involved in a recessive neurometabolic disease (Gaucher's disease). Screening of PD patients for *GBA* mutations showed a higher number of heterozygous mutations carriers as compared to healthy controls. Mutations have been found in about 2–4% of Caucasian PD patients and less than 1% of controls [33].

The *LRRK2* Gene: Mapping and Cloning

Although the discovery of mutations in the *SNCA*, *PRKN*, *PINK1*, and *DJ-1* genes clearly contributed to our understanding of the pathogenesis of PD, they were identified in a limited number of PD cases, often with early-onset or pathologically atypical features.

A new locus for PD, termed PARK8, was identified in a large family with autosomal dominant PD, known as the "Sagamihara family" from the region in Japan where the family originated from [34]. The clinical features in affected individuals of the kindred were reported to resemble very closely classical PD, with an average of symptoms onset at 51 ± 6 years. A pattern of "pure nigral degeneration" without Lewy bodies (LB) was found at autopsy in six PD patients examined, another carrier of the disease haplotype developed multiple system atrophy type P-like pathology, and one showed classical LB pathology [35]. In this family, a genome-wide linkage scan yielded significant evidence for linkage of PD to the centromeric region of chromosome 12 (12p11.2-q13.1). The haplotype analysis suggested an incomplete penetrance of the mutation [34, 35]. In 2004 the linkage to PARK8 was confirmed in two Caucasian families, "family A" (a German–Canadian kindred) and "family D" (from Western Nebraska) with dominantly inherited neurodegeneration [36], and thereafter in several Basque PD [37] families suggesting PARK8 to be a relatively common locus and refining the critical region. A wide clinical–pathological spectrum was shown in these families, including typical PD but also dementia and amyotrophy, diffuse LB and tau pathology, nigral degeneration without inclusions, and atypical, ubiquitin-positive inclusions [38].

In 2004 two independent groups, by positional cloning, identified mutations in a gene at that time annotated as DKFZp434H2111, which cosegregated with PD in several PARK8-linked pedigrees [3, 4]. The gene was renamed *LRRK2* (leucine-rich repeat kinase 2) and the encoded protein LRRK2 or dardarin (from the Basque term *dardara*, meaning tremor, since resting tremor was a consistent clinical feature of the Basque patients who carried *LRRK2* mutations).

Subsequently, early in 2005, several groups identified a single *LRRK2* mutation (c.G6055A) leading to a G2019S substitution in the encoded protein, which was present in familial and sporadic PD with unprecedented high frequency [39–42]. The following years have seen an explosion of research into the *LRRK2* gene in PD and related disorders. The I2020T mutation was detected as the cause of disease in the original "Sagamihara family" [43].

The G2019S Mutation

Prevalence of G2019S Across Populations

G2019S is particularly important among the PD-causing mutations in *LRRK2*. This mutation was identified by several groups as a common cause of the disease, being detected initially in ~5–6% of large cohorts of familial PD in Europe and the USA [39, 41] and in ~1–2% of sporadic PD from the UK [40]. Due to the unprecedented high frequency in familial and late-onset classical parkinsonism, which in the past would have been identified as "idiopathic PD," this specific mutation has been extensively studied worldwide (Fig. 1.1).

So far, large screenings revealed that the frequency of G2019S is population specific. The G2019S mutation has been reported at the highest frequency (up to 37%) among familial PD cases of North African descent and in familial Ashkenazi Jewish patients (23%) [44, 45]. Similar frequencies were replicated in independent studies on PD cases from Tunisia [46–48] and in Ashkenazi Jews [48–51]. Remarkably, the frequency of this mutation was considerably high among sporadic cases (41% North Africans and 13% Ashkenazi Jews) and rarely identified in healthy controls too (3% North Africans and 1.3% Ashkenazi Jews). Other studies reported the presence of the G2019S among 1–2% of healthy North Africans, Ashkenazi, and

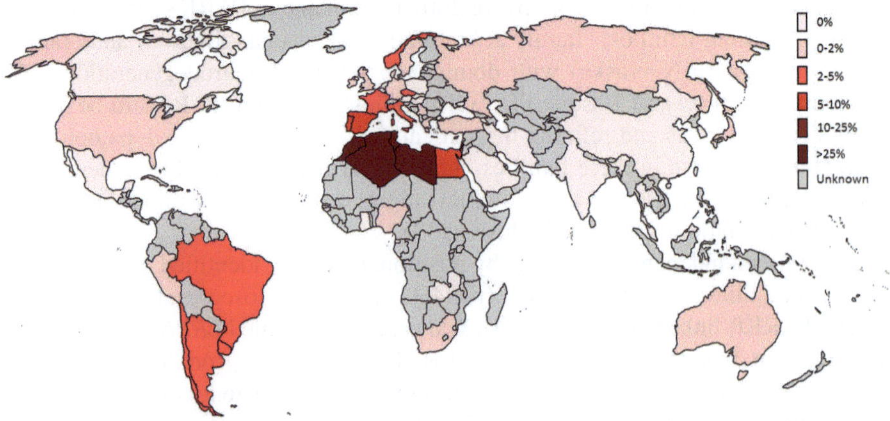

Fig. 1.1 Rough estimates of worldwide G2019S prevalence in PD patients (familial and sporadic)

Sephardic Jewish subjects [49, 50, 52, 53]. So far the G2019S mutation has not been found in sub-Saharan Africans, with exceptions of South Africans where the mutation was present in subjects with European and Jewish ancestry only [54–57]. Little is known about the prevalence in Middle Eastern populations. G2019S is rare in Turkey [58] and has not been identified so far in Yemenite Jews [51] and in Iran [59].

In Western Europe there is a south–north gradient of frequency. The G2019S is found in 9–16% of familial and 3–4% of sporadic PD patients in Portugal [60, 61]; it accounts for 6–16% of familial and 2–6% of sporadic PD in different regions of Spain: Catalonia [62], Cantabria [63, 64], Asturias [65], Galicia [63], and Basque regions (patients without Basque ancestry) [66], while it is less common in patients of Basque origin (1–2%) [66].

In Italy, the G2019S mutation has been reported up to 6–7% of familial and ~1–2% of sporadic cases [39, 67–71]. Similarly, in France the mutation accounts for ~3.5% of familial and ~1.9% of sporadic cases [44, 72–75]. Two independent screening in Sardinians, an isolated population, reported a frequency of ~1.5% in both familial and sporadic cases [76, 77]. Interestingly, the mutation that appeared to be common in the western Mediterranean basin is instead very rare in Greece and Crete [78–81].

A slightly lower frequency was reported in UK screenings of PD patients of Caucasian ethnicity (2.5% familial and 0.3–1.6% sporadic) [40, 82, 83] and also in populations of Celtic and Baltic origin (Ireland 1.1% of familial PD [42, 84], Norway ~1.5% of familial PD [85], and Sweden 1.4% of sporadic cases [86]). Mutation analyses in more than 300 familial and 1200 sporadic PD in Germany suggested a very low frequency of this mutation (0.8% of familial cases [87, 88] 0.2–0.9% of sporadic) [87–89], as well as in Belgium [90], the Netherlands [91], Denmark [92], and Austria [93].

In Poland, Serbia, Hungary, Czech Republic, and Slovakia, the G2019S appeared to be rare (or found in a single subject) [87, 94–98]. On the contrary, four studies have been performed in Russia, where the mutation accounts for 4–7% of familial and 1% of sporadic cases [99–102]. However, subjects were included from a mixed ethnic background, since at least two PD families and one sporadic case reported their ethnic origin as Ashkenazi Jews [101].

An analogous observation can be done when analyzing patients from the USA, where the frequency of the G2019S in Caucasian PD reaches 2–3.5% in familial and 0.5–1.6% in sporadic cases [46, 49, 103–109]; it seems to be rare among American Indians and Afro-Americans (but the sample size for these two ethnic groups is still insufficient to make firm conclusions) [108, 109], whereas it was reported to be higher when patients of Ashkenazi Jewish ancestry were included [49]. In Canadian PD patients, the G2019S is rare/absent [110, 111].

Four different populations of South America, where the Spanish, Portuguese, and Italian ethnic backgrounds are strong, have been studied for this mutation. In Uruguay [112], Chile [113], and Argentina [114], G2019S accounts for 3.5–5.5% of familial and 2.9–4.2% of sporadic cases. The 5.45% of a PD cohort from Argentina was found to carry the mutation, all of them being of Jewish ancestry. While in Peru, the G2019S appears to be rare [112]. Controversial results came from large screening in Brazil (from 3 to 6.8% in familial and 0 to 1.7% in sporadic cases), probably due to the high degree of ethnic heterogeneity within the study cohorts [115–117].

G2019S is rare/absent among Chinese patients with familial and sporadic PD [118–121], as well as in Korea [122, 123] and in India [121, 124]. So far, only three patients have been reported with this mutation in Japan (0.7% of sporadic cases) [48, 125].

Finally, the mutation is present in Australia, among PD patients with European ancestry (2–6% of familial, 0.4% of sporadic PD) [126, 127], while it has not been identified in Australian Aboriginal.

Taken together, these data show that a single *LRRK2* mutation represents the most frequent known genetic determinant of PD. The frequency of the G2019S mutation varies widely across populations, indicating that ethnicity is an important factor. For some populations, independent studies on the prevalence of the mutation are already available, and often, the reported results are consistent. These observations imply that most neurologists who treat patients with movement disorders will see patients with *LRRK2*-related PD that may be addressed to genetic testing. This estimation could be even higher if we include the other *LRRK2* definitely pathogenic mutations.

Origins of the G2019S Mutation

So far three different haplotypes have been identified in patients carrying the G2019S mutation.

Haplotype 1

The first studies on unrelated carriers of the G2019S of European or Middle Eastern–North African origin revealed that all shared the same haplotype, consistent with a common founder [42, 44, 45, 70, 128, 129].

Subsequently, the same haplotype has been identified among subjects carrying the G2019S mutation from Italy (independent subset) [71], France [47], Germany [87], Russia [99] Sardinia [76, 77], Spain [65], Portugal [61, 70], Brazil [70], Chile [113], Uruguay and Peru [112], and Australia [126].

According to a general rule in population genetics, the geographic center of the origin of a mutation corresponds to the area where that mutation is most frequent [130]. The highest prevalence of the G2019S mutation has been reported in Berbers [52], followed by North African Arabs, Ashkenazi, and Sephardic Jews. The frequency data combined with the identification of a common haplotype among these populations support the hypothesis that the mutation of haplotype I originated in North Africa or in the Middle East and then spread to other countries following the patterns of migration.

Further studies provided important insights on the estimated age of the common founder for the haplotype 1 carriers. Analyzing the haplotypes of European and Ashkenazi Jews [129] and Tunisian G2019S carriers [131], the age estimated of the common ancestor (using the 30-year intergeneration interval) was 2250 (95% CI 1650–3120) and 3120 (95% CI 2340–4620) years ago, respectively. A third study, on Ashkenazi Jews only, estimated a more recent founder approximately 1525–1830 years ago (150–450 A.D.) [132]. This estimation would fit with the absence of the G2019S in Yemenite Jews [51]. The Yemenite Jews evolved completely separate from all of the other Jewish populations. Most of them arrived in Yemen in the early second century A.D. (~160 A.D.). Finally, a multicentric study proposed a consensus of haplotype 1 origin, estimating the founding mutational event in Ashkenazi Jews ancestors in a period ranging from 4500 to 9100 years. In this scenario, being the Ashkenazi Jews history more recent (at most 2000 years old), it is possible that the G2019S have arisen at least 4000 years ago in the Near East and then Ashkenazi ancestors may have kept the mutation through the different diasporas. Thereafter, the mutation may have been reintroduced by gene flow from Ashkenazi Jews to other European and North African populations [133].

Haplotype 2

A different G2019S haplotype was identified in three families from Western Europe, which appeared to share a more recent founder than haplotype 1. The geographic origin of this haplotype is less certain [129].

Haplotype 3

The third haplotype has been found in Japanese patients carrying the G2019S mutation [125]. This haplotype differs across the markers closest to the mutation, which would suggest an independent origin of the mutation in Japanese and European populations rather than a single ancient founder. Interestingly, the haplotype 3 has also been observed in a single sporadic Turkish patient [134]. This may be the result of a common ancestry (plausibly explained by the large centuries-long migration of the Turkic people across Central Asia) or coincidental presence of Japanese ancestors.

Incomplete Penetrance of G2019S

Incomplete penetrance was already suspected for the mutations underlying the PARK8 locus at the time of the linkage studies. Most of the penetrance analyses have been performed on the frequent G2019S mutation.

Analyses performed on Ashkenazi Jews from the USA revealed a lifetime penetrance of 31.8% [45]. A slighter lower penetrance (24–26% at 80 years) was estimated in independent groups of US Ashkenazi Jews [49, 135].

The International *LRRK2* Consortium performed a penetrance study on the largest dataset of G2019S carriers. By analyzing a large sample of PD patients, they calculated a 28% risk of PD at 59 years, 51% at 69 years, and 74% at 79 years for *LRRK2* G2019S carriers without differences in penetrance by sex or ethnic group [136]. Interestingly a penetrance study in Tunisian G2019S PD cases, after stratifying by homozygous ($n = 23$) and heterozygous carriers, reported a penetrance consistently higher in homozygotes in each age group. Considering possible biases in estimating penetrance only from families, this finding, if true, would indicate a gene dosage effect, although the age of onset was not dissimilar between the two groups [46]. However, subsequent studies collecting clinical data of homozygous carriers showed no phenotype differences between heterozygous and homozygous carriers ruling out a gene dosage effect [44, 137, 138].

The reduced penetrance of this frequent mutation is in keeping with the *LRRK2* G2019S being the most important genetic determinant, known so far, of sporadic PD. Penetrance can also be expressed in terms of risk (calculated as odd ratio) to develop the disease. For an Ashkenazi Jew who carries the G2019S, the risk of developing PD increases ~18-fold [45]. By analyzing the G2019S in North Africans, a lifetime odds ratio for developing PD of 48.6 (CI 11.2–211.0) [44] has been calculated.

Nevertheless, additional studies in different populations are warranted before G2019S genetic counseling can be implemented, since the precise estimation of the penetrance in some countries is still controversial.

Dissimilar results across the abovementioned Ashkenazi Jews from the USA and other G2019S carriers might be influenced by different methodological approaches (e.g. including only patients with both parents genotyped, excluding patients with *GBA* mutations, etc.) or by additional genetic or nongenetic factors that can act as modifiers.

The analysis of candidate genes involved in neurodegeneration as potential genetic modifiers of *LRRK2* has been reported. The first to be explored was *PRKN*, since patients who simultaneously harbored *PRKN* mutations and *LRRK2* G2019S have been mentioned in several studies [61, 73, 99, 139–141]. However, the clinical and cosegregation analysis of patients carrying heterozygous *PRKN* mutations and the G2019S revealed that the combination of the two does not influence the symptoms or the age at disease onset [142].

Polymorphic variations in the microtubule-associated protein tau (*MAPT*) have been proposed to be significantly associated with age of disease onset in individuals with *LRRK2* mutations [143]. Moreover, SNCA variants have been found as determinant of age of onset in G2019S carriers [144]. It is a common observation among neurologists of the different penetrance of *LRRK2* mutations in affected families, implying the great importance of genetic modifiers. Further analyses, especially on large samples and families carrying the G2019S, are warranted to identify genetic factors that can act as modifiers of *LRRK2* mutations.

The R1441 Mutational Hot Spot

The *LRRK2* R1441 residue is the second most common spot of pathogenic *LRRK2* mutations, after G2019S. Three non-synonymous substitutions (R1441C, R1441G, and R1441H) and the synonymous R1441R have been reported in several patients.

R1441C: The Second Most Frequent Pathogenic *LRRK2* Mutation

This mutation (c. 4321C > T) represents the second known most common mutation of the *LRRK2* gene. The R1441C was identified as causative mutation of the PARK8-linked "family D" (Western Nebraska) [4]. Cosegregation was reported also in smaller PD families from Germany [4, 87], Italy [69], Belgium [90], the USA [42, 145], and Iran [59]. The mutation has also been reported in a few other families, but additional affected relatives were not available for cosegregation analysis [62, 69, 90, 146]. The R1441C is also found among sporadic cases and has been reported in patients from Italy [70], Sardinia [77], Russia (Slavic origin) [100], China [147], and Belgium [90]. The variant was absent in large cohorts of ethnically matched controls (>1000 German, 530 Italian, 208 Sardinian, 400 Chinese, 178 Belgian, and 300 American). Interestingly, the R1441C has been found to be more common than G2019S in southern Italy [148].

Haplotype analysis of *LRRK2* R1441C carriers from 20 families of different geographical areas revealed in total four classes of haplotypes. Only for the two major haplotypes, the phase could be established [149]. A first haplotype was identified in the Italian carriers, as well as in German, Spanish, and American patients. A second haplotype was present in the American family D (Western Nebraska) and in Belgian R1441C families. A German and an Irish patient shared a third haplotype for which phase could not be unambiguously determined. Finally, a Chinese proband carried alleles that could not be assigned to any of three previous haplotype classes.

The phenotype associated with this mutation is similar to that of classic PD [149]. The mutation exhibits incomplete penetrance, which could explain its presence in sporadic cases, but calculations performed so far must be interpreted with caution as only a small number of R1441C mutation carriers have been identified until now.

R1441G: A Founder Pathogenic Mutation in the Basques

The *LRRK2* R1441G (c. 4321C > G) was initially described in patients with autosomal dominant late-onset PD in PARK8-linked families of Basque ethnicity [3]. The Basques are a homogeneous ethnic group who historically were isolated by linguistic and geographical barriers. The first report on the frequency of this mutation in Basque PD (~8% of familial cases) [3] and the absence in other large populations screened (except for a US patient reported to be of Hispanic descent [50]) suggested that this variant was population specific. Further studies investigated the prevalence of this mutation in Basque. One group detected the R1441G in 16.4% and 4.0% of familial and sporadic Basque PD, respectively [150], while a more recent study reported a prevalence of 46% in familial Basque patients and 2.5% of sporadic cases [66]. It has also been identified at lower frequencies in patients from nearby provinces in Spain who did not report Basque ancestry (6% of non-Basques living in the Basque countries [66], 2.7% in Asturias [151], 0.7% in Catalonia [62], two families from the neighboring region of Navarre, and one from La Rioja [63]), while it is rare in Cantabria [64]. Haplotype analysis on R1441G carriers from Basque and neighborhood regions [63, 150, 152] indicates that this mutation occurred in a single common ancestor, which in one study was estimated to have lived 1350 (95% CI, 1020–1740) years ago [152]. Since the Basque population has a history of emigration to Europe and North, Central, and South Americas, it would not be surprising to find isolated cases in those countries. However, a single case from Uruguay and a family from Japan carrying the R1441G have been reported with a different haplotype than the Basque, suggesting in these cases independent mutational events [112, 153].

R1441H and R1441R: Uncommon but also Likely Pathogenic

This variant, c.G4322A on *LRRK2* cDNA, occurs immediately adjacent to the two previously reported pathogenic mutations, c.C4321T (R1441C) and c.C4321G (R1441G), resulting in a different substitution of the same amino acid residue (R1441H).

R1441H has been described in a US PD family, but only the proband and an unaffected sibling were available for testing [146]. It was also reported in PD families from Crete [81], Portugal [61], and Taiwan [84], all not large enough to demonstrate definitive cosegregation with the disease.

Haplotype analysis of the abovementioned R1441H carriers showed diversity suggesting a number of independent founders [154]. Subsequently, the R1441H mutation has been identified in two cases from Australia, both of British origin and with a possible common haplotype, although in these cases the phase was not assessed [126]. A further proof in favor of a pathogenic role of this variant came from the identification of R1441H in two slightly larger French families [72].

R1441H was not found in 281 Americans, 300 Cretans, 200 Portugueses, 174 Europeans, and a set of 1000 control samples (600 North Americans, 200 Taiwaneses, 200 Norwegians, 200 Irish, and 200 Spanish). Moreover several studies screened by sequence the *LRRK2* exon 31 in a large sample of healthy controls (>3000 Caucasian [3, 4, 69, 90]) in order to check for the R1441C and R1441G, and none reported mutation in the adjacent nucleotide.

The clinical presentation of affected R1441H carriers appears to be similar to typical Parkinson's disease with an age at onset range of 32–66 years. All display levodopa-responsive parkinsonism; however, the disease in one of the siblings from the Greek R1441H family appeared to transition into a progressive supranuclear palsy-like disorder [81].

To further highlight the nature of codon 1441 as a mutational hot spot, two groups reported a R1441R (c.C4323T) in a sporadic PD patient [101] from Russia and a PD patient with ascertained LB pathology who additionally developed dementia and dysautonomia (PDD) [155]. As for the R1441H, we can indirectly assume that the variant is rare in the Caucasian population, since sequencing controls for the other mutations at the same codon did not reveal any R1441R carriers. This variant is predicted to lead to a synonymous substitution, which would suggest a nonpathogenic role. Moreover, being the nucleotide change close to the splice site, cDNA analysis from the brain of the PDD patient was performed and did not reveal any aberrations on the *LRRK2* transcript [155]. Taken together these results suggest that R1441R is likely to represent a rare but nonpathogenic polymorphism.

Mutations in *LRRK2* are associated with pleomorphic pathology, although the Lewy bodies (LB)-positive pathology is the most common pattern, particularly for the G2019S mutation [38, 40, 82, 156, 157].

In a large screen of 405 LB-positive brains, eight (~2%) have been found to be carriers of the G2019S mutation, including four with brainstem type, three with transitional type, and one with diffuse LB pathology. In two G2019S-positive brains, Alzheimer-type pathology was also present, and it was of enough severity to make a concomitant pathological diagnosis of Alzheimer's disease [157].

A further study on 80 brains with PD or LB dementia screened for the G2019S mutation, and three were found to be carriers. Typical brainstem-type LB-positive pathology was found in one, while the Lewy body variant of Alzheimer's disease was diagnosed in the second. The third brain showed only cell loss in the substantia nigra and locus coeruleus, but no α-synuclein inclusions were detected. There were only rare tau-positive tangles and occasional plaques. No other ubiquitin-positive inclusions were present either [156].

In family D (Western Nebraska), all R1441C carriers examined showed substantia nigra neuronal loss. Two cases had LB pathology, one brainstem type, and the other one diffuse type. The third case had "nonspecific" substantia nigra degeneration with ubiquitin-positive neuronal inclusions. The final case had PSP-like changes with tau-immunoreactive neuronal and glial lesions [4].

The neuropathological examination of R1441G Basque carriers displayed "nonspecific" nigral degeneration in the substantia nigra without α-synuclein, tau, or ubiquitin inclusions [158].

Japanese cases with the I2020T mutation were found to display hyperphosphorylated tau aggregates [159]. Therefore, despite LBs represent the predominant feature in neuropathological studies, the overall *LRRK2*-associated pathology has revealed great variability, probably recapitulating the heterogeneity of PD itself, which can be a more complex disease than what we thought until now.

The Other LRRK2 *Variants: Which Are Pathogenic?*

Besides the most recurrent G2019S and R1441C/R1441G/R1441H, more than 50 different *LRRK2* sequence variants have been reported in familial and sporadic PD cases so far; moreover, few novel *LRRK2* substitutions have been found in healthy control subjects only (Fig. 1.2).

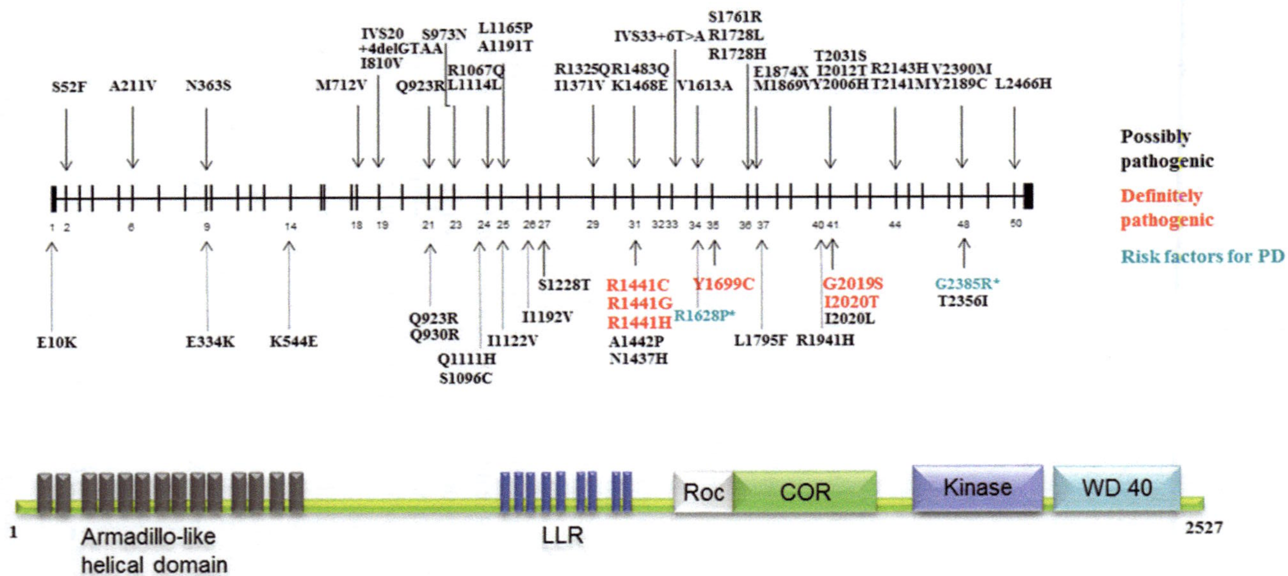

Fig. 1.2 Schematic representation of the *LRRK2* gene, the dardarin protein, and its known functional domains. *LRRK2* 83 coding variants and three putative splice variants are grouped according to evidence of pathogenicity (see main text)

The Y1699C and I2020T mutations are considered as definitely pathogenic. The Y1669C was identified in two independent large families, the Lincolnshire kindred [3, 82] (family PL) of European ancestry, and "family A" (German–Canadian) [4, 38]. The I2020T was identified in "family 32" [4] and "T10738" [89], both of German ancestry. Additionally the same mutation was identified segregating in the large PARK8-linked Sagamihara kindred [43] and in two smaller Japanese families coming from the neighborhood of the Sagamihara region [48].

The role of several other variants remains unclear, since often no family members were available to assess cosegregation and a limited number of ethnically matched controls were screened. Overall, the criteria that may be applied to consider the pathogenicity of the *LRRK2* variants should consider several standpoints: frequency in healthy subjects, cosegregation in families, confirmation in independent studies, and pathogenic consequences on cellular and animal models.

Association Studies on *LRRK2*

In the past few years, many groups put special effort in search of common risk factors for complex diseases. Among these, PD and other neurodegenerative disorders have been extensively studied. However, even using high-throughput techniques allowing to genotype hundreds of thousands of SNPs and covering the whole genome in cases and controls (genome-wide association studies, GWA), no reproducible risk loci have been reported so far.

One caveat is that the GWA approach can be problematic because the massive number of statistical tests performed presents an unprecedented potential for false-positive results.

After the discovering of mutations in the *LRRK2* gene, several studies aimed to explore whether common variant of this gene could represent a risk factor for PD.

Two association studies on *LRRK2* have been performed in Caucasians. The first enrolled 340 PD patients and 608 controls from Germany. 121 SNPs (81 tagging SNPs) were genotyped attempting to represent the complete DNA variation of the *LRRK2* gene [160]. The second study analyzed four common coding SNPs (L953L, R1398H, G1624G, and T2397M) in 250 controls and 121 unrelated PD, mostly with early-onset and positive family history [141]. Neither of these studies revealed any evidences of association between PD and the *LRRK2* SNPs at both allelic and genotypic levels.

In 2005, one study performed in Singapore yielded a significant association. A set of 21 tagging SNPs covering the *LRRK2* gene were genotyped in 466 sporadic PD and 374 control individuals all of Chinese ancestry. The authors identified a common haplotype that was highly overrepresented within cases ($p = 0.005$) and, when present in two copies, significantly increased the risk of PD (OR = 5.5, 95% C.I. = 2.1–14.0, $P = 0.0001$) [161]. However, no *LRRK2* variants within the risk haplotype were reported as the biologically relevant factors.

The G2385R Variant

The *LRRK2* G2385R represents the first common genetic risk factor for PD in the Asian population. This variant was first reported in a small PD family from Taiwan [84]. Evidence for cosegregation with PD in that family was limited due to the small pedigree size; however, the mutation was reported to be absent in 200 ethnically matched controls and, therefore, interpreted as putatively pathogenic. At that time very limited data were available on the nature and frequency of *LRRK2* mutations and on the polymorphism content of the gene in patients from Asia.

Several groups conducted a mutational screening of three known PD-causing mutations (I2012T, G2019S, and I2020T) which appeared to be very rare or absent in Asian PD patients [43, 118, 120, 121]. A sequence of the whole *LRRK2* in Chinese Han patients revealed four coding variants (A419V, P755L, M1869V which were novel substitutions, and the G2385R) that were tested for association with PD in 608 Chinese Han cases and 373 ethnically matched controls.

The heterozygosity for the G2385R variant was significantly higher among PD cases than controls (10% vs 4% *p* 0012). This suggested that the G2385R variant, or another variant in linkage disequilibrium, is associated with PD in the Taiwanese population.

Since then, several association studies on Asian populations from Taiwan, Singapore, Mainland China, Korea, and Malaysia replicated this finding with a similar size effect. Interestingly the association was also reported in Japanese PD patients and controls, giving a risk of developing PD increased of ~twofold [125, 166] (Table 1.2).

Two groups performed a haplotype analysis of G2385R carriers in a cohort of Chinese Han from Taiwan [165, 168]. A single common haplotype shared by carriers has been identified, likely originated from a single ancestor who lived approximately 4800 years ago. Also all Japanese G2385R carriers shared the same haplotype, with a set of markers (D12S2516, D12S2519, and D12S2521) which overlapped with the Chinese haplotype. This might suggest that the G2385R of Chinese Han and Japanese ancestry has arisen from a common ancestor [125].

The R1628P Variant

The *LRRK2* R1628P has been identified in a multicentric study which combined 1986 Chinese individuals from three independent centers in Taiwan and Singapore and so far represents the second most frequent genetic risk factor for PD in Asia [184]. This variant was approximately twice as frequent in affected individuals as control subjects (~6% of PD and ~3.5% of controls, odds ratio 1.84, 95% C.I.: 1.20–2.83, nominal p value = 0.006) [184].

Table 1.2 Association studies on Asian populations (from Taiwan, Singapore, China, Korea, Japan, and Malaysia) showing the G2385R variant as significantly associated with Parkinson's disease

Geographical location	Ethnicity	PD	Controls	OR (95% CI)	References
Taiwan	Chinese Han	61/608 (10%)	18/373 (4.8%)	2.20 (1.28–3.78)	[162]
Singapore	Chinese	37/495 (7.5%)	18/494 (3.6%)	2.14 (1.20–3.81)	[163]
Taiwan	Chinese Han	27/305 (9%)	1/176 (0.5%)	17.00 (2.29–126.20)	[164]
Taiwan	Chinese	34/410 (9.3%)	13/335 (3.9%)	2.24 (1.16–4.32)	[165]
Japan	Japanese	52/448 (11.6%)	22/457 (4.8%)	2.60 (1.55–4.35)	[166]
Singapore	Malay	2/98 (2%)	2/173 (1.2%)	1.75 (0.25–12.85)	[167]
	Indian	0/66 (0%)	0/133 (0%)		
Shanghai	Chinese Han	14/235 (6%)	0/214 (0%)	28.08 (1.66–473.72)	[168]
Mainland of China	Chinese Han	71/600 (11.8%)	11/334 (3.3%)	3.94 (2.06–7.55)	[169]
Japan and USA	Japanese	69/601 (11.5%)	101/1628 (6.2%)	1.96 (1.42–2.70)	[170]
Japan	Japanese	30/229 (13.1%)	23/358 (6.4%)	2.06	[171]
Korea	Koreans	82/923 (8.9%)	21/422 (5%)	1.83	[172]
Asia	Taiwanese	369 (NA)	300 (NA)	1.62	[173]
	Korean	844 (NA)	587 (NA)	1.87	
	Japanese	173 (NA)	95 (NA)	1.44	
Malaysia	Malaysian	695 (NA)	507 (NA)	2.22	[174]
Total		479/5018 (9.6%)	230/5097 (4.5%)	2.23 (1.89–2.62) *p* value < 0.0001	

This finding was replicated in two independent Chinese Han cohorts from Singapore [185] and Taiwan [186]. On the contrary, the R1628P is rare in Japan and in non-Chinese Asians [170, 184, 187].

Haplotype analysis strongly indicates that carriers of the R1628P variant share an extended haplotype, indicative of a founder effect [184]. The mutation has been estimated to arise ~2500 years ago and, in contrast to the older G2385R, has remained confined to subjects of Chinese Han ethnicity.

Like for the G2385R, the clinical phenotype of the affected R1628P carriers is that of typical late-onset L-dopa-responsive PD [184, 186, 187].

Taken together, these studies indicate for the first time that common population specific genetic risk factors for PD exist. The association of both *LRRK2* variants with PD in Asia has been extensively confirmed in independent dataset of patients. These findings open several opportunities of studies for researchers and clinicians. Discovering how those variants can increase the risk of death of dopaminergic neurons might provide important insight into the pathogenesis of the disease. Other interesting prospects can be provided in clinical practice, for example, studying the effect of neuroprotective drugs in large cohorts of asymptomatic carriers of these two *LRRK2* variants, in order to explore whether the risk of developing PD would decrease in the treated subjects.

In conclusion, the *LRRK2* gene displays a high polymorphic content in terms of single nucleotide substitutions. No deletions or duplications have been identified until now. Variants identified in patients are located in almost all exons. However, most of them still lack a definite proof of pathogenicity (Tables 1.3). This has direct

Table 1.3 *LRRK2* genetic variants associated with Parkinson's disease that are possibly pathogenic, but need more evidence to be definitely associated with the disease

Possibly pathogenic *LRRK2* variants			
cDNA change	Protein change	Protein domain	References
28G > A	E10K	LRRK2 repeats	[175]
155C > T	S52F	LRRK2 repeats	[72]
632C > T	A211V	LRRK2 repeats	[176]
1000G > A	E334K	LRRK2 repeats	[175]
1088A > G	N363S	LRRK2 repeats	[72]
1630A > G	K544E	LRRK2 repeats	[176]
2134A > G	M712V	LRRK2 repeats	[106]
2242119_2242122delGTAA	–	LRRK2 repeats	[104]
2769G > C	Q923R	LRRK2 repeats	[115]
2789A > G	Q930R	LRRK2 repeats	[89]
2918G > A	S973N	LRRK2 repeats	[93]
3200G > A	R1067Q	LRRK2 repeats	[177]
3287C > G	S1096C	LRRK2 repeats	[89]
3333G > T	Q1111H	LRRK2 repeats	[175]
3364A > G	I1122V	LRRK2 repeats	[4]
3574A > G	I1192V	LRRK2 repeats	[175]
3451G > A	A1191T	LRRK2 repeats	[178]
3494 T > C	L1165P	LRRK2 repeats	[179]
3287G > C	S1228T	LRRK2 repeats	[89]
3974G > A	R1325Q		[90]
4111A > G	I1371V		[69]
4309A > C	N1437H	Roc domain	[180]
4324G > C	A1442P	Roc domain	[126]
4402A > G	K1468E	Roc domain	[90]
4448G > A	R1483Q	Roc domain	[90]
45,361 + 3A > G	–		[146]

(continued)

Table 1.3 (continued)

Possibly pathogenic *LRRK2* variants			
cDNA change	Protein change	Protein domain	References
48,271 + 6 T > A	–		[177]
4838 T > C	V1613A	COR	[100]
5183G > T	R1728L	COR	[106]
5183G > A	R1728H	COR	[106]
5281A > C	S1761R	COR	[181]
5385G > T	L1795F	COR	[175]
5605A > G	M1869V		[69]
5620G > T	E1874X		[69]
5822G > A	R1941H	Kinase	[82]
6016 T > C	Y2006H	Kinase	[73]
6035 T > C	I2012T	Kinase	[90]
Unknown	I2020L	Kinase	[182]
6091A > T	T2031S	Kinase	[73]
6422C > T	T2141M		[106]
6428G > A	R2143H		[106]
6566A > G	Y2189C	WD40	[90]
7168G > A	V2390M	WD40	[183]
7397 T > A	L2466H	WD40	[106]
7067C > T	T2356I	WD40	[82]

Table 1.4 *LRRK2* variants that are definitely associated with Parkinson's disease (upper panel: causative mutations, lower panel: risk variants)

dbSNP rs #	cDNA change	Protein change	Protein domain	References
Definitely pathogenic LRRK2 mutations				
	4321C > G	R1441G	Roc domain	[3]
rs33939927	4321C > T	R1441C	Roc domain	[4]
rs34995376	4322G > A	R1441H	Roc domain	[146]
rs35801418	5096A > G	Y1699C	COR domain	[3, 4]
rs34637584	6055G > A	G2019S	Kinase domain	[39–41]
rs35870237	6059 T > C	I2020T	Kinase domain	[4]
LRRK2 variants associated with increased risk				
rs33949390	4883G > C	R1628P	COR domain	[184]
rs34778348	7153G > A	G2385R		[162]

practical consequences for the genetic studies. *LRRK2* is a large gene containing 51 exons. A time-/cost-saving strategy to perform the mutational analysis could be to first screen for the frequent G2019S mutation. If negative, other validated mutations (R1441G/R1441C/R1441H, I2020T, and Y1699C) can be tested next (Tables 1.4). Where a considerable number of affected family members are available for testing, an option is to screen the entire *LRRK2* gene, which raises the possibility of discovering one of the above-reported doubtful variants, or even a novel mutation that could be tested for cosegregation in order to verify its pathogenic role. Concerning the significance of these data for the genetic counseling, it is worth to consider that screening the whole coding region or single variants of uncertain role in unselected cases is still a matter of debate, since the identification of any variant would result in more questions than answers for both clinicians and patients. *LRRK2* mutations penetrance is a key piece of information for a proper genetic counseling. Only a minority of *LRRK2* mutation carriers will develop the disease, making the predictive genetic testing more similar to *BRCA* test for breast cancer than to presymptomatic test in Huntington's disease. Dominant transmission involves more subjects and generations inside the family. The involvement of the offspring and the absence of neuroprotective therapy make the offer of predictive/presymptomatic genetic tests in neurodegenerative disease controversial. However, whenever presymptomatic testing is offered, detailed information and counseling at a center with expertise in this area are required [188, 189].

Conflict of Interest The author declares no conflicts of interest.

References

1. Polymeropoulos MH, Lavedan C, Leroy E, Ide SE, Dehejia A, Dutra A et al (1997) Mutation in the alpha-synuclein gene identified in families with Parkinson's disease. Science 276(5321):2045–2047
2. Singleton AB, Farrer M, Johnson J, Singleton A, Hague S, Kachergus J et al (2003) alpha-Synuclein locus triplication causes Parkinson's disease. Science 302(5646):841
3. Paisan-Ruiz C, Jain S, Evans EW, Gilks WP, Simon J, van der Brug M et al (2004) Cloning of the gene containing mutations that cause PARK8-linked Parkinson's disease. Neuron 44(4):595–600
4. Zimprich A, Biskup S, Leitner P, Lichtner P, Farrer M, Lincoln S et al (2004) Mutations in LRRK2 cause autosomal-dominant parkinsonism with pleomorphic pathology. Neuron 44(4):601–607
5. Zimprich A, Benet-Pages A, Struhal W, Graf E, Eck SH, Offman MN et al (2011) A mutation in VPS35, encoding a subunit of the retromer complex, causes late-onset Parkinson disease. Am J Hum Genet 89(1):168–175
6. Vilarino-Guell C, Wider C, Ross OA, Dachsel JC, Kachergus JM, Lincoln SJ et al (2011) VPS35 mutations in Parkinson disease. Am J Hum Genet 89(1):162–167
7. Kitada T, Asakawa S, Hattori N, Matsumine H, Yamamura Y, Minoshima S et al (1998) Mutations in the parkin gene cause autosomal recessive juvenile parkinsonism. Nature 392(6676):605–608
8. Bonifati V, Rizzu P, van Baren MJ, Schaap O, Breedveld GJ, Krieger E et al (2003) Mutations in the DJ-1 gene associated with autosomal recessive early-onset parkinsonism. Science 299(5604):256–259
9. Valente EM, Abou-Sleiman PM, Caputo V, Muqit MM, Harvey K, Gispert S et al (2004) Hereditary early-onset Parkinson's disease caused by mutations in PINK1. Science 304(5674):1158–1160
10. Ramirez A, Heimbach A, Grundemann J, Stiller B, Hampshire D, Cid LP et al (2006) Hereditary parkinsonism with dementia is caused by mutations in ATP13A2, encoding a lysosomal type 5 P-type ATPase. Nat Genet 38(10):1184–1191
11. Paisan-Ruiz C, Bhatia KP, Li A, Hernandez D, Davis M, Wood NW et al (2009) Characterization of PLA2G6 as a locus for dystonia-parkinsonism. Ann Neurol 65(1):19–23

12. Di Fonzo A, Dekker MC, Montagna P, Baruzzi A, Yonova EH, Correia Guedes L et al (2009) FBXO7 mutations cause autosomal recessive, early-onset parkinsonian-pyramidal syndrome. Neurology 72(3):240–245
13. Shojaee S, Sina F, Banihosseini SS, Kazemi MH, Kalhor R, Shahidi GA et al (2008) Genome-wide linkage analysis of a Parkinsonian-pyramidal syndrome pedigree by 500 K SNP arrays. Am J Hum Genet 82(6):1375–1384
14. Edvardson S, Cinnamon Y, Ta-Shma A, Shaag A, Yim YI, Zenvirt S et al (2012) A deleterious mutation in DNAJC6 encoding the neuronal-specific clathrin-uncoating co-chaperone auxilin, is associated with juvenile parkinsonism. PLoS One 7(5), e36458
15. Quadri M, Fang M, Picillo M, Olgiati S, Breedveld GJ, Graafland J et al (2013) Mutation in the SYNJ1 gene associated with autosomal recessive, early-onset Parkinsonism. Hum Mutat 34(9):1208–1215
16. Leroy E, Boyer R, Auburger G, Leube B, Ulm G, Mezey E et al (1998) The ubiquitin pathway in Parkinson's disease. Nature 395(6701):451–452
17. Strauss KM, Martins LM, Plun-Favreau H, Marx FP, Kautzmann S, Berg D et al (2005) Loss of function mutations in the gene encoding Omi/HtrA2 in Parkinson's disease. Hum Mol Genet 14(15):2099–2111
18. Lautier C, Goldwurm S, Durr A, Giovannone B, Tsiaras WG, Pezzoli G et al (2008) Mutations in the GIGYF2 (TNRC15) gene at the PARK11 locus in familial Parkinson disease. Am J Hum Genet 82(4):822–833
19. Chartier-Harlin MC, Dachsel JC, Vilarino-Guell C, Lincoln SJ, Lepretre F, Hulihan MM et al (2011) Translation initiator EIF4G1 mutations in familial Parkinson disease. Am J Hum Genet 89(3):398–406
20. Vilarino-Guell C, Rajput A, Milnerwood AJ, Shah B, Szu-Tu C, Trinh J et al (2014) DNAJC13 mutations in Parkinson disease. Hum Mol Genet 23(7):1794–1801
21. Gasser T, Muller-Myhsok B, Wszolek ZK, Oehlmann R, Calne DB, Bonifati V et al (1998) A susceptibility locus for Parkinson's disease maps to chromosome 2p13. Nat Genet 18(3):262–265
22. Li YJ, Scott WK, Hedges DJ, Zhang F, Gaskell PC, Nance MA et al (2002) Age at onset in two common neurodegenerative diseases is genetically controlled. Am J Hum Genet 70(4):985–993
23. Pankratz N, Nichols WC, Uniacke SK, Halter C, Rudolph A, Shults C et al (2002) Genome screen to identify susceptibility genes for Parkinson disease in a sample without parkin mutations. Am J Hum Genet 71(1):124–135
24. Singleton AB, Farrer MJ, Bonifati V (2013) The genetics of Parkinson's disease: progress and therapeutic implications. Mov Disord 28(1):14–23
25. Gonzalez-Perez A, Gayan J, Marin J, Galan JJ, Saez ME, Real LM et al (2009) Whole-genome conditional two-locus analysis identifies novel candidate genes for late-onset Parkinson's disease. Neurogenetics 10(3):173–181
26. Pankratz N, Wilk JB, Latourelle JC, DeStefano AL, Halter C, Pugh EW et al (2009) Genomewide association study for susceptibility genes contributing to familial Parkinson disease. Hum Genet 124(6):593–605
27. Sutherland GT, Halliday GM, Silburn PA, Mastaglia FL, Rowe DB, Boyle RS et al (2009) Do polymorphisms in the familial Parkinsonism genes contribute to risk for sporadic Parkinson's disease? Mov Disord 24(6):833–838
28. Skipper L, Wilkes K, Toft M, Baker M, Lincoln S, Hulihan M et al (2004) Linkage disequilibrium and association of MAPT H1 in Parkinson disease. Am J Hum Genet 75(4):669–677
29. Tobin JE, Latourelle JC, Lew MF, Klein C, Suchowersky O, Shill HA et al (2008) Haplotypes and gene expression implicate the MAPT region for Parkinson disease: the GenePD study. Neurology 71(1):28–34
30. Zabetian CP, Hutter CM, Factor SA, Nutt JG, Higgins DS, Griffith A et al (2007) Association analysis of MAPT H1 haplotype and subhaplotypes in Parkinson's disease. Ann Neurol 62(2):137–144

31. Saiki M, Baker A, Williams-Gray CH, Foltynie T, Goodman RS, Taylor CJ et al (2010) Association of the human leucocyte antigen region with susceptibility to Parkinson's disease. J Neurol Neurosurg Psychiatry 81(8):890–891
32. Hamza TH, Zabetian CP, Tenesa A, Laederach A, Montimurro J, Yearout D et al (2010) Common genetic variation in the HLA region is associated with late-onset sporadic Parkinson's disease. Nat Genet 42(9):781–785
33. Aharon-Peretz J, Rosenbaum H, Gershoni-Baruch R (2004) Mutations in the glucocerebrosidase gene and Parkinson's disease in Ashkenazi Jews. N Engl J Med 351(19):1972–1977
34. Funayama M, Hasegawa K, Kowa H, Saito M, Tsuji S, Obata F (2002) A new locus for Parkinson's disease (PARK8) maps to chromosome 12p11.2-q13.1. Ann Neurol 51(3):296–301
35. Hasegawa K, Stoessl AJ, Yokoyama T, Kowa H, Wszolek ZK, Yagishita S (2009) Familial parkinsonism: study of original Sagamihara PARK8 (I2020T) kindred with variable clinicopathologic outcomes. Parkinsonism Relat Disord 15(4):300–306
36. Zimprich A, Muller-Myhsok B, Farrer M, Leitner P, Sharma M, Hulihan M et al (2004) The PARK8 locus in autosomal dominant parkinsonism: confirmation of linkage and further delineation of the disease-containing interval. Am J Hum Genet 74(1):11–19
37. Paisan-Ruiz C, Saenz A, Lopez de Munain A, Marti I, Martinez Gil A, Marti-Masso JF et al (2005) Familial Parkinson's disease: clinical and genetic analysis of four Basque families. Ann Neurol 57(3):365–372
38. Wszolek ZK, Pfeiffer RF, Tsuboi Y, Uitti RJ, McComb RD, Stoessl AJ et al (2004) Autosomal dominant parkinsonism associated with variable synuclein and tau pathology. Neurology 62(9):1619–1622
39. Di Fonzo A, Rohe CF, Ferreira J, Chien HF, Vacca L, Stocchi F et al (2005) A frequent LRRK2 gene mutation associated with autosomal dominant Parkinson's disease. Lancet 365(9457):412–415
40. Gilks WP, Abou-Sleiman PM, Gandhi S, Jain S, Singleton A, Lees AJ et al (2005) A common LRRK2 mutation in idiopathic Parkinson's disease. Lancet 365(9457):415–416
41. Nichols WC, Pankratz N, Hernandez D, Paisan-Ruiz C, Jain S, Halter CA et al (2005) Genetic screening for a single common LRRK2 mutation in familial Parkinson's disease. Lancet 365(9457):410–412
42. Kachergus J, Mata IF, Hulihan M, Taylor JP, Lincoln S, Aasly J et al (2005) Identification of a novel LRRK2 mutation linked to autosomal dominant parkinsonism: evidence of a common founder across European populations. Am J Hum Genet 76(4):672–680
43. Funayama M, Hasegawa K, Ohta E, Kawashima N, Komiyama M, Kowa H et al (2005) An LRRK2 mutation as a cause for the parkinsonism in the original PARK8 family. Ann Neurol 57(6):918–921
44. Lesage S, Durr A, Tazir M, Lohmann E, Leutenegger AL, Janin S et al (2006) LRRK2 G2019S as a cause of Parkinson's disease in North African Arabs. N Engl J Med 354(4):422–423
45. Ozelius LJ, Senthil G, Saunders-Pullman R, Ohmann E, Deligtisch A, Tagliati M et al (2006) LRRK2 G2019S as a cause of Parkinson's disease in Ashkenazi Jews. N Engl J Med 354(4):424–425
46. Ishihara L, Gibson RA, Warren L, Amouri R, Lyons K, Wielinski C et al (2007) Screening for Lrrk2 G2019S and clinical comparison of Tunisian and North American Caucasian Parkinson's disease families. Mov Disord 22(1):55–61
47. Lesage S, Ibanez P, Lohmann E, Pollak P, Tison F, Tazir M et al (2005) G2019S LRRK2 mutation in French and North African families with Parkinson's disease. Ann Neurol 58(5):784–787
48. Tomiyama H, Li Y, Funayama M, Hasegawa K, Yoshino H, Kubo S et al (2006) Clinicogenetic study of mutations in LRRK2 exon 41 in Parkinson's disease patients from 18 countries. Mov Disord 21(8):1102–1108
49. Clark LN, Wang Y, Karlins E, Saito L, Mejia-Santana H, Harris J et al (2006) Frequency of LRRK2 mutations in early- and late-onset Parkinson disease. Neurology 67(10):1786–1791

50. Deng H, Le W, Guo Y, Hunter CB, Xie W, Huang M et al (2006) Genetic analysis of LRRK2 mutations in patients with Parkinson disease. J Neurol Sci 251(1–2):102–106
51. Djaldetti R, Hassin-Baer S, Farrer MJ, Vilarino-Guell C, Ross OA, Kolianov V et al (2008) Clinical characteristics of Parkinson's disease among Jewish Ethnic groups in Israel. J Neural Transm 115(9):1279–1284
52. Change N, Mercier G, Lucotte G (2008) Genetic screening of the G2019S mutation of the LRRK2 gene in Southwest European, North African, and Sephardic Jewish subjects. Genet Test 12(3):333–339
53. Saunders-Pullman R, Lipton RB, Senthil G, Katz M, Costan-Toth C, Derby C et al (2006) Increased frequency of the LRRK2 G2019S mutation in an elderly Ashkenazi Jewish population is not associated with dementia. Neurosci Lett 402(1–2):92–96
54. Bardien S, Marsberg A, Keyser R, Lombard D, Lesage S, Brice A et al (2010) LRRK2 G2019S mutation: frequency and haplotype data in South African Parkinson's disease patients. J Neural Transm 117(7):847–853
55. Yonova-Doing E, Atadzhanov M, Quadri M, Kelly P, Shawa N, Musonda ST et al (2012) Analysis of LRRK2, SNCA, Parkin, PINK1, and DJ-1 in Zambian patients with Parkinson's disease. Parkinsonism Relat Disord 18(5):567–571
56. Cilia R, Sironi F, Akpalu A, Cham M, Sarfo FS, Brambilla T et al (2012) Screening LRRK2 gene mutations in patients with Parkinson's disease in Ghana. J Neurol 259(3):569–570
57. Okubadejo N, Britton A, Crews C, Akinyemi R, Hardy J, Singleton A et al (2008) Analysis of Nigerians with apparently sporadic Parkinson disease for mutations in LRRK2, PRKN and ATXN3. PLoS One 3(10), e3421
58. Hanagasi HA, Lohmann E, Dursun B, Honore A, Lesage S, Dogu O et al (2011) LRRK2 mutations are uncommon in Turkey. Eur J Neurol 18(10), e137
59. Shojaee S, Sina F, Farboodi N, Fazlali Z, Ghazavi F, Ghorashi SA et al (2009) A clinic-based screening of mutations in exons 31, 34, 35, 41, and 48 of LRRK2 in Iranian Parkinson's disease patients. Mov Disord 24(7):1023–1027
60. Bras JM, Guerreiro RJ, Ribeiro MH, Januario C, Morgadinho A, Oliveira CR et al (2005) G2019S dardarin substitution is a common cause of Parkinson's disease in a Portuguese cohort. Mov Disord 20(12):1653–1655
61. Ferreira JJ, Guedes LC, Rosa MM, Coelho M, van Doeselaar M, Schweiger D et al (2007) High prevalence of LRRK2 mutations in familial and sporadic Parkinson's disease in Portugal. Mov Disord 22(8):1194–1201
62. Gaig C, Ezquerra M, Marti MJ, Munoz E, Valldeoriola F, Tolosa E (2006) LRRK2 mutations in Spanish patients with Parkinson disease: frequency, clinical features, and incomplete penetrance. Arch Neurol 63(3):377–382
63. Gonzalez-Fernandez MC, Lezcano E, Ross OA, Gomez-Esteban JC, Gomez-Busto F, Velasco F et al (2007) Lrrk2-associated parkinsonism is a major cause of disease in Northern Spain. Parkinsonism Relat Disord 13(8):509–515
64. Infante J, Rodriguez E, Combarros O, Mateo I, Fontalba A, Pascual J et al (2006) LRRK2 G2019S is a common mutation in Spanish patients with late-onset Parkinson's disease. Neurosci Lett 395(3):224–226
65. Mata IF, Ross OA, Kachergus J, Huerta C, Ribacoba R, Moris G et al (2006) LRRK2 mutations are a common cause of Parkinson's disease in Spain. Eur J Neurol 13(4):391–394
66. Gorostidi A, Ruiz-Martinez J, Lopez de Munain A, Alzualde A, Marti Masso JF (2009) LRRK2 G2019S and R1441G mutations associated with Parkinson's disease are common in the Basque Country, but relative prevalence is determined by ethnicity. Neurogenetics 10(2):157–159
67. Civitelli D, Tarantino P, Nicoletti G, Ciro Candiano IC, Annesi F, De Marco EV et al (2007) LRRK2 G6055A mutation in Italian patients with familial or sporadic Parkinson's disease. Clin Genet 71(4):367–370
68. De Rosa A, Criscuolo C, Mancini P, De Martino M, Giordano IA, Pappata S et al (2009) Genetic screening for LRRK2 gene G2019S mutation in Parkinson's disease patients from Southern Italy. Parkinsonism Relat Disord 15(3):242–244

69. Di Fonzo A, Tassorelli C, De Mari M, Chien HF, Ferreira J, Rohe CF et al (2006) Comprehensive analysis of the LRRK2 gene in sixty families with Parkinson's disease. Eur J Hum Genet 14(3):322–331
70. Goldwurm S, Di Fonzo A, Simons EJ, Rohe CF, Zini M, Canesi M et al (2005) The G6055A (G2019S) mutation in LRRK2 is frequent in both early and late onset Parkinson's disease and originates from a common ancestor. J Med Genet 42(11), e65
71. Marongiu R, Ghezzi D, Ialongo T, Soleti F, Elia A, Cavone S et al (2006) Frequency and phenotypes of LRRK2 G2019S mutation in Italian patients with Parkinson's disease. Mov Disord 21(8):1232–1235
72. Lesage S, Condroyer C, Lannuzel A, Lohmann E, Troiano A, Tison F et al (2009) Molecular analyses of the LRRK2 gene in European and North-African autosomal dominant Parkinson's disease. J Med Genet 46(7):458–464
73. Lesage S, Janin S, Lohmann E, Leutenegger AL, Leclere L, Viallet F et al (2007) LRRK2 exon 41 mutations in sporadic Parkinson disease in Europeans. Arch Neurol 64(3):425–430
74. Lesage S, Leclere L, Lohmann E, Borg M, Ruberg M, Durr A et al (2007) Frequency of the LRRK2 G2019S mutation in siblings with Parkinson's disease. Neurodegener Dis 4(2–3):195–198
75. Funalot B, Nichols WC, Perez-Tur J, Mercier G, Lucotte G (2006) Genetic screening for two LRRK2 mutations in French patients with idiopathic Parkinson's disease. Genet Test 10(4):290–293
76. Cossu G, van Doeselaar M, Deriu M, Melis M, Molari A, Di Fonzo A et al (2007) LRRK2 mutations and Parkinson's disease in Sardinia: a Mediterranean genetic isolate. Parkinsonism Relat Disord 13(1):17–21
77. Floris G, Cannas A, Solla P, Murru MR, Tranquilli S, Corongiu D et al (2009) Genetic analysis for five LRRK2 mutations in a Sardinian parkinsonian population: importance of G2019S and R1441C mutations in sporadic Parkinson's disease patients. Parkinsonism Relat Disord 15(4):277–280
78. Kalinderi K, Fidani L, Bostantjopoulou S, Katsarou Z, Kotsis A (2007) The G2019S LRRK2 mutation is uncommon amongst Greek patients with sporadic Parkinson's disease. Eur J Neurol 14(10):1088–1090
79. Papapetropoulos S, Adi N, Shehadeh L, Bishopric N, Singer C, Argyriou AA et al (2008) Is the G2019S LRRK2 mutation common in all southern European populations? J Clin Neurosci 15(9):1027–1030
80. Papapetropoulos S, Argyriou AA, Mitsi G, Chroni E (2007) Re: the G2019S LRRK2 mutation is uncommon amongst Greek patients with familial Parkinson's disease. Eur J Neurol 14(11), e6
81. Spanaki C, Latsoudis H, Plaitakis A (2006) LRRK2 mutations on Crete: R1441H associated with PD evolving to PSP. Neurology 67(8):1518–1519
82. Khan NL, Jain S, Lynch JM, Pavese N, Abou-Sleiman P, Holton JL et al (2005) Mutations in the gene LRRK2 encoding dardarin (PARK8) cause familial Parkinson's disease: clinical, pathological, olfactory and functional imaging and genetic data. Brain 128(Pt 12):2786–2796
83. Williams-Gray CH, Goris A, Foltynie T, Brown J, Maranian M, Walton A et al (2006) Prevalence of the LRRK2 G2019S mutation in a UK community based idiopathic Parkinson's disease cohort. J Neurol Neurosurg Psychiatry 77(5):665–667
84. Mata IF, Kachergus JM, Taylor JP, Lincoln S, Aasly J, Lynch T et al (2005) Lrrk2 pathogenic substitutions in Parkinson's disease. Neurogenetics 6(4):171–177
85. Toft M, Haugarvoll K, Ross OA, Farrer MJ, Aasly JO (2007) LRRK2 and Parkinson's disease in Norway. Acta Neurol Scand Suppl 187:72–75
86. Carmine Belin A, Westerlund M, Sydow O, Lundstromer K, Hakansson A, Nissbrandt H et al (2006) Leucine-rich repeat kinase 2 (LRRK2) mutations in a Swedish Parkinson cohort and a healthy nonagenarian. Mov Disord 21(10):1731–1734

87. Hedrich K, Winkler S, Hagenah J, Kabakci K, Kasten M, Schwinger E et al (2006) Recurrent LRRK2 (Park8) mutations in early-onset Parkinson's disease. Mov Disord 21(9):1506–1510
88. Moller JC, Rissling I, Mylius V, Hoft C, Eggert KM, Oertel WH (2008) The prevalence of the G2019S and R1441C/G/H mutations in LRRK2 in German patients with Parkinson's disease. Eur J Neurol 15(7):743–745
89. Berg D, Schweitzer K, Leitner P, Zimprich A, Lichtner P, Belcredi P et al (2005) Type and frequency of mutations in the LRRK2 gene in familial and sporadic Parkinson's disease*. Brain 128(Pt 12):3000–3011
90. Nuytemans K, Rademakers R, Theuns J, Pals P, Engelborghs S, Pickut B et al (2008) Founder mutation p.R1441C in the leucine-rich repeat kinase 2 gene in Belgian Parkinson's disease patients. Eur J Hum Genet 16(4):471–479
91. Macedo MG, Verbaan D, Fang Y, van Rooden SM, Visser M, Anar B et al (2009) Genotypic and phenotypic characteristics of Dutch patients with early onset Parkinson's disease. Mov Disord 24(2):196–203
92. Bech S, Norremolle A, Winge K, Hasholt L, Tommerup N, Svenstrup K et al (2011) The lrrk2 p.Gly2019Ser mutation is uncommon in a Danish cohort with various neurodegenerative disorders. Parkinsonism Relat Disord 17(5):398–399
93. Haubenberger D, Bonelli S, Hotzy C, Leitner P, Lichtner P, Samal D et al (2007) A novel LRRK2 mutation in an Austrian cohort of patients with Parkinson's disease. Mov Disord 22(11):1640–1643
94. Bialecka M, Hui S, Klodowska-Duda G, Opala G, Tan EK, Drozdzik M (2005) Analysis of LRRK 2 G 2019 S and I 2020 T mutations in Parkinson's disease. Neurosci Lett 390(1):1–3
95. Jankovic MZ, Kresojevic ND, Dobricic VS, Markovic VV, Petrovic IN, Novakovic IV et al (2015) Identification of novel variants in LRRK2 gene in patients with Parkinson's disease in Serbian population. J Neurol Sci 353(1–2):59–62
96. Bognar C, Baldovic M, Benetin J, Kadasi L, Zatkova A (2013) Analysis of Leucine-rich repeat kinase 2 (LRRK2) and Parkinson protein 2 (parkin, PARK2) genes mutations in Slovak Parkinson disease patients. Gen Physiol Biophys 32(1):55–66
97. Balicza P, Bereznai B, Takats A, Klivenyi P, Dibo G, Hidasi E et al (2012) The absence of the common LRRK2 G2019S mutation in 120 young onset Hungarian Parkinon's disease patients. Ideggyogy Sz 65(7–8):239–242
98. Fiala O, Pospisilova L, Prochazkova J, Matejckova M, Martasek P, Novakova L et al (2010) Parkin mutations and phenotypic features in Czech patients with early-onset Parkinson's disease. Neuro Endocrinol Lett 31(2):187–192
99. Illarioshkin SN, Shadrina MI, Slominsky PA, Bespalova EV, Zagorovskaya TB, Bagyeva G et al (2007) A common leucine-rich repeat kinase 2 gene mutation in familial and sporadic Parkinson's disease in Russia. Eur J Neurol 14(4):413–417
100. Pchelina SN, Yakimovskii AF, Emelyanov AK, Ivanova ON, Schwarzman AL, Singleton AB (2008) Screening for LRRK2 mutations in patients with Parkinson's disease in Russia: identification of a novel LRRK2 variant. Eur J Neurol 15(7):692–696
101. Pchelina SN, Yakimovskii AF, Ivanova ON, Emelianov AK, Zakharchuk AH, Schwarzman AL (2006) G2019S LRRK2 mutation in familial and sporadic Parkinson's disease in Russia. Mov Disord 21(12):2234–2236
102. Shadrina MI, Illarioshkin SN, Bagyeva G, Bespalova EV, Zagorodskaia TB, Slominskii PA et al (2007) A PARK8 form of Parkinson's disease: a mutational analysis of the LRRK2 gene in Russian population. Zh Nevrol Psikhiatr Im S S Korsakova 107(3):46–50
103. Chen-Plotkin AS, Yuan W, Anderson C, McCarty Wood E, Hurtig HI, Clark CM et al (2008) Corticobasal syndrome and primary progressive aphasia as manifestations of LRRK2 gene mutations. Neurology 70(7):521–527
104. Johnson J, Paisan-Ruiz C, Lopez G, Crews C, Britton A, Malkani R et al (2007) Comprehensive screening of a North American Parkinson's disease cohort for LRRK2 mutation. Neurodegener Dis 4(5):386–391

105. Kay DM, Zabetian CP, Factor SA, Nutt JG, Samii A, Griffith A et al (2006) Parkinson's disease and LRRK2: frequency of a common mutation in U.S. movement disorder clinics. Mov Disord 21(4):519–523
106. Paisan-Ruiz C, Nath P, Washecka N, Gibbs JR, Singleton AB (2008) Comprehensive analysis of LRRK2 in publicly available Parkinson's disease cases and neurologically normal controls. Hum Mutat 29(4):485–490
107. Papapetropoulos S, Singer C, Ross OA, Toft M, Johnson JL, Farrer MJ et al (2006) Clinical heterogeneity of the LRRK2 G2019S mutation. Arch Neurol 63(9):1242–1246
108. Patra B, Parsian AJ, Racette BA, Zhao JH, Perlmutter JS, Parsian A (2009) LRRK2 gene G2019S mutation and SNPs [haplotypes] in subtypes of Parkinson's disease. Parkinsonism Relat Disord 15(3):175–180
109. Scholz S, Mandel RJ, Fernandez HH, Foote KD, Rodriguez RL, Barton E et al (2006) LRRK2 mutations in a clinic-based cohort of Parkinson's disease. Eur J Neurol 13(12):1298–1301
110. Dupre N, Riviere JB, Myers RH, Provencher P, Pourcher E, Emond F et al (2007) LRRK2 is not a significant cause of Parkinson's disease in French-Canadians. Can J Neurol Sci 34(3):333–335
111. Grimes DA, Racacho L, Han F, Panisset M, Bulman DE (2007) LRRK2 screening in a Canadian Parkinson's disease cohort. Can J Neurol Sci 34(3):336–338
112. Mata IF, Cosentino C, Marca V, Torres L, Mazzetti P, Ortega O et al (2009) LRRK2 mutations in patients with Parkinson's disease from Peru and Uruguay. Parkinsonism Relat Disord 15(5):370–373
113. Perez-Pastene C, Cobb SA, Diaz-Grez F, Hulihan MM, Miranda M, Venegas P et al (2007) Lrrk2 mutations in South America: a study of Chilean Parkinson's disease. Neurosci Lett 422(3):193–197
114. Gatto EM, Parisi V, Converso DP, Poderoso JJ, Carreras MC, Marti-Masso JF et al (2013) The LRRK2 G2019S mutation in a series of Argentinean patients with Parkinson's disease: clinical and demographic characteristics. Neurosci Lett 537:1–5
115. Camargos ST, Dornas LO, Momeni P, Lees A, Hardy J, Singleton A et al (2009) Familial Parkinsonism and early onset Parkinson's disease in a Brazilian movement disorders clinic: phenotypic characterization and frequency of SNCA, PRKN, PINK1, and LRRK2 mutations. Mov Disord 24(5):662–666
116. Munhoz RP, Wakutani Y, Marras C, Teive HA, Raskin S, Werneck LC et al (2008) The G2019S LRRK2 mutation in Brazilian patients with Parkinson's disease: phenotype in monozygotic twins. Mov Disord 23(2):290–294
117. Pimentel MM, Moura KC, Abdalla CB, Pereira JS, de Rosso AL, Nicaretta DH et al (2008) A study of LRRK2 mutations and Parkinson's disease in Brazil. Neurosci Lett 433(1):17–21
118. Fung HC, Chen CM, Hardy J, Hernandez D, Singleton A, Wu YR (2006) Lack of G2019S LRRK2 mutation in a cohort of Taiwanese with sporadic Parkinson's disease. Mov Disord 21(6):880–881
119. Lin CH, Tzen KY, Yu CY, Tai CH, Farrer MJ, Wu RM (2008) LRRK2 mutation in familial Parkinson's disease in a Taiwanese population: clinical, PET, and functional studies. J Biomed Sci 15(5):661–667
120. Lu CS, Simons EJ, Wu-Chou YH, Fonzo AD, Chang HC, Chen RS et al (2005) The LRRK2 I2012T, G2019S, and I2020T mutations are rare in Taiwanese patients with sporadic Parkinson's disease. Parkinsonism Relat Disord 11(8):521–522
121. Tan EK, Shen H, Tan LC, Farrer M, Yew K, Chua E et al (2005) The G2019S LRRK2 mutation is uncommon in an Asian cohort of Parkinson's disease patients. Neurosci Lett 384(3):327–329
122. Cho JW, Kim SY, Park SS, Kim HJ, Ahn TB, Kim JM et al (2007) The G2019S LRRK2 mutation is rare in Korean patients with Parkinson's disease. Can J Neurol Sci 34(1):53–55
123. Choi JM, Woo MS, Ma HI, Kang SY, Sung YH, Yong SW et al (2008) Analysis of PARK genes in a Korean cohort of early-onset Parkinson disease. Neurogenetics 9(4):263–269

124. Punia S, Behari M, Govindappa ST, Swaminath PV, Jayaram S, Goyal V et al (2006) Absence/rarity of commonly reported LRRK2 mutations in Indian Parkinson's disease patients. Neurosci Lett 409(2):83–88
125. Zabetian CP, Morino H, Ujike H, Yamamoto M, Oda M, Maruyama H et al (2006) Identification and haplotype analysis of LRRK2 G2019S in Japanese patients with Parkinson disease. Neurology 67(4):697–699
126. Huang Y, Halliday GM, Vandebona H, Mellick GD, Mastaglia F, Stevens J et al (2007) Prevalence and clinical features of common LRRK2 mutations in Australians with Parkinson's disease. Mov Disord 22(7):982–989
127. Mellick GD, Siebert GA, Funayama M, Buchanan DD, Li Y, Imamichi Y et al (2009) Screening PARK genes for mutations in early-onset Parkinson's disease patients from Queensland, Australia. Parkinsonism Relat Disord 15(2):105–109
128. Hernandez DG, Paisan-Ruiz C, McInerney-Leo A, Jain S, Meyer-Lindenberg A, Evans EW et al (2005) Clinical and positron emission tomography of Parkinson's disease caused by LRRK2. Ann Neurol 57(3):453–456
129. Zabetian CP, Hutter CM, Yearout D, Lopez AN, Factor SA, Griffith A et al (2006) LRRK2 G2019S in families with Parkinson disease who originated from Europe and the Middle East: evidence of two distinct founding events beginning two millennia ago. Am J Hum Genet 79(4):752–758
130. Watterson GA, Guess HA (1977) Is the most frequent allele the oldest? Theor Popul Biol 11(2):141–160
131. Warren L, Gibson R, Ishihara L, Elango R, Xue Z, Akkari A et al (2008) A founding LRRK2 haplotype shared by Tunisian, US, European and Middle Eastern families with Parkinson's disease. Parkinsonism Relat Disord 14(1):77–80
132. Bar-Shira A, Hutter CM, Giladi N, Zabetian CP, Orr-Urtreger A (2009) Ashkenazi Parkinson's disease patients with the LRRK2 G2019S mutation share a common founder dating from the second to fifth centuries. Neurogenetics 10(4):355–358
133. Lesage S, Patin E, Condroyer C, Leutenegger AL, Lohmann E, Giladi N et al (2010) Parkinson's disease-related LRRK2 G2019S mutation results from independent mutational events in humans. Hum Mol Genet 19(10):1998–2004
134. Pirkevi C, Lesage S, Condroyer C, Tomiyama H, Hattori N, Ertan S et al (2009) A LRRK2 G2019S mutation carrier from Turkey shares the Japanese haplotype. Neurogenetics 10(3):271–273
135. Marder K, Wang Y, Alcalay RN, Mejia-Santana H, Tang MX, Lee A et al (2015) Age-specific penetrance of LRRK2 G2019S in the Michael J. Fox Ashkenazi Jewish LRRK2 Consortium. Neurology 85(1):89–95
136. Healy DG, Falchi M, O'Sullivan SS, Bonifati V, Durr A, Bressman S et al (2008) Phenotype, genotype, and worldwide genetic penetrance of LRRK2-associated Parkinson's disease: a case–control study. Lancet Neurol 7(7):583–590
137. Ishihara L, Warren L, Gibson R, Amouri R, Lesage S, Durr A et al (2006) Clinical features of Parkinson disease patients with homozygous leucine-rich repeat kinase 2 G2019S mutations. Arch Neurol 63(9):1250–1254
138. Brice A (2005) Genetics of Parkinson's disease: LRRK2 on the rise. Brain 128(Pt 12):2760–2762
139. Marras C, Klein C, Lang AE, Wakutani Y, Moreno D, Sato C et al (2010) LRRK2 and Parkin mutations in a family with parkinsonism-Lack of genotype-phenotype correlation. Neurobiol Aging 31(4):721–722
140. Dachsel JC, Mata IF, Ross OA, Taylor JP, Lincoln SJ, Hinkle KM et al (2006) Digenic parkinsonism: investigation of the synergistic effects of PRKN and LRRK2. Neurosci Lett 410(2):80–84
141. Paisan-Ruiz C, Lang AE, Kawarai T, Sato C, Salehi-Rad S, Fisman GK et al (2005) LRRK2 gene in Parkinson disease: mutation analysis and case control association study. Neurology 65(5):696–700

142. Solla P, Cannas A, Floris G, Murru MR, Corongiu D, Tranquilli S et al (2010) Parkin Exon rearrangements and sequence variants in LRRK2 mutations carriers: analysis on a possible modifier effect on LRRK2 penetrance. Parkinsons Dis 2010:537698
143. Golub Y, Berg D, Calne DB, Pfeiffer RF, Uitti RJ, Stoessl AJ et al (2009) Genetic factors influencing age at onset in LRRK2-linked Parkinson disease. Parkinsonism Relat Disord 15(7):539–541
144. Botta-Orfila T, Ezquerra M, Pastor P, Fernandez-Santiago R, Pont-Sunyer C, Compta Y et al (2012) Age at onset in LRRK2-associated PD is modified by SNCA variants. J Mol Neurosci 48(1):245–247
145. Latourelle JC, Sun M, Lew MF, Suchowersky O, Klein C, Golbe LI et al (2008) The Gly2019Ser mutation in LRRK2 is not fully penetrant in familial Parkinson's disease: the GenePD study. BMC Med 6:32
146. Zabetian CP, Samii A, Mosley AD, Roberts JW, Leis BC, Yearout D et al (2005) A clinic-based study of the LRRK2 gene in Parkinson disease yields new mutations. Neurology 65(5):741–744
147. Tan EK, Skipper L, Chua E, Wong MC, Pavanni R, Bonnard C et al (2006) Analysis of 14 LRRK2 mutations in Parkinson's plus syndromes and late-onset Parkinson's disease. Mov Disord 21(7):997–1001
148. Criscuolo C, De Rosa A, Guacci A, Simons EJ, Breedveld GJ, Peluso S et al (2011) The LRRK2 R1441C mutation is more frequent than G2019S in Parkinson's disease patients from southern Italy. Mov Disord 26(9):1733–1736
149. Haugarvoll K, Rademakers R, Kachergus JM, Nuytemans K, Ross OA, Gibson JM et al (2008) Lrrk2 R1441C parkinsonism is clinically similar to sporadic Parkinson disease. Neurology 70(16 Pt 2):1456–1460
150. Simon-Sanchez J, Marti-Masso JF, Sanchez-Mut JV, Paisan-Ruiz C, Martinez-Gil A, Ruiz-Martinez J et al (2006) Parkinson's disease due to the R1441G mutation in Dardarin: a founder effect in the Basques. Mov Disord 21(11):1954–1959
151. Mata IF, Taylor JP, Kachergus J, Hulihan M, Huerta C, Lahoz C et al (2005) LRRK2 R1441G in Spanish patients with Parkinson's disease. Neurosci Lett 382(3):309–311
152. Mata IF, Hutter CM, Gonzalez-Fernandez MC, de Pancorbo MM, Lezcano E, Huerta C et al (2009) Lrrk2 R1441G-related Parkinson's disease: evidence of a common founding event in the seventh century in Northern Spain. Neurogenetics 10(4):347–353
153. Hatano T, Funayama M, Kubo S, Mata IF, Oji Y, Mori A et al (2014) Identification of a Japanese family with LRRK2 p.R1441G-related Parkinson's disease. Neurobiol Aging 35(11):2656.e17–2656.e23
154. Ross OA, Spanaki C, Griffith A, Lin CH, Kachergus J, Haugarvoll K et al (2009) Haplotype analysis of Lrrk2 R1441H carriers with parkinsonism. Parkinsonism Relat Disord 15(6):466–467
155. Gaig C, Ezquerra M, Marti MJ, Valldeoriola F, Munoz E, Llado A et al (2008) Screening for the LRRK2 G2019S and codon-1441 mutations in a pathological series of parkinsonian syndromes and frontotemporal lobar degeneration. J Neurol Sci 270(1–2):94–98
156. Giasson BI, Covy JP, Bonini NM, Hurtig HI, Farrer MJ, Trojanowski JQ et al (2006) Biochemical and pathological characterization of Lrrk2. Ann Neurol 59(2):315–322
157. Ross OA, Toft M, Whittle AJ, Johnson JL, Papapetropoulos S, Mash DC et al (2006) Lrrk2 and Lewy body disease. Ann Neurol 59(2):388–393
158. Marti-Masso JF, Ruiz-Martinez J, Bolano MJ, Ruiz I, Gorostidi A, Moreno F et al (2009) Neuropathology of Parkinson's disease with the R1441G mutation in LRRK2. Mov Disord 24(13):1998–2001
159. Kalia LV, Lang AE, Hazrati LN, Fujioka S, Wszolek ZK, Dickson DW et al (2015) Clinical correlations with Lewy body pathology in LRRK2-related Parkinson disease. JAMA Neurol 72(1):100–105
160. Biskup S, Mueller JC, Sharma M, Lichtner P, Zimprich A, Berg D et al (2005) Common variants of LRRK2 are not associated with sporadic Parkinson's disease. Ann Neurol 58(6):905–908

161. Skipper L, Li Y, Bonnard C, Pavanni R, Yih Y, Chua E et al (2005) Comprehensive evaluation of common genetic variation within LRRK2 reveals evidence for association with sporadic Parkinson's disease. Hum Mol Genet 14(23):3549–3556
162. Di Fonzo A, Wu-Chou YH, Lu CS, van Doeselaar M, Simons EJ, Rohe CF et al (2006) A common missense variant in the LRRK2 gene, Gly2385Arg, associated with Parkinson's disease risk in Taiwan. Neurogenetics 7(3):133–138
163. Tan EK (2006) Identification of a common genetic risk variant (LRRK2 Gly2385Arg) in Parkinson's disease. Ann Acad Med Singapore 35(11):840–842
164. Fung HC, Chen CM, Hardy J, Singleton AB, Wu YR (2006) A common genetic factor for Parkinson disease in ethnic Chinese population in Taiwan. BMC Neurol 6:47
165. Farrer MJ, Stone JT, Lin CH, Dachsel JC, Hulihan MM, Haugarvoll K et al (2007) Lrrk2 G2385R is an ancestral risk factor for Parkinson's disease in Asia. Parkinsonism Relat Disord 13(2):89–92
166. Funayama M, Li Y, Tomiyama H, Yoshino H, Imamichi Y, Yamamoto M et al (2007) Leucine-rich repeat kinase 2 G2385R variant is a risk factor for Parkinson disease in Asian population. Neuroreport 18(3):273–275
167. Tan EK, Zhao Y, Tan L, Lim HQ, Lee J, Yuen Y et al (2007) Analysis of LRRK2 Gly2385Arg genetic variant in non-Chinese Asians. Mov Disord 22(12):1816–1818
168. Li C, Ting Z, Qin X, Ying W, Li B, Guo Qiang L et al (2007) The prevalence of LRRK2 Gly2385Arg variant in Chinese Han population with Parkinson's disease. Mov Disord 22(16):2439–2443
169. An XK, Peng R, Li T, Burgunder JM, Wu Y, Chen WJ et al (2008) LRRK2 Gly2385Arg variant is a risk factor of Parkinson's disease among Han-Chinese from mainland China. Eur J Neurol 15(3):301–305
170. Zabetian CP, Yamamoto M, Lopez AN, Ujike H, Mata IF, Izumi Y et al (2009) LRRK2 mutations and risk variants in Japanese patients with Parkinson's disease. Mov Disord 24(7):1034–1041
171. Miyake Y, Tsuboi Y, Koyanagi M, Fujimoto T, Shirasawa S, Kiyohara C et al (2010) LRRK2 Gly2385Arg polymorphism, cigarette smoking, and risk of sporadic Parkinson's disease: a case–control study in Japan. J Neurol Sci 297(1–2):15–18
172. Kim JM, Lee JY, Kim HJ, Kim JS, Shin ES, Cho JH et al (2010) The LRRK2 G2385R variant is a risk factor for sporadic Parkinson's disease in the Korean population. Parkinsonism Relat Disord 16(2):85–88
173. Ross OA, Soto-Ortolaza AI, Heckman MG, Aasly JO, Abahuni N, Annesi G et al (2011) Association of LRRK2 exonic variants with susceptibility to Parkinson's disease: a case–control study. Lancet Neurol 10(10):898–908
174. Gopalai AA, Lim SY, Chua JY, Tey S, Lim TT, Mohamed Ibrahim N et al (2014) LRRK2 G2385R and R1628P mutations are associated with an increased risk of Parkinson's disease in the Malaysian population. Biomed Res Int 2014:867321
175. Nichols WC, Elsaesser VE, Pankratz N, Pauciulo MW, Marek DK, Halter CA et al (2007) LRRK2 mutation analysis in Parkinson disease families with evidence of linkage to PARK8. Neurology 69(18):1737–1744
176. Xiromerisiou G, Hadjigeorgiou GM, Gourbali V, Johnson J, Papakonstantinou I, Papadimitriou A et al (2007) Screening for SNCA and LRRK2 mutations in Greek sporadic and autosomal dominant Parkinson's disease: identification of two novel LRRK2 variants. Eur J Neurol 14(1):7–11
177. Skipper L, Shen H, Chua E, Bonnard C, Kolatkar P, Tan LC et al (2005) Analysis of LRRK2 functional domains in nondominant Parkinson disease. Neurology 65(8):1319–1321
178. Schlitter AM, Woitalla D, Mueller T, Epplen JT, Dekomien G (2006) The LRRK2 gene in Parkinson's disease: mutation screening in patients from Germany. J Neurol Neurosurg Psychiatry 77(7):891–892
179. Covy JP, Yuan W, Waxman EA, Hurtig HI, Van Deerlin VM, Giasson BI (2009) Clinical and pathological characteristics of patients with leucine-rich repeat kinase-2 mutations. Mov Disord 24(1):32–39

180. Aasly JO, Vilarino-Guell C, Dachsel JC, Webber PJ, West AB, Haugarvoll K et al (2010) Novel pathogenic LRRK2 p.Asn1437His substitution in familial Parkinson's disease. Mov Disord 25(13):2156–2163
181. Lorenzo-Betancor O, Samaranch L, Ezquerra M, Tolosa E, Lorenzo E, Irigoyen J et al (2012) LRRK2 haplotype-sharing analysis in Parkinson's disease reveals a novel p.S1761R mutation. Mov Disord 27(1):146–151
182. Cilia R, Siri C, Rusconi D, Allegra R, Ghiglietti A, Sacilotto G et al (2014) LRRK2 mutations in Parkinson's disease: confirmation of a gender effect in the Italian population. Parkinsonism Relat Disord 20(8):911–914
183. Clarimon J, Pagonabarraga J, Paisan-Ruiz C, Campolongo A, Pascual-Sedano B, Marti-Masso JF et al (2008) Tremor dominant parkinsonism: clinical description and LRRK2 mutation screening. Mov Disord 23(4):518–523
184. Ross OA, Wu YR, Lee MC, Funayama M, Chen ML, Soto AI et al (2008) Analysis of Lrrk2 R1628P as a risk factor for Parkinson's disease. Ann Neurol 64(1):88–92
185. Tan EK, Tan LC, Lim HQ, Li R, Tang M, Yih Y et al (2008) LRRK2 R1628P increases risk of Parkinson's disease: replication evidence. Hum Genet 124(3):287–288
186. Lu CS, Wu-Chou YH, van Doeselaar M, Simons EJ, Chang HC, Breedveld GJ et al (2008) The LRRK2 Arg1628Pro variant is a risk factor for Parkinson's disease in the Chinese population. Neurogenetics 9(4):271–276
187. Tan EK, Tang M, Tan LC, Wu YR, Wu RM, Ross OA et al (2008) Lrrk2 R1628P in non-Chinese Asian races. Ann Neurol 64(4):472–473
188. Goldwurm S, Zini M, Mariani L, Tesei S, Miceli R, Sironi F et al (2007) Evaluation of LRRK2 G2019S penetrance: relevance for genetic counseling in Parkinson disease. Neurology 68(14):1141–1143
189. McInerney-Leo A (2005) Genetic testing in Parkinson's disease. Mov Disord 20(7): 908–909

Chapter 2
Clinical Features of *LRRK2* Carriers with Parkinson's Disease

Meir Kestenbaum and Roy N. Alcalay

Abstract *LRRK2* mutations are present in 1% of all sporadic Parkinson's disease (PD) cases and 5% of all familial PD cases. Several mutations in the *LRRK2* gene are associated with PD, the most common of which is the Gly2019Ser mutation. In the following review, we summarize the demographics and motor and non-motor symptoms of *LRRK2* carriers with PD, as well as symptoms in non-manifesting carriers. The clinical features of *LRRK2*-associated PD are often indistinguishable from those of idiopathic PD on an individual basis. However, *LRRK2* PD patients are likely to have less non-motor symptoms compared to idiopathic PD patients, including less olfactory and cognitive impairment. *LRRK2*-associated PD patients are less likely to report REM sleep behavior disorder (RBD) than noncarriers. In addition, it is possible that carriers are more prone to cancer than noncarriers with PD, but larger studies are required to confirm this observation. Development of more sensitive biomarkers to identify mutation carriers at risk of developing PD, as well as biomarkers of disease progression among *LRRK2* carriers with PD, is required. Such biomarkers would help evaluate interventions, which may prevent PD among non-manifesting carriers, or slow down disease progression among carriers with PD.

Keywords LRRK2 • Parkinson's disease • Clinical features

Introduction

Since 2004 [1], when the association between mutations in the *LRRK2* gene and Parkinson's disease (PD) was first described, numerous studies have been conducted in different populations to define the clinical phenotype of *LRRK2*-associated PD (here and thereafter *LRRK2* PD). Several mutations in the *LRRK2* gene have been shown to be associated with PD, the most common of them being the Gly2019Ser mutation. While tremendous effort has been invested in clinically phenotyping

---

M. Kestenbaum • R.N. Alcalay (✉)
Department of Neurology, Columbia University Medical Center, New York, NY, USA
e-mail: rna2104@columbia.edu

H.J. Rideout (ed.), *Leucine-Rich Repeat Kinase 2 (LRRK2)*,
Advances in Neurobiology 14, DOI 10.1007/978-3-319-49969-7_2

LRRK2 mutation carriers, there are significant limitations to the existing literature. Many of the published studies are single-site studies with small number of participants (and therefore may be underpowered), and the vast majority of the studies are cross-sectional rather than longitudinal.

Although the clinical features of *LRRK2* PD are, in most cases, indistinguishable from idiopathic PD (i.e., PD noncarriers) on an individual basis, recent studies have reported differences in motor and non-motor features between idiopathic PD and *LRRK2* PD. Specifically, *LRRK2* carriers may have slightly slower PD progression, and carriers have fewer non-motor symptoms, including less cognitive impairment and lower frequency of REM sleep behavior disorder and hyposmia when compared to noncarriers with PD. The demographics and motor and non-motor symptoms of *LRRK2* PD carriers, as well as symptoms in non-manifesting carriers, are described in the following review.

LRRK2 Epidemiology

Ethnic Distribution

PD is a common neurodegenerative disease affecting 1–2% of people older than 65 years. The etiology of PD has not been fully elucidated, but genetic factors clearly play a role in its pathogenesis. Along with *GBA* mutations, *LRRK2* mutations are to date the most common dominantly inherited (with reduced penetrance) genetic risk factor implicated in late-onset familial and sporadic PD [2]. Several mutations in the *LRRK2* gene have been shown to be associated with PD, the most common of which is the Gly2019Ser mutation [3]. Other mutations and variants in the *LRRK2* gene that have been associated with PD are Arg1441Gly, Arg1441Cys, Arg1441His, Ile2020Thr, Tyr1699Cys [3], Arg1628Pro, and Gly2385Arg [4]. The largest analysis to date [3], in which data from 24 populations worldwide were combined, showed that out of 19,376 unrelated PD patients, approximately 1% of PD patients without a family history of disease and 4% of PD with an affected first-degree relative were carriers of the *LRRK2* Gly2019Ser mutation. The frequency of *LRRK2* mutation carriers among PD patients varies greatly in different populations and is presented in Table 2.1. *LRRK2* mutations are most prevalent among North African Berbers. Lesage et al. [5] reported that 7/17 (41%) of North African families from Morocco, Algeria, and Tunisia with familial PD were carriers of the *LRRK2* Gly2019Ser mutation compared to 5/174 (2.9%) families of European origin, mostly French. Ishihara et al. [6] screened for the *LRRK2* mutation in Tunisian familial PD patients and reported that 38 out of 91 families (42%) were carriers of the *LRRK2* Gly2019Ser mutation compared to only 1 out of 39 (2.6%) North American Caucasian familial PD patients.

The *LRRK2* Gly2019Ser mutation is also common among Ashkenazi Jews (AJ). Orr-Urtreger et al. [7] reported that the frequency of *LRRK2* mutations in 344 AJ PD

Table 2.1 Frequency of *LRRK2* mutation carriers among different PD patient populations

Reference	Mutation	Ethnicity	Number of participants	Sporadic PD	Familial PD	Comments
Lesage [5]	Gly2019Ser	North African (Algerian, Tunisian, Moroccan)	198		41%	
Ishihara [6]	Gly2019Ser	Tunisian	659		42%	
Hulihan [11]	Gly2019Ser	Tunisian	609	30%		
Orr-Urtreger [7]	Gly2019Ser	Ashkenazi Jewish	126	10.6%	26%	
Alcalay [8]	Gly2019Ser	Ashkenazi Jewish	488	19.9%		Excluding *GBA* carriers
Ferreira [12]	Gly2019Ser, Arg1441Gly, Arg1441Cys, Arg1441His	Portuguese	138	3.7%	16.1%	
Pulkes [13]	Arg1628Pro	Thai	958	11%		
Sierra [14]	Gly2019Ser	Spanish	493	8.7%		
Cilia [15]	Gly2019Ser, Arg1441Gly, Arg1441Cys, Arg1441His, Ile2020Thr	Italian	2976	1.6%		
Yescas [16]	Gly2019Ser, Arg1441His, Arg1441Cys	Mexican	319	0.9%		
Sanyal [17]	Gly2019Ser, Arg1441Gly, Arg1441Cys, Arg1441His, Tyr1699Cys, Ile2020Thr, Ile2012Thr	East Indian	320	0%		
Chien [18]	Gly2019Ser	Brazilian	200	0%		
Zhang [9]	Arg1628Pro	Chinese	1059	7.2%		
Kim [10]	Gly2385Arg Arg1628Pro	Korean Korean	1345	8.9% 0.8%		
Johnson [19]	Gly2019Ser	North American	79		2.5%	

(continued)

Table 2.1 (continued)

Reference	Mutation	Ethnicity	Number of participants	Sporadic PD	Familial PD	Comments
Saunders-Pullman [20]	Gly2019Ser	North American (Hispanics, non-Hispanic, non-Jewish Caucasians)	104	2.9%		
Cilia [15]	Gly2019Ser, Arg1441Gly, Arg1441Cys, Arg1441His, Ile2020Thr	Ghana	100		0%	

patients is 14.8%. The frequency ranged from 10.6% of sporadic AJ PD patients to 26% of familial PD, but only 2.7% of non-AJ cases [7]. Alcalay et al. [8] reported that 97 of 488 (19.9%) AJ with PD, excluding carriers of *GBA* mutations, recruited in three medical centers in New York and Tel Aviv were carriers of a *LRRK2* mutation. This cohort contained both familial and nonfamilial PD subjects, but *LRRK2* carriers had a twofold risk of having a first-degree relative with PD compared to noncarriers (39.1% vs. 20.5%, respectively). In other populations worldwide, mutations in the *LRRK2* gene are less common. However, *LRRK2* variants with low PD penetrance are common in East Asian populations. In a cohort of 600 Chinese PD patients, Zhang et al. [9] reported that approximately 7% of PD patients are carriers of the Arg1628Pro *LRRK2* variant compared to 2.4% of controls. Kim et al. [10] reported a twofold increase in Gly2385Arg *LRRK2* variant carrier rate (approximately 9%) in a cohort of 923 Korean PD patients compared to controls.

Gender Distribution

While idiopathic PD is more common in men, gender distribution is more evenly distributed in *LRRK2* PD. A recently published meta-analysis described 50 studies of *LRRK2* PD where information on sex was available. A total of 1080 *LRRK2* PD patients were identified. The male to female ratio was 1.02:1.00 (50.6% men and 49.4% women), concluding that *LRRK2* PD lacks a sex effect [21]. Some studies report equal gender distribution among *LRRK2* carriers [8], while others report that *LRRK2* mutations were more common in men [7, 22, 23]. In contrast, Marras et al. [24] reported more women than men were carriers of the *LRRK2* mutation. High female frequency of *LRRK2* carriers compared to noncarriers was also reported in an Italian cohort of 2523 unrelated consecutive PD patients [15]. In conclusion, as opposed to idiopathic PD, *LRRK2* PD is probably *not* more common in men. Whether mutations are slightly more penetrant in women or whether PD risk among carriers is similar across genders remains to be studied.

Age at Motor Onset

Many studies have reported data on the age of onset of motor symptoms in *LRRK2* PD. Disease age of onset is highly variable and has been reported in patients in their 30s as well as after the age of 80 [25]. There are conflicting reports on the effect of *LRRK2* mutations on age of disease onset. In some studies, age of disease onset is similar in *LRRK2* PD and in idiopathic PD and has been shown to be between 52 and 60 years [8, 24–26]; other studies report younger age of disease onset in *LRRK2* carriers [27]. Healy et al. [3] reported an average age of disease onset of 58.1 years in *LRRK2* carriers, which was lower by approximately 3 years than in cases of idiopathic PD at the Queen Square Brain Bank. In this study, 8% of *LRRK2* mutation carriers developed symptoms of PD before the age of 40 (young-onset PD), compared to 4% of idiopathic PD patients.

Penetrance

LRRK2 mutations are associated with autosomal dominant PD with incomplete penetrance. The penetrance varies among different mutations and variants. For example, penetrance is reportedly very low with the Gly2385Arg variant and very high among carriers of the Ile2020Thr mutation. The penetrance of the Gly2019Ser mutation is very controversial, and different studies have reported penetrance estimates ranging from 24 to 100% risk (see Table 2.2) [3]. The reason for such a wide range of penetrance estimation is unclear. Some of the disparity in penetrance estimation may reflect differences in population ethnicity, biases in patient recruitment (familial PD vs. unrelated PD patients' relatives), differences in statistical methods, and differences between *LRRK2* mutations. A recently published paper by Marder et al. [28] estimated the penetrance of PD among Gly2019Ser mutation carriers, in

Table 2.2 *LRRK2* Gly2019Ser penetrance estimates for Parkinson's disease

Author	Penetrance at age 60	Penetrance at age 80	Methods of calculating penetrance
Marder [28]		26%	Kin-cohort method
Lesage [5]	33%	100%	Two highly selected familial autosomal dominant
Healy [3]	28%	74%	In 133 families, mainly familial PD cases
Latourelle [29]	30%	55%	Group of selected familial PD
Goldwurm [30]	15%	32%	Family members of unrelated PD patients
Goldwurm [31]	12%	33%	Kin-cohort study of 24 families of unrelated PD patients
Clark [26]	12%	24%	28 families of unrelated PD patients
Trinh [32]	50%	100%	Kin-cohort analysis

2270 relatives of PD patients from New York and Israel, at 26% by age of 80 using the kin-cohort method. This risk was almost threefold higher for the *LRRK2* carriers than in relatives predicted to be noncarriers of the Gly2019Ser mutation. Healy et al. [3] reported PD risk of 28% at age 59 and 74% at the age of 79. Currently, it is not possible to predict who among unaffected Gly2019Ser carriers will develop PD in the future.

Homozygous Carriers

Mutations in the *LRRK2* gene can cause autosomal dominant PD. Nearly all reported *LRRK2* PD cases have been heterozygote carriers of one of the mutations associated with PD. In our review we found 32 reported cases of homozygote carriers of the Gly2019Ser mutation. The phenotype of homozygous carriers is similar to that of heterozygote carriers, with regard to age of onset, Unified Parkinson's Disease Rating Scale (UPDRS), Montreal Cognitive Assessment (MoCA), Geriatric Depression Scale (GDS), and non-motor symptoms [8, 33].

Motor Symptoms

Motor Phenotype

Given the heterogeneity of PD, on an individual level, it is often impossible to distinguish a *LRRK2* carrier from a noncarrier. Carriers have a similar response to medical and surgical treatment and have similar rates of side effects and complications [3, 15, 24, 34, 35]. The largest report to date regarding the clinical features of 356 patients with *LRRK2* Gly2019Ser PD stated that the core features were asymmetric parkinsonism with tremor, bradykinesia, and rigidity that responded to dopamine replacement therapy. According to this study, *LRRK2* PD is indistinguishable from idiopathic PD [3], and at some point during the course of the disease, 93% of patients will develop all the cardinal motor symptoms of PD, including rest tremor, rigidity, and bradykinesia. Some reports suggest that tremor as a presenting symptom in *LRRK2* PD was more common than in idiopathic PD, affecting 63–64% compared to 39–52%, respectively [3, 24]. This observation was suggested also by a study focusing on tremor-dominant PD patients [36]. In contrast, other studies report that tremor is less frequent in *LRRK2* PD compared to noncarriers [8, 22]. Alcalay et al. [22] described the motor phenotype of 691 PD patients with early-onset PD (age of onset younger than 51 years). Of these, 26 cases were Gly2019Ser carriers. Gly2019Ser carriers were more likely to manifest the postural instability and gait difficulty (PIGD) phenotype than the tremor-predominant (TD) motor phenotype. This observation was similarly seen in another study [8] reporting the phenotype of 553 AJ PD patients, of whom 140 were *LRRK2* carriers. *LRRK2* carriers were more likely to have a PIGD phenotype, report their first symptom in the lower extremities, and have persistent levodopa response for more than 5 years. Given the high prevalence of *GBA* mutations among AJ PD patients, in both studies, participants were screened for *GBA* mutations, and *GBA* carriers were excluded from the analysis. *LRRK2* mutation status was not associated with performance on the UPDRS part III (motor part) or presence of dyskinesia. A recent study reported in a Chinese PD population that the *LRRK2* Gly2385Arg variant was associated with motor fluctuations only in women [37]. Another study [38] compared gait and mobility in 50 patients with Gly2019Ser *LRRK2* PD to 50 noncarrier PD patients and found that *LRRK2* PD cases had greater gait variability and less consistent walking patterns than idiopathic PD. Gait parameters, assessed in three walking conditions (usual walking, dual tasking, and fast walking), were quantified by an accelerometer: In all three walking conditions, gait variability was larger, and the walking pattern was less consistent among *LRRK2* PD compared to idiopathic PD. The *LRRK2* carriers also took longer to complete the timed up and go (TUG) task and were more likely to report having fallen in the previous year. This study reported that *LRRK2* carriers were more associated with the PIGD subtype. Marras et al. [24] also reported gait disorder to be more common in *LRRK2* PD compared to idiopathic PD.

Time to first fall (as a marker of motor disease severity) was reported to be longer in patients with a mutation in *LRRK2* and was longer than in the Queen Square Brain Bank (QSBB) series [3].

Response to Treatment and Development of Motor Complications

The response to treatment with levodopa and development of motor complications in *LRRK2* are generally similar to idiopathic PD, although there is controversy in the literature. Almost all *LRRK2* carriers are reported to have good response of symptoms to levodopa similar to idiopathic PD patients [3, 24]. Drug-induced dyskinesia incidence is similar in *LRRK2* carriers and idiopathic PD, but the time to onset of dyskinesia in *LRRK2* carriers was longer by almost 3 years on average (8.4 years in carriers compared to 5.6 years in noncarriers) [3]. Yahalom et al. [34] reported that the prevalence of levodopa-induced dyskinesia (LID) and the mean duration of therapy from levodopa initiation to the development of LID do not differ between *LRRK2* mutation carriers and noncarriers in a cohort of 349 patients from Israel. Craig et al. [39] further demonstrated that the motor features in *LRRK2* PD and idiopathic PD do not differ in UPDRS scores, frequency of motor symptoms, or levodopa equivalent dose of treatment.

Rate of Disease Progression

Rate of disease progression in *LRRK2* patients was also described; however, the vast majority of studies are either retrospective or cross-sectional. A retrospective study, which compared the motor progression of AJ PD patients with and without *LRRK2* Gly2019Ser mutation, reported no difference in the time of progression to Hoehn and Yahr (HY) stage 3 (i.e., motor instability) [27]. There are still no published longitudinal studies describing rate of disease progression. Such studies are needed and will provide valuable information.

Deep Brain Stimulation

In the past 20 years, deep brain stimulation (DBS) has become an important treatment option for PD patients with motor complications and medication refractory tremor. In their review, Healy et al. [3] commented that 22 patients with *LRRK2* PD underwent stereotactic functional neurosurgery. Eighteen of them underwent subthalamic DBS, three had pallidotomy, and one had thalamotomy. The mean time from PD onset to surgery was 11.4 years, and the indications were usually motor fluctuations and dyskinesia (similar to idiopathic PD patients). It is estimated that *LRRK2* carriers are good candidates for DBS given the slower cognitive decline and the reports on dyskinesia [40]. Because of the frequency of Gly2019Ser mutations and the fact that only a fraction of PD patients undergo DBS, the reports on DBS treatment in carriers are limited. Two small studies have been published. One study including 13 *LRRK2* carriers and 26 noncarriers reported no difference in DBS outcomes after follow-up of 3 years post-DBS surgery [35]. The second study, including 15 carriers and 12 noncarriers, reported similar Hoehn and Yahr outcomes, but better improvement on the UPDRS-III (motor exam) among *LRRK2* carriers when patients were "off" medication and "on" DBS [41].

Cognition

Several studies compared the cognitive function of *LRRK2* carriers to idiopathic PD. Studies of cognition are prone to epidemiological challenges, as cognitive performance is highly correlated with level of education, age, and longer PD duration. Differences between carriers and noncarriers are subtle and not always identified by screening tests such as the Montreal Cognitive Assessment (MoCA) and Mini-Mental State Exam (MMSE) tests.

In contrast, the following studies support the notion of milder cognitive changes in *LRRK2* PD patients compared to idiopathic PD patients:

A study assessing the neuropsychological performance in *LRRK2* Gly2019Ser PD patients and noncarrier PD patients from three movement disorder centers in New York and Tel Aviv reported that carriers performed better than noncarriers in attention (Stroop word reading test), executive function (Stroop interference test), and language (category fluency) [46]. Srivatsal et al. [47] also reported that *LRRK2* mutation carriers have better performance on certain cognitive tests, as well as lower rates of dementia compared to idiopathic PD patients. This study was conducted in eight sites and 1447 participants underwent a battery of seven cognitive tests. Twenty-nine of them had *LRRK2* mutations (24 with Gly2019Ser mutation and 5 with Arg1441Cys mutation). Mutation carriers demonstrated better performance in the MMSE and Letter-Number Sequencing test and had significantly lower frequency of dementia (4% compared to 19.6% in noncarriers). The motor UPDRS scores were also better in the mutation carriers. This study suggests that *LRRK2* carriers might have an overall milder disease course, although these findings require replication.

In addition, Somme et al. [48] described that *LRRK2* PD patients showed less impairment on scales for general cognition (Mattis Dementia Rating Scale) and episodic verbal memory (Rey's auditory verbal learning test, immediate recall, and delayed recall) and had less apathy and hallucinations compared to idiopathic PD.

Healy et al. [3] reported that the proportion of PD patients that develop cognitive impairment within 2 years of symptom onset is less than half in the *LRRK2* carriers than in idiopathic PD. Aasly et al. [49] also reported the *LRRK2* carriers in a Norwegian cohort have only mild cognitive changes even after many years of disease. Cognitive studies are summarized in Table 2.3

Depression/Anxiety

Findings on depression and anxiety among carriers with PD have been inconsistent. Several studies reported no difference in the Geriatric Depression Scale (GDS) between *LRRK2* carriers and noncarrier PD patients [8, 38]. Alcalay et al. [8] reported similar GDS scores in 97 *LRRK2* PD compared to 391 idiopathic PD in a cohort of AJ. Craig et al. [39] also reported no difference in the frequency of depression and anxiety between carriers and noncarriers. Shanker et al. [44] reported no difference in lifetime affective disorders between *LRRK2* carriers and noncarriers, but did report a trend toward a greater risk of premorbid mood disorder in the carriers.

On the other hand, a study comparing 25 *LRRK2* PD patients to 84 idiopathic PD patients from four movement disorder centers in Germany, Canada, the USA, and Brazil reported higher Beck Depression Inventory scores in *LRRK2* carriers but no difference in the State-Trait Anxiety Inventory (STAI) [24]. Belarbi et al. [45] also reported higher frequency of depression in the *LRRK2* carriers than in noncarriers. Goldwurm et al. [25] reported that 69% of *LRRK2* carriers in their cohort were depressed according to the Hamilton Depression Rating Scale scores.

Table 2.3 Studies assessing differences between *LRRK2* PD and idiopathic PD on cognitive tasks

Reference	Cognitive task	Findings
Cognitive screens		
Alcalay et al. [8]	MoCA	No significant differences
Alcalay et al. [22, 42]	MMSE	No significant differences
Trinh et al. [32]	MMSE, MoCA, FAB	No significant differences
Lesage et al. [5]	MMSE	Mildly worse performance among carriers
Neuropsychological testing		
Ben Sassi et al. [43]	MMSE, MoCA, FAB	No significant differences
Shanker et al. [44]	FAB	No significant differences
Mirelman et al. [38]	MoCA, trail making tests A and B, verbal fluency, digit span, and Stroop test	No significant differences
Belarbi et al. [45]	Neuropsychological battery	No significant differences in cognitive performance. Higher rates of depression and hallucinations among carriers
Alcalay et al. [46]	Neuropsychological battery	*LRRK2* Gly2019Ser PD patients performed better than idiopathic PD patients on attention (Stroop word reading test), executive function (Stroop interference test), and language (category fluency) tasks.
Srivatsal et al. [47]	Neuropsychological battery, MMSE, clinical diagnosis of dementia	*LRRK2* Gly2019Ser PD patients performed better than idiopathic PD patients on the MMSE and Letter-Number Sequencing test and had significantly lower frequency of dementia (4% compared to 19.6% in noncarriers)
Somme et al. [48]	Neuropsychological battery and the Mattis Dementia Rating Scale	*LRRK2* PD patients showed less impairment on scales for general cognition (Mattis Dementia Rating Scale) and episodic verbal memory (Rey's auditory verbal learning test, immediate recall, and delayed recall)

Olfaction

The majority of studies reported less olfactory disturbance in *LRRK2* PD compared to idiopathic PD. A study from four movement disorder centers in Europe and the USA comparing olfaction with the Brief Smell Identification Test (B-SIT) reported significantly better olfactory identification in *LRRK2* PD subjects compared to idiopathic PD [24]. Saunders-Pullman et al. [50] also reported that *LRRK2* PD patients had significantly better University of Pennsylvania Smell Identification

Test (UPSIT) scores than idiopathic PD patients, by an average of six points, but had lower UPSIT scores than healthy controls. An additional study by Saunders-Pullman et al. [51] recently reaffirmed that *LRRK2* PD had better olfactory scores compared to idiopathic PD. Silveira-Moriyama et al. [52] reported higher scores on the "Sniffin' sticks" and better odor identification in *LRRK2* PD compared to idiopathic PD, but lower scores compared to controls. Johansen et al. [53] showed that sporadic PD had significantly lower scores in olfaction (assessed by B-SIT) compared with *LRRK2* PD. Craig et al. [39] showed that UPSIT scores in *LRRK2* Gly2019Ser PD were higher than those in idiopathic PD by an average of five points.

However, some studies reported similarly impaired olfaction in *LRRK2* PD and idiopathic PD in mean UPSIT scores [54] and in the "Sniffin' sticks" test [32]. Silveira-Moriyama et al. [54] reported that 19 *LRRK2* PD patients did not differ in their UPSIT scores from 145 idiopathic PD patients in a cohort from Portugal and the UK and that *LRRK2* PD patients' UPSIT scores were lower than those of healthy controls. Valldeoriola et al. [55] reported that UPSIT scores in 14 *LRRK2* PD patients do not differ from those of 14 idiopathic PD patients in a Spanish cohort. Healy et al. [3] reported that abnormal olfaction was common and found in 51% of *LRRK2* PD patients after an average disease duration of 5 years.

Sleep Disorders

Sleep disorders are frequently seen in *LRRK2* PD. Goldwurm et al. [25] reported that 56% of *LRRK2* PD patients had sleep disorders mainly manifesting as repeated awakenings and insomnia. Healy et al. [3] reported that 69% of *LRRK2* Gly2019Ser patients had sleep disturbances, but there was no significant difference from idiopathic PD patients. The most common sleep disturbances were insomnia and sleep fragmentation, but rapid eye movement (REM) sleep behavior disorder (RBD) and restless legs syndrome were also noted. When *LRRK2* carriers were compared to noncarriers in three different cohorts [8, 32, 56], carriers were less likely to report RBD. This observation was not observed by others [39, 45].

Autonomic Dysfunction

Most reports on autonomic dysfunction did not observe a difference between carriers and noncarriers. Gaig et al. [39] reported that dysautonomic symptoms in *LRRK2* Gly2019Ser PD were not significantly different from those in idiopathic PD in a cohort of 1251 Spanish PD patients. They described a cohort of 33 *LRRK2* PD patients and compared them to idiopathic PD patients. Hao et al. [57] and Li et al. [4] found non-motor symptoms to be very common in PD patients, but did not find a difference in two different studies using the non-motor symptom questionnaire

(NMSQ) in a Chinese cohort comparing *LRRK2* PD (with Arg1628Pro or Gly2385Arg variants) and idiopathic PD. Alcalay et al. also reported no difference in NMS questionnaires in *LRRK2* carriers and noncarriers [8] in a cohort of AJ PD patients. Trinh et al. [32] reported no differences between *LRRK2* carriers and noncarriers in the Scales for Outcomes in Parkinson's Disease-Autonomic (SCOPA-AUT) questionnaires in a large cohort of Tunisian PD patients. Healy et al. [3] reported that 48% of *LRRK2* carriers had constipation, 28% were affected with urinary symptoms (mainly frequency and urge incontinence), and 11% of men reported erectile dysfunction; however, there was no comparison to idiopathic PD in this study.

Vision

Marras et al. reported worse performance on the 100-hue test of color discrimination in 25 *LRRK2* PD patients compared with 84 idiopathic PD patients from four movement disorder centers in Europe, Brazil, Canada, and the USA [24].

Cancer

Cancer frequency among idiopathic PD patients is probably lower than in the general population with the exception of melanomas [58, 59] and possibly breast cancer. Epidemiological studies of cancer frequency among *LRRK2* carriers face many challenges, including small numbers, mechanism by which cancer history is collected, and selection bias (those with cancer may have died and are not available to provide history). In contrast to idiopathic PD patients, the Gly2019Ser *LRRK2* mutation may not be overrepresented in patients with melanoma [60]. A study from three movement disorder centers in Israel reported a higher prevalence (odds ratio of 3.38) of non-skin cancers in 79 AJ *LRRK2* PD patients (carriers of the Gly2019Ser mutation) compared to 401 noncarrier PD patients [61]. Cancer history was obtained by personal interview and reviewing patients' files. The most common non-skin cancers were breast and prostate cancers. The study did not include a control group, so it remains unclear whether cancer frequency among *LRRK2* carriers is higher than in controls without PD. Another study reported a higher (almost threefold) prevalence of non-skin cancers (primarily breast and prostate cancers) as seen by chart reviews in 32 Gly2019Ser PD mutation carriers compared to 132 idiopathic PD in a cohort of 163 AJ [62]. Sixty-seven percent of the *LRRK2* carriers in that study were diagnosed with cancer before the onset of PD, whereas only 40% of noncarriers developed their first non-skin cancer before onset of PD.

Agalliu et al. [63] reported in a recently published multinational study from five centers in Europe, Israel, and the USA that Gly2019Ser *LRRK2* mutation carriers have an overall increased risk of cancer compared to noncarriers, especially for hormone-related cancer and breast cancer in women. Gly2019Ser *LRRK2* mutation carriers had a 1.62-fold risk of developing non-skin cancers and a 2.3-fold risk of developing breast cancer compared to noncarriers. The cancer history was self-reported by the patients and confirmed by medical record review and tumor registry databases.

Ruiz-Martinez et al. [64] reported on cancer prevalence using population-based cancer registry in 732 Spanish PD patients with Arg1441Gly or Gly2019Ser mutations and in idiopathic PD. In this study cancer prevalence did not differ between PD Gly2019Ser carriers, PD Arg1441Gly carriers, and PD noncarriers, with the exception of a high prevalence of hematological cancers in the Arg1441Gly group.

The underlying biological mechanism that links the *LRRK2* Gly2019Ser mutation and cancer remains largely unknown. Proposed mechanisms include mitogenic kinase activity (given that Gly2019Ser mutations increase LRRK2 activity) and possible modification of the immune system. The gain-of-function carcinogenic theory is supported by the findings that amplification and overexpression of the *LRRK2* gene have been reported in papillary renal and thyroid carcinomas [65].

Pathology

The neuropathology of *LRRK2* PD was previously reviewed [66, 67]. Briefly, the most common pathology in autopsies of Gly2019Ser carriers is Lewy body pathology. However, as in the case of the clinical studies of *LRRK2* mutation carriers, the literature may be biased because of study design. First, the vast majority of Gly2019Ser autopsies were obtained from patients with neurodegenerative disorders in spite of an estimated penetrance of 30%. This is because brain banks are skewed toward collection from patients with neurodegeneration rather than healthy controls. Among autopsies of patients with parkinsonism, there may be additional bias. On one hand, in many cases, only Lewy body brain banks were genotyped in order to estimate Gly2019Ser mutation frequency in these banks; on the other hand, there is likely a publication bias where Gly2019Ser autopsies with Lewy body pathology are less likely to be reported as compared with brains with unusual pathology. For example, the Columbia University Brain Bank has six Gly2019Ser autopsies. All manifest Lewy body pathology, but one of them also contains tau inclusion bodies consistent with progressive supranuclear palsy pathology. Most of the non-Lewy body brain bank was not genotyped for the Gly2019Ser mutation. When the clinical features of those with Lewy body pathology were compared to those with other pathology (e.g., neuronal loss in substantia nigra), Kalia et al. [67] reported that the presence of Lewy bodies was associated with non-motor features of PD, while cases without Lewy body were more likely to manifest a more pure motor deficit. Carriers of mutations other than Gly2019Ser are less likely to have Lewy bodies, which were present only in 35% of the reported autopsies (8 out of

23). No autopsy with the common Asian variant Gly2385Arg has been reported to date.

In summary, pathological features of different *LRRK2* mutations may vary considerably. The majority of G2019S carriers with PD had Lewy body containing neurons, which may also correlate with the presence (or lack thereof) of non-motor symptoms. Autopsies of mutation carriers without PD are sparse and may be extremely informational for our understanding of the pathogenesis of the *LRRK2* mutations and their incomplete penetrance.

Symptoms in Non-manifesting LRRK2 *Carriers*

A few groups studied non-manifesting *LRRK2* mutation carriers. Study design often includes analyses of families with *LRRK2* mutations, where family member carriers and noncarriers are compared. To date, there are no studies which can clinically distinguish carriers without PD from their noncarrier family members.

A recently published study [68] assessed non-motor symptoms in 256 family relatives of AJ PD patients (non-manifesting Gly2019Ser *LRRK2* mutation carriers and noncarriers of the Gly2019Ser mutation). Non-manifesting carriers had higher non-motor symptom score on the non-motor symptom (NMS) questionnaire (questions relating to constipation and urinary urgency) than noncarriers. Differences between groups were more pronounced with older subject age and also included anxiety and daytime sleepiness. No differences between groups were found in motor scores (UPDRS part III), cognitive function (MoCA), or olfaction (UPSIT) [68].

Saunders-Pullman et al. reported that non-manifesting carriers of the *LRRK2* mutation had lower UPSIT scores compared to healthy controls but higher UPSIT scores compared to *LRRK2* PD patients [50]. Another study by Saunders-Pullman et al. [51] reported that non-manifesting *LRRK2* carriers had higher UPSIT scores than healthy controls, but their olfaction did not differ from first-degree family members of idiopathic PD patients.

Mirelman et al. assessed gait in 52 first-degree relatives of PD patients, in whom 25 were non-manifesting *LRRK2* mutation carriers and 27 were noncarriers. Mutation carriers had subtle gait changes. Although the groups did not differ in gait speed, stride time, or stride length, mutation carriers had altered gait variability (measure of gait consistency and stability) during fast walking and dual tasking [69]. The groups did not differ in the UPDRS part III scores.

Thaler et al. [70] reported that non-manifesting *LRRK2* carriers demonstrated poorer performance on computerized measure of executive functioning (Stoop test score and response time) compared with that of noncarriers. The groups did not differ in their MoCA scores, GDS scores, or UPDRS part III motor scores.

The only notable difference Marras et al. [24] reported between non-manifesting *LRRK2* mutation carriers and noncarriers was a marginally higher frequency of postural and action tremor in the former.

Conclusions and Future Directions

In this review we highlighted the current data available on the demographics, motor and non-motor symptoms of *LRRK2* PD carriers, and symptoms in non-manifesting carriers. *LRRK2* is the most common dominantly inherited genetic factor implicated in late-onset familial and sporadic PD identified to date. Several mutations in the *LRRK2* gene have been shown to be associated with PD. The clinical phenotype of *LRRK2* PD is variable, similar to idiopathic PD, but is probably slightly milder, with less non-motor involvement compared to idiopathic PD patients. There is considerable variability in the literature regarding the epidemiology, penetrance, and symptoms of *LRRK2* PD. This heterogeneity may be partially explained by phenotypic variations of the different mutations in the *LRRK2* gene and the differences in experimental methods applied to identify motor and non-motor symptoms. Development of more sensitive biomarkers for identifying and monitoring motor and non-motor symptom progression is needed. Longitudinal studies and pathological data of *LRRK2* PD will likely help to shed light on the disease characteristics of this intriguing subgroup of PD patients.

Acknowledgement: Dr. Kestenbaum's fellowship at Columbia University was funded by the Parkinson's Disease Foundation. The authors would like to thank Mr. Chris Liong for reviewing and editing the manuscript.

Conflict of Interest The authors declare no conflicts of interest.

References

1. Zimprich A, Biskup S, Leitner P et al (2004) Mutations in LRRK2 cause autosomal-dominant parkinsonism with pleomorphic pathology. Neuron 44(4):601–607
2. Bekris LM, Mata IF, Zabetian CP (2010) The genetics of Parkinson disease. J Geriatr Psychiatry Neurol 23(4):228–242
3. Healy DG, Falchi M, O'Sullivan SS et al (2008) Phenotype, genotype, and worldwide genetic penetrance of LRRK2-associated Parkinson's disease: a case–control study. Lancet Neurol 7(7):583–590
4. Li DW, Gu Z, Wang C et al (2015) Non-motor symptoms in Chinese Parkinson's disease patients with and without LRRK2 G2385R and R1628P variants. J Neural Transm 122(5):661–667
5. Lesage S, Ibanez P, Lohmann E et al (2005) G2019S LRRK2 mutation in French and North African families with Parkinson's disease. Ann Neurol 58(5):784–787
6. Ishihara L, Gibson RA, Warren L et al (2007) Screening for Lrrk2 G2019S and clinical comparison of Tunisian and North American Caucasian Parkinson's disease families. Mov Disord 22(1):55–61
7. Orr-Urtreger A, Shifrin C, Rozovski U et al (2007) The LRRK2 G2019S mutation in Ashkenazi Jews with Parkinson disease: is there a gender effect? Neurology 69(16):1595–1602
8. Alcalay RN, Mirelman A, Saunders-Pullman R et al (2013) Parkinson disease phenotype in Ashkenazi Jews with and without LRRK2 G2019S mutations. Mov Disord 28(14): 1966–1971

9. Zhang Z, Burgunder JM, An X et al (2009) LRRK2 R1628P variant is a risk factor of Parkinson's disease among Han-Chinese from mainland China. Mov Disord 24(13): 1902–1905
10. Kim JM, Lee JY, Kim HJ et al (2010) The LRRK2 G2385R variant is a risk factor for sporadic Parkinson's disease in the Korean population. Parkinsonism Relat Disord 16(2):85–88
11. Hulihan MM, Ishihara-Paul L, Kachergus J et al (2008) LRRK2 Gly2019Ser penetrance in Arab-Berber patients from Tunisia: a case–control genetic study. Lancet Neurol 7(7): 591–594
12. Ferreira JJ, Guedes LC, Rosa MM et al (2007) High prevalence of LRRK2 mutations in familial and sporadic Parkinson's disease in Portugal. Mov Disord 22(8):1194–1201
13. Pulkes T, Papsing C, Thakkinstian A et al (2014) Confirmation of the association between LRRK2 R1628P variant and susceptibility to Parkinson's disease in the Thai population. Parkinsonism Relat Disord 20(9):1018–1021
14. Sierra M, Gonzalez-Aramburu I, Sanchez-Juan P et al (2011) High frequency and reduced penetrance of LRRK2 G2019S mutation among Parkinson's disease patients in Cantabria (Spain). Mov Disord 26(13):2343–2346
15. Cilia R, Siri C, Rusconi D et al (2014) LRRK2 mutations in Parkinson's disease: confirmation of a gender effect in the Italian population. Parkinsonism Relat Disord 20(8):911–914
16. Yescas P, Lopez M, Monroy N et al (2010) Low frequency of common LRRK2 mutations in Mexican patients with Parkinson's disease. Neurosci Lett 485(2):79–82
17. Sanyal J, Sarkar B, Ojha S, Banerjee TK, Ray BC, Rao VR (2010) Absence of commonly reported leucine-rich repeat kinase 2 mutations in Eastern Indian Parkinson's disease patients. Genet Test Mol Biomarkers 14(5):691–694
18. Chien HF, Figueiredo TR, Hollaender MA et al (2014) Frequency of the LRRK2 G2019S mutation in late-onset sporadic patients with Parkinson's disease. Arq Neuropsiquiatr 72(5):356–359
19. Johnson J, Paisan-Ruiz C, Lopez G et al (2007) Comprehensive screening of a North American Parkinson's disease cohort for LRRK2 mutation. Neurodegener Dis 4(5):386–391
20. Saunders-Pullman R, Cabassa J, San Luciano M et al (2011) LRRK2 G2019S mutations may be increased in puerto ricans. Mov Disord 26(9):1772–1773
21. Gan-Or Z, Leblond CS, Mallett V, Orr-Urtreger A, Dion PA, Rouleau GA (2015) LRRK2 mutations in Parkinson disease; a sex effect or lack thereof? a meta-analysis. Parkinsonism Relat Disord 21:778–782
22. Alcalay RN, Mejia-Santana H, Tang MX et al (2009) Motor phenotype of LRRK2 G2019S carriers in early-onset Parkinson disease. Arch Neurol 66(12):1517–1522
23. Nabli F, Ben Sassi S, Amouri R, Duda JE, Farrer MJ, Hentati F (2015) Motor phenotype of LRRK2-associated Parkinson's disease: a tunisian longitudinal study. Mov Disord 30(2): 253–258
24. Marras C, Schule B, Munhoz RP et al (2011) Phenotype in parkinsonian and nonparkinsonian LRRK2 G2019S mutation carriers. Neurology 77(4):325–333
25. Goldwurm S, Zini M, Di Fonzo A et al (2006) LRRK2 G2019S mutation and Parkinson's disease: a clinical, neuropsychological and neuropsychiatric study in a large Italian sample. Parkinsonism Relat Disord 12(7):410–419
26. Clark LN, Wang Y, Karlins E et al (2006) Frequency of LRRK2 mutations in early—and late-onset Parkinson disease. Neurology 67(10):1786–1791
27. Yahalom G, Orlev Y, Cohen OS et al (2014) Motor progression of Parkinson's disease with the leucine-rich repeat kinase 2 G2019S mutation. Mov Disord 29(8):1057–1060
28. Marder K, Wang Y, Alcalay RN et al (2015) Age-specific penetrance of LRRK2 G2019S in the Michael J. Fox Ashkenazi Jewish LRRK2 Consortium. Neurology 85(1):89–95
29. Latourelle JC, Sun M, Lew MF et al (2008) The Gly2019Ser mutation in LRRK2 is not fully penetrant in familial Parkinson's disease: the GenePD study. BMC Med 6:32
30. Goldwurm S, Zini M, Mariani L et al (2007) Evaluation of LRRK2 G2019S penetrance: relevance for genetic counseling in Parkinson disease. Neurology 68(14):1141–1143

31. Goldwurm S, Tunesi S, Tesei S et al (2011) Kin-cohort analysis of LRRK2-G2019S penetrance in Parkinson's disease. Mov Disord 26(11):2144–2145
32. Trinh J, Amouri R, Duda JE et al (2014) Comparative study of Parkinson's disease and leucine-rich repeat kinase 2 p.G2019S parkinsonism. Neurobiol Aging 35(5):1125–1131
33. Ishihara L, Warren L, Gibson R et al (2006) Clinical features of Parkinson disease patients with homozygous leucine-rich repeat kinase 2 G2019S mutations. Arch Neurol 63(9): 1250–1254
34. Yahalom G, Kaplan N, Vituri A et al (2012) Dyskinesias in patients with Parkinson's disease: effect of the leucine-rich repeat kinase 2 (LRRK2) G2019S mutation. Parkinsonism Relat Disord 18(9):1039–1041
35. Greenbaum L, Israeli-Korn SD, Cohen OS et al (2013) The LRRK2 G2019S mutation status does not affect the outcome of subthalamic stimulation in patients with Parkinson's disease. Parkinsonism Relat Disord 19(11):1053–1056
36. Clarimon J, Pagonabarraga J, Paisan-Ruiz C et al (2008) Tremor dominant parkinsonism: clinical description and LRRK2 mutation screening. Mov Disord 23(4):518–523
37. Gao C, Pang H, Luo XG, Ren Y, Shang H, He ZY (2013) LRRK2 G2385R variant carriers of female Parkinson's disease are more susceptible to motor fluctuation. J Neurol 260(11): 2884–2889
38. Mirelman A, Heman T, Yasinovsky K et al (2013) Fall risk and gait in Parkinson's disease: the role of the LRRK2 G2019S mutation. Mov Disord 28(12):1683–1690
39. Gaig C, Vilas D, Infante J et al (2014) Nonmotor symptoms in LRRK2 G2019S associated Parkinson's disease. PLoS One 9(10), e108982
40. Pal G, Hall D, Ouyang B, Phelps J, Alcalay RN (2016) Genetic and clinical predictors of deep brain stimulation in young-onset Parkinson's disease. Mov Disord Clin Pract 3:285–287, In press
41. Sayad M, Zouambia M, Chaouch M et al (2016) Greater improvement in LRRK2 G2019S patients undergoing subthalamic nucleus deep brain stimulation compared to non-mutation carriers. BMC Neurosci 17(1):6
42. Alcalay RN, Mejia-Santana H, Tang MX et al (2010) Self-report of cognitive impairment and mini-mental state examination performance in PRKN, LRRK2, and GBA carriers with early onset Parkinson's disease. J Clin Exp Neuropsychol 32(7):775–779
43. Ben Sassi S, Nabli F, Hentati E et al (2012) Cognitive dysfunction in Tunisian LRRK2 associated Parkinson's disease. Parkinsonism Relat Disord 18(3):243–246
44. Shanker V, Groves M, Heiman G et al (2011) Mood and cognition in leucine-rich repeat kinase 2 G2019S Parkinson's disease. Mov Disord 26(10):1875–1880
45. Belarbi S, Hecham N, Lesage S et al (2010) LRRK2 G2019S mutation in Parkinson's disease: a neuropsychological and neuropsychiatric study in a large Algerian cohort. Parkinsonism Relat Disord 16(10):676–679
46. Alcalay RN, Mejia-Santana H, Mirelman A et al (2015) Neuropsychological performance in LRRK2 G2019S carriers with Parkinson's disease. Parkinsonism Relat Disord 21(2): 106–110
47. Srivatsal S, Cholerton B, Leverenz JB et al (2015) Cognitive profile of LRRK2-related Parkinson's disease. Mov Disord 30(5):728–733
48. Somme JH, Molano Salazar A, Gonzalez A et al (2015) Cognitive and behavioral symptoms in Parkinson's disease patients with the G2019S and R1441G mutations of the LRRK2 gene. Parkinsonism Relat Disord 21(5):494–499
49. Aasly JO, Toft M, Fernandez-Mata I et al (2005) Clinical features of LRRK2-associated Parkinson's disease in central Norway. Ann Neurol 57(5):762–765
50. Saunders-Pullman R, Stanley K, Wang C et al (2011) Olfactory dysfunction in LRRK2 G2019S mutation carriers. Neurology 77(4):319–324
51. Saunders-Pullman R, Mirelman A, Wang C et al (2014) Olfactory identification in LRRK2 G2019S mutation carriers: a relevant marker? Ann Clin Transl Neurol 1(9):670–678

52. Silveira-Moriyama L, Munhoz RP, de J Carvalho M et al (2010) Olfactory heterogeneity in LRRK2 related Parkinsonism. Mov Disord 25(16):2879–2883
53. Johansen KK, Waro BJ, Aasly JO (2014) Olfactory dysfunction in sporadic Parkinson's disease and LRRK2 carriers. Acta Neurol Scand 129(5):300–306
54. Silveira-Moriyama L, Guedes LC, Kingsbury A et al (2008) Hyposmia in G2019S LRRK2-related parkinsonism: clinical and pathologic data. Neurology 71(13):1021–1026
55. Valldeoriola F, Gaig C, Muxi A et al (2011) 123I-MIBG cardiac uptake and smell identification in parkinsonian patients with LRRK2 mutations. J Neurol 258(6):1126–1132
56. Pont-Sunyer C, Iranzo A, Gaig C et al (2015) Sleep disorders in parkinsonian and nonparkinsonian LRRK2 mutation carriers. PLoS One 10(7), e0132368
57. Hao M, Pan N, Zhang Q, Wang X (2014) Mutant of leucine-rich repeat kinase 2 is not associated with non-motor symptoms in Chinese Parkinson's disease patients. Int J Clin Exp Med 7(8):2253–2257
58. Inzelberg R, Rabey JM, Melamed E et al (2011) High prevalence of malignant melanoma in Israeli patients with Parkinson's disease. J Neural Transm 118(8):1199–1207
59. Bajaj A, Driver JA, Schernhammer ES (2010) Parkinson's disease and cancer risk: a systematic review and meta-analysis. Cancer Causes Control 21(5):697–707
60. Hassin-Baer S, Laitman Y, Azizi E et al (2009) The leucine rich repeat kinase 2 (LRRK2) G2019S substitution mutation. Association with Parkinson disease, malignant melanoma and prevalence in ethnic groups in Israel. J Neurol 256(3):483–487
61. Inzelberg R, Cohen OS, Aharon-Peretz J et al (2012) The LRRK2 G2019S mutation is associated with Parkinson disease and concomitant non-skin cancers. Neurology 78(11):781–786
62. Saunders-Pullman R, Barrett MJ, Stanley KM et al (2010) LRRK2 G2019S mutations are associated with an increased cancer risk in Parkinson disease. Mov Disord 25(15):2536–2541
63. Agalliu I, San Luciano M, Mirelman A et al (2015) Higher frequency of certain cancers in LRRK2 G2019S mutation carriers with Parkinson disease: a pooled analysis. JAMA Neurol 72(1):58–65
64. Ruiz-Martinez J, de la Riva P, Rodriguez-Oroz MC et al (2014) Prevalence of cancer in Parkinson's disease related to R1441G and G2019S mutations in LRRK2. Mov Disord 29(6):750–755
65. Looyenga BD, Furge KA, Dykema KJ et al (2011) Chromosomal amplification of leucine-rich repeat kinase-2 (LRRK2) is required for oncogenic MET signaling in papillary renal and thyroid carcinomas. Proc Natl Acad Sci U S A 108(4):1439–1444
66. Poulopoulos M, Levy OA, Alcalay RN (2012) The neuropathology of genetic Parkinson's disease. Mov Disord 27(7):831–842
67. Kalia LV, Lang AE, Hazrati LN et al (2015) Clinical correlations with Lewy body pathology in LRRK2-related Parkinson disease. JAMA Neurol 72(1):100–105
68. Mirelman A, Alcalay RN, Saunders-Pullman R et al (2015) Nonmotor symptoms in healthy Ashkenazi Jewish carriers of the G2019S mutation in the LRRK2 gene. Mov Disord 30:981–986
69. Mirelman A, Gurevich T, Giladi N, Bar-Shira A, Orr-Urtreger A, Hausdorff JM (2011) Gait alterations in healthy carriers of the LRRK2 G2019S mutation. Ann Neurol 69(1):193–197
70. Thaler A, Mirelman A, Gurevich T et al (2012) Lower cognitive performance in healthy G2019S LRRK2 mutation carriers. Neurology 79(10):1027–1032

Part II
Fundamentals of LRRK2 Biology

Chapter 3
LRRK2 Phosphorylation

R. Jeremy Nichols

Abstract Mutations in the leucine-rich repeat kinase 2 (LRRK2) gene were discovered in 2004 and have been found to be the most frequently mutated gene in Parkinson's disease. LRRK2 is a large multi-domain protein with a functional GTPase and kinase domain. The signal transduction pathways in which LRRK2 is dysfunctional in the disease state are only now being resolved, but we do know that LRRK2 is, itself, a substrate of multiple kinases and phosphatases and exists in variable phosphorylated states. Autophosphorylation of LRRK2 can impact GTPase and pathological outcomes. LRRK2 serines (910/935/955/973) are differentially phosphorylated in pathogenic PD mutations and after LRRK2 kinase inhibition. The phosphorylation status of LRRK2 can therefore provide key insight into the mechanisms of kinase dysfunction during disease. This chapter will describe the identification of LRRK2 phosphorylation sites and how phosphoregulation of LRRK2 reveals its own kinase activity and regulates its ubiquitination and localization in vitro, in cells, and in tissues.

Keywords Parkinson's disease • LRRK2 • Synuclein • Posttranslational modification • Kinase • Phosphatase • Phosphorylation • Dephosphorylation • Ubiquitin ligase • Deubiquitinase • Ubiquitination • Deubiquitination • Signal transduction

Introduction

Parkinson's Disease

Parkinson's disease (PD) is a progressive neurodegenerative disease affecting 1–2 % of the population over 65 years of age. There are approximately 60,000 newly diagnosed patients per year in the USA. It is estimated that the prevalence of PD cases worldwide will double by the year 2030 [1, 2]. The increasing disability caused by the progression of disease burdens the patients, their caregivers, as well as society, and without new treatments for disease modifying or prevention of onset, this cost will increase. Classical clinical features of PD include resting tremor,

R.J. Nichols (✉)
The Parkinson's Institute, Sunnyvale, CA 94089, USA
e-mail: jnichols@parkinsonsinstitute.org

H.J. Rideout (ed.), *Leucine-Rich Repeat Kinase 2 (LRRK2)*,
Advances in Neurobiology 14, DOI 10.1007/978-3-319-49969-7_3

bradykinesia, postural instability, and rigidity. PD also exhibits a wide variety of non-motor features such as autonomic dysfunction and dementia. Although the pattern of neuronal loss in PD is well characterized from postmortem studies, the molecular mechanisms of neurodegenerative cell death are still being elucidated. The majority of PD patients suffer from idiopathic disease with no clear etiology. However, approximately 5 % of patients present with familial PD. Furthermore, exposure to a number of environmental toxicants has been shown to increase the risk of PD [3–5]. The protein targets of environmental insults or the products of mutated genes are important therapeutic targets for drug development aimed at disease modification or prevention, a huge unmet need in the PD field.

Mutations in the leucine-rich repeat kinase 2 (LRRK2) gene are the most common cause of inherited PD [6–10], and genome-wide association studies have also identified LRRK2 as a risk factor for sporadic PD [11, 12]. With the association in both familial and idiopathic disease, it is likely that the molecular basis of LRRK2 dysfunction-based PD is similar between the two. Further, the clinical phenotypes of PD caused by LRRK2 mutations are largely indistinguishable from idiopathic disease (reviewed in [13]). Protein kinases mediate the transfer of phosphate to the hydroxyl group of Ser/Thr/Tyr residues of substrate proteins, which, in turn, alters downstream function of those phosphorylated proteins [14, 15]. The activity of kinases themselves is typically mediated by an upstream phosphorylation event, thereby transducing signals from one protein to another. The normal and pathological roles of LRRK2 are not fully understood; however, a consensus is forming around roles in vesicle trafficking and neurotransmitter release along with neuronal outgrowth. This has been solidified by the recent discovery that LRRK2 phosphorylates a subset of Rab GTPases to regulate their interaction with guanine nucleotide dissociation inhibitors [16]. This chapter summarizes the current understanding of the phosphoregulation of LRRK2 and how it is deregulated in pathogenic conditions.

LRRK2 Protein Domain Structure

The LRRK2 gene encodes a large multi-domain protein of 2527 amino acids, with both GTPase and kinase enzymatic domains, Fig. 3.1. The amino terminus [1-1287aa] is dispensable for kinase and GTPase activity [17], but participates in

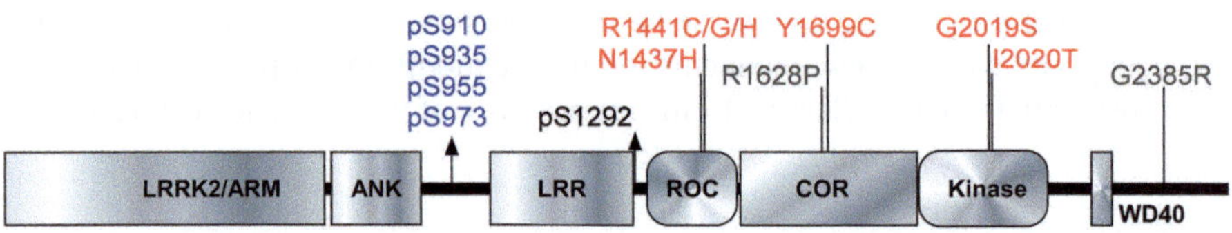

Fig. 3.1 Domain architecture of LRRK2. Domain architecture of the LRRK2 protein is shown with sites of LRRK2 phosphorylation by upstream kinases in blue and the in vivo relevant autophosphorylation site in black. Pathogenic PD-related mutations in LRRK2 are shown in *red*, while risk factor mutations are shown in *gray*

regulation of LRRK2, containing a phosphorylation cluster (described below), an armadillo-like and ankyrin-like repeats, and the namesake leucine-rich repeat domain. The minimal catalytic fragment for kinase activity encompasses the remainder of the protein [1326-2527aa] [17]. This includes an active GTPase domain, termed Ras of complex proteins (Roc), which is juxtaposed to the C-terminal of ROC (COR) domain in a class of enzymes termed ROCO proteins, of which LRRK2 is a member [18, 19]. The adjacent kinase domain bears similarity to mixed lineage kinases which are typically involved in kinase signaling cascades. The carboxy terminus contains a WD40 domain and is essential for kinase activity, where deletion of the last seven amino acids inhibits activity [17, 20, 21].

LRRK2 is likely a dimer with several intramolecular interaction interfaces throughout the multiple domains of the protein, which lends itself to various potential regulatory mechanisms. The GTPase domain of LRRK2 can be purified as an active monomer that binds and hydrolyzes GTP [22]; however, it is thought that the RocCOR domain mediates dimerization of LRRK2 [23–25]. In one proposed model, LRRK2 activation is a result of GTPase activation by nucleotide-dependent dimerization [26]. This model competes with the identification of GTPase-activating proteins (GAPs) and guanine nucleotide exchange factors (GEFs) that interact with LRRK2 to potentially regulate the activity of the LRRK2 GTPase domain, i.e., ArfGAP1 and ARHGEF7 [27–29]. Some studies have shown that the mutations within the GTPase and ROC domain alter LRRK2 kinase activity. If GAPs or GEFs regulate LRRK2 GTPase activity, there is likely an impact on kinase activity [27].

LRRK2 PD-Related Mutations and Effects on Enzyme Activity

LRRK2 activity is derived from functional inputs from throughout the enzyme itself; these could be intramolecular or intermolecular in the case of LRRK2 dimers/oligomers or protein-protein interactions. Many of the LRRK2 missense mutations found to segregate with disease encode substitutions in the catalytic core of the GTPase and kinase domains [N1437H, R1441C/G, Y1699C, G2019S, and I2020T] (Fig. 3.1) [13, 30, 31]. The most common mutation in inherited and idiopathic PD encodes a G2019S substitution and is located in subdomain VII of the kinase domain "DFG" motif [14, 15, 32]. This mutation has been consistently found to increase kinase activity at least two- to threefold [33]. Structural studies with the kinase domain of ROCO4, a *Dictyostelium discoideum* ROCO protein homolog of LRRK2, revealed that an analogous Ser substitution in the Gly of the DFG motif results in a salt bridge between Ser2019 and Q1918 (as predicted in the human orthologue) stabilizing a potential active state of the kinase [26, 34]. I2020T displays decreased activity in some assays and increased activity in others. Ray et al. also noted differences in I2020T activity due to different substrate-dependent readouts of kinase activity [35].

It is hypothesized that the increased kinase activity of LRRK2 PD-related mutations results in hyperphosphorylation, and thus aberrant function, of a yet to be identified substrate. Kinase-inactive mutations of LRRK2 blunt or do not exhibit some detrimental phenotypes of mutant LRRK2 expression such as neurite shortening [36]. It is therefore plausible that inhibiting LRRK2 with small-molecule kinase inhibitors to reduce the kinase activity of LRRK2 would be a potential benefit to patients carrying these mutations in LRRK2. Further, since patients with LRRK2 mutations present clinically, and in most cases pathologically, similar to idiopathic PD, it is also possible that modulation of LRRK2 kinase activity could benefit idiopathic PD as well [13]. One difficulty in relating the observed in vitro change in enzyme activity to downstream pathology is the lack of an in vivo LRRK2 substrate that is validated across laboratories and for which there are methods to detect this substrate phosphorylation. This prevents correlation of pathological outcomes with altered substrate phosphorylation in model systems and PD patient samples.

There are also some discrepancies in the reported effects of mutations in the RocCOR domain on LRRK2 GTPase and kinase activity [37]. The RocCOR GTPase domain mutations exhibit increased GTP-binding activity using immobilized GTP agarose [38] or thin-layer chromatography with LRRK2 purified from ^{32}P-orthophosphate metabolically labeled cells [39, 40]. In careful kinetic analysis, R1441H shows slowed GTP hydrolysis and increased affinity for GTP [22], which could account for the apparent increased association with GTP agarose in similar LRRK2 mutants in other studies. In many signaling cascades, activated GTPases in the GTP-bound state bind and activate kinases. Since the LRRK2 polypeptide contains both modules of this type of signaling cascade, and altered GTPase activity of the Roc domain has the potential to change kinase activity, it is likely that there is a similar mechanism of intramolecular regulation of LRRK2 kinase function. R1441C mutation was found to increase activity by 20 % using purified recombinant full-length LRRK2 R1441C [41]. LRRK2 R1441G/C and Y1699C mutations both increase the kinase activity of LRRK2 in cells [41, 42]. The prolonged GTP-bound "active state" is a potential explanation for the increased kinase activity of LRRK2 in these PD mutants.

The extra catalytic amino and carboxy terminal domains of LRRK2 also play a role in the activity and function of LRRK2. The carboxy terminus can interact with the catalytic domains of LRRK2, and deletion of the carboxy terminus or substitution of the risk factor mutation Gly2385 to Arg decreases kinase activity of LRRK2 [17, 20, 21, 43]. The amino terminus of LRRK2 is heavily phosphorylated in cells and regulates several aspects of LRRK2 biology (discussed below), and deletion of this domain enables purification of active kinase [17].

LRRK2 Phosphorylation

LRRK2 Autophosphorylation

Identification of LRRK2 Autophosphorylation Sites

Many kinases undergo autophosphorylation in vitro; however, the relevance to physiological function has yet to be defined for many of these enzymes. One classic example of kinase autophosphorylation as a regulatory mechanism in neuronal transmission is with Ca++/calmodulin-dependent kinase which is involved in long-term potentiation. CaMKII autophosphorylation of a regulatory domain leads to activation of the enzyme, while inhibition of protein phosphatase1 leads to persistent activation of the protein [44–47]. LRRK2 autophosphorylation has been examined in several studies directed at identifying LRRK2 autophosphorylation sites using in vitro autophosphorylation kinase assays or *trans*-phosphorylation of purified LRRK2 fragments. Autophosphorylation of LRRK2 is located throughout the protein, with several sites of modification found within the catalytic core of the enzyme. Table 3.1 lists these sites and the literature sources from which these were reported. In Greggio et al. 2009, phosphosites were mapped in *trans*-phosphorylation reactions using LRRK2 aa970-2527 with bacterially expressed LRRK2 fragments where they found Thr1024, Thr1343, T1491, and T2031 as sites of LRRK2 autophosphorylation [48]. Kamikawaji also identified Thr1491, as well as Ser1403, Thr1404, and Thr1410, located within the ROC domain and Thr1967 and Thr1969 in the kinase domain [39]. LRRK2 Thr1348, Thr1349, and Thr1357 were reported as major autophosphorylation sites, while mutation of T1410, T1491, and T1503 resulted in minimal loss of phosphorylation pattern in two-dimensional thin-layer chromatography analysis, indicating these are low-stoichiometry sites [49]. In positional scanning peptide library screens to search for an optimal LRRK2 substrate recognition sequence, it was observed that LRRK2 prefers Thr residues in the context of peptide substrates [50, 51]; this correlates well with the fact that most autophosphorylation sites are Thr residues (Table 3.1). The stoichiometry of LRRK2 autophosphorylation sites is low within cells, and this is reflected in the lack of mass spectrometric identification or immunological reagents to detect these phosphosites in lysates of cells. An alternate explanation is that the LRRK2 protein analyzed is in a basal state of low kinase activity. The Ser1292 phosphorylation site of LRRK2 was isolated in Gloeckner et al. [52] and in Sheng et al. [42], and this site does appear to undergo autophosphorylation in cells. Anti-pSer1292 antibodies can detect modification of this site and reveal changes in LRRK2 kinase activity in LRRK2 N1437H, R1441C/G/H, Y1699C, and G2019S in cells and tissues [41, 42].

In some kinase signal transduction cascades, in response to an inducer, upstream activating kinases phosphorylate downstream kinases within the activation loop (or T-loop) of the enzyme. The T-loop lies between subdomains VII and VIII of the kinase domain and can be single or multiple Ser/Thr/Tyr residues [14, 15]. This phosphorylation causes a conformational change flipping the T-loop into an active conformation and turning on the enzyme, thus transmitting the signal downstream. In LRRK2, there are three potential sites of T-loop phosphorylation, namely, Thr2031, Ser2032, and Ser2035, which were reported to be sites of LRRK2 autophosphorylation [53]. There have been no clear activators of LRRK2 kinase activity found thus far, though Dzamko et al. observed a slight increase in LRRK2 kinase activity after LPS treatment [54]. Reactive oxygen stress with hydrogen peroxide has also been found to increase LRRK2 autophosphorylation [53]. It is therefore possible that the enzymatic studies with LRRK2 have been performed on kinase with basal levels of unstimulated activity. LRRK2 was found to autophosphorylate at Thr2031 by mass spectrometry [42, 48, 50]. However, there is little evidence of

Table 3.1 Phosphosites of LRRK2

LRRK2 P-site	Peptide sequence	Gloeckner et al. [52]	Nichols et al. [58]	Sheng et al. [42]	Greggio et al. [48]	Kamikawaji et al. [39]	Kamikawaji et al. [49]	Webber et al. [55]	Pungaliya et al. [50]
pS3	MAsGSCQGC	Auto							
pS5	MASGsCQGCEE	Auto							
p424	ASANALStLLEQNV	Auto							
pT524	PEESREDtEFHHKL	Auto							
pT776	QDVRKALtISIGKG	Auto							
pT826	GPLFPDKtSNLRKQ	Auto							
pT833	tSNLRKQtNIAStL	Auto							
pT838	KQtNIAStLARMVI	Auto							
pS850	VIRYQMKsAVEEGT	Upstream							
pS858	AVEEGTAsGSDGNF	Upstream							
pS860	EEGTASGsDGNFSE	Upstream	Upstream						
pS865	SGSDGNFsEDVLSK	Upstream							
pS895	AQSDDLDsEGSEGS	Upstream							
pS898	DDLDSEGsEGSFLV	Upstream							
pS908	SFLVKKKsNSISVG	Upstream							
pS910	LVKKKSNsISVGEF	Upstream	Upstream						
pS935	NLQRHSNsLGPIFD	Upstream	Upstream						
pS971	LQSHMRHsDSISSL	Upstream							
pS973	SHMRHSDsISSLAS	Upstream	Upstream						
pS975	MRHSDSIsSLASER	Upstream							
pS976	RHSDSISsLASERE	Upstream	Upstream						
pS979	DSISSLAsEREYIT	Upstream							
pT1024	ELHQNALtSFPQQL				Auto				
pS1124	NKISGICsPLRLKE	Auto							
pS1292	PNEMGKLsKIWDLP	Auto		Auto					

pT1343	KLMIVGNtGSGKTT	Auto		Auto	Auto			Auto	
pS1345	MIVGNTGsGKTTLL	Auto			Auto				
pT1348	GNtGSGKtTLLQQL				Auto		Auto	Auto	
pT1349	NTGSGKTtLLQQLM				Auto				
pT1357	LLQQLMKtKKSDLG			Auto			Auto	Auto	Auto
pT1368	DLGMQSAtVGIDVK	Auto		Auto					Auto
pS1403	AGREEFYsTHPHFM					Auto		Auto	
pT1404	GREEFYStHPHFMT					Auto			
pT1410	STHPHFMtQRALYL	Auto		Auto		Auto			
pS1443	FNIKARAsSSPVIL			Auto					
pS1444	NIKARASsSPVILV								Auto
pS1445	IKARASSsPVILVG								Auto
pT1452	SPVILVGtHLDVSD	Auto		Auto					
pT1470	KACMSKItKELLNK			Auto					
pT1491	DYHFVNAtEESDALA	Auto		Auto	Auto	Auto		Auto	
pT1503	ALAKLRKtIINESLN	Auto		Auto		Auto		Auto	
pS1508	RKTIINEsLNFKIRD			Auto					
pT1512	KIMAQILtVKVEGCP			Auto					Auto
pS1913	KIFNKHTsLRLLRQE								Auto
pT1967	QQDKASLtRTLQHRI					Auto			
pT1969	DKASLTRtLQHRIAL					Auto			
pT2031	CCRMGIKtSEGTPGF	Auto		Auto	Auto				
pS2032	CRMGIKTsEGTPGFR			Auto	Auto				Auto
pT2035	GIKTSEGtPGFRAPE			Auto	Auto				Auto
pT2483	NRKNTEGtQKQKEIQ	Auto							
pT2524	LAEKMRRtSVE								Auto

The sites of LRRK2 phosphorylation are listed with the primary amino acid sequence. Literature sources are indicated and if the site was identified from in vitro autophosphorylation reactions (Auto) or found on the native protein as likely upstream kinase phosphorylation sites (Upstream) are indicated. The amino acid sequence of the phosphorylation sites are given with the site of modification in lowercase

LRRK2 autoactivation in vitro. Mutation of T2031S increased LRRK2 kinase activity similar to G2019S and mutation of Thr2031, Ser2032, and Ser2035 to Ala repressed kinase activity [21, 53] showing that the regulation of these sites as true T-loop activation sites remains unresolved.

Effects of Autophosphorylation on LRRK2 Function

In the absence of a validated downstream substrate for LRRK2, autophosphorylation is a potential tool to understand LRRK2 kinase activity in experimental or pathological conditions. Autophosphorylation has been shown to impact downstream function of LRRK2 in vitro. An Ala substitution of Thr1503 decreases GTPase activity and kinase activity [55]. Substitution of Thr1367 to Ala decreases the kinase activity of LRRK2 [39]. Thr1348 and Thr1349 were characterized as major autophosphorylation sites of LRRK2, and mutation of these residues to Asp (to mimic phosphorylation) decreases kinase activity [49]. However, phosphorylation of the Roc domain by LRRK2, partially through Thr1343 and Thr1348, could enhance GTPase activity [56]. Autophosphorylation of Ser1292 has recently been observed within cells as an indicator of LRRK2 kinase activity [42]. However, mutation of Ser1292 does not affect kinase activity in vitro [41, 42]. Interestingly, a S1292A mutation reduces the effects of LRRK2 PD-related mutations on primary neurite outgrowth in rat cortical primary neurons [42]. Ser1292 phosphorylation has also been shown to contribute to the potential role of LRRK2 in regulating lysosome size and function in primary astrocytes. This could be related to findings from prior expression studies in Plowey et al. showing that LRRK2 G2019S mutant induces neurite shortening concomitant with an increase in autophagic vesicle formation, which could be due to LRRK2 pSer1292 effects on lysosomal size [57]. Cumulatively, autophosphorylation can alter in vitro and in vivo activity of LRRK2, representing intramolecular regulation of the enzyme.

LRRK2 Phosphorylation by Upstream Kinases

Identification of LRRK2 Upstream Kinase Phosphorylation Sites

LRRK2 is constitutively phosphorylated in cells, but the fraction of LRRK2 that is modified is currently unknown. LRRK2 was first described to be phosphorylated in cells in West et al. [38], reporting phosphorylated residues Ser910 and Ser935 [38]. Gloeckner et al. first described the phosphorylation of LRRK2 as constitutive by using a comparative quantitative mass spectrometry on strep-tagged LRRK2 isolated from cells labeled with heavy and light amino acids using a SILAC (stable isotope labeling in culture) approach [52]. They were able to distinguish constitutive phosphorylation sites present in active and inactive LRRK2 [K1906M] purified from cells from those sites induced by autophosphorylation in vitro. The presence of phosphorylated residues on kinase-inactive LRRK2 indicates that these are not likely autophosphorylation sites. Endogenous LRRK2 immunoprecipitated from Swiss3T3 cells was characterized by mass spectrometry to verify that Ser860, Ser910, and Ser935 and S973/S976 were observed on native protein [58]. Ectopically expressed LRRK2 was metabolically labeled in HEK293 cells with ^{32}P-orthrophosphate in Lobbestael et al., directly demonstrating that LRRK2 is phosphorylated in cells [59]. Further evidence that Ser910 and Ser935 and S973/S976 are not sites of autophosphorylation was provided by showing that dephosphorylated LRRK2 was not able to rephosphorylate these sites in vitro [41, 58, 60, 61]. Table 3.1 also describes LRRK2 phosphorylation sites derived from the literature and publically available phosphosite curation databases that are likely modified by upstream kinases.

Phosphorylation-Dependent 14-3-3 Binding to LRRK2

A functional significance of the Ser910/Ser935 phosphorylation sites stemmed from a SILAC-based protein-protein interaction study with LRRK2, which found 14-3-3 proteins highly enriched in LRRK2 immunoprecipitates. 14-3-3 proteins can bind phosphopeptides as dimers, engaging two phosphosites at once and can be separated by approximately 20-25aa [62–65]. 14-3-3 can also bind carboxy terminal phosphopeptides [66] as well as bind proteins in a non-phosphoamino acid-dependent manner [67]. 14-3-3 was shown to bind to Ser910/Ser935, where mutation of either site disrupts phosphorylation of the other site and also ablates 14-3-3 interaction. This was contemporaneously shown by Li et al., also finding that LRRK2 bound 14-3-3 proteins through Ser935 [68]. Mutation of Ser910/Ser935, but not Ser955/Ser973, results in loss of 14-3-3 binding [60, 69]. Interestingly the absence of binding 14-3-3 results in the localization of some fraction of LRRK2 to skein-like structures and cytoplasmic accumulations [61, 70, 71]. 14-3-3 proteins were also found to be LRRK2 interactors in an unbiased screen for LRRK2-interacting proteins using proteoarrays [72]. 14-3-3 not only binds phosphorylated LRRK2, but overexpression of 14-3-3 proteins may alter LRRK2 kinase activity and neurotoxicity in cells [73].

Regulation of LRRK2 Phosphorylation

PD Mutations and Kinase Inhibition Change LRRK2 Phosphorylation

PD-Related Mutations Alter LRRK2 Phosphorylation

PD-related mutations in LRRK2 not only impact the enzymatic activity of LRRK2 but also alter the phosphorylation status of an autophosphorylation site and the upstream kinase sites in cells. An important recent finding was that serines 910, 935,

955, and 973 are dephosphorylated in pathogenic PD mutations found within the RocCOR domain of the protein [N1437H, R1441C/G/H, Y1699C] and in the kinase domain of LRRK2 [I2020T] [41, 58–60]. The mechanism of how altered GTPase activity [R1441C/G/H and Y1699C] or kinase activity [I2020T] causes the dephosphorylation of LRRK2 is unclear; it is however possible that the mutations change enzymatic activity or conformation to impart the deregulation of these sites through changing access to a phosphatase. Interestingly the G2019S mutation of LRRK2, though exhibiting increased kinase activity, neither increases nor decreases the phosphorylation status of LRRK2. This is in contrast to the RocCOR mutations that also increase kinase activity while concomitantly decreasing Ser910/Ser935/Ser955/Ser973 phosphorylation. The PD risk factor LRRK2 mutation encoding G2385R also decreases the phosphorylation of LRRK2 at Ser910/Ser935 [43].

Kinase Inhibition Induces LRRK2 Dephosphorylation and Loss of 14-3-3 Binding

Another aspect of the dynamic regulation of LRRK2 is that Ser910/Ser935/Ser955/Ser973 is rapidly dephosphorylated in cells and tissues after inhibition of LRRK2 with small molecules [60, 61, 74] similar to N1437H, R1441C/G/H, Y1699C, and I2020T. This similarly results in the loss of 14-3-3 interaction and relocalization of the protein to skein-like structures or accumulations in the cytoplasm of cells. The relocalization phenotype has only been observed on expressed protein; however, it is coincident with loss of 14-3-3 association which has been observed on endogenous R1441C protein in mouse knock-in tissues and on endogenous LRRK2 in mouse tissues or cells treated with LRRK2 inhibitors [58]. The cytoplasmic accumulations of LRRK2 could be associated with microtubules, vesicles, or aggregates of LRRK2 protein itself. The change in phosphorylation, interacting partners, and localization of LRRK2 is specific to inhibition of the kinase domain of LRRK2. The A2016T substitution desensitizes LRRK2 to inhibitors by creating a steric block to inhibitor engagement of the kinase domain without major changes to kinase activity [51]. When this mutant is exposed to selective LRRK2 inhibitors in cells, LRRK2 remains phosphorylated at Ser910/Ser935/Ser955/Ser973 and Ser1292 in cells, confirming that dephosphorylation is a result of direct inhibition of LRRK2. Loss of phosphorylation at Ser935 was recently adapted to a high-throughput amenable detection method using a time-resolved fluorescence energy transfer assay to identify regulators of LRRK2 phosphorylation [75], where several LRRK2 inhibitor molecules were identified.

When LRRK2 is dephosphorylated, it is blocked from being trafficked to a secretory pathway. LRRK2 can be found in exosomes from cell culture model systems and in urine from patient samples [76]. It may be that the dephosphorylation of LRRK2 sequesters the protein in subcellular locale that prevents involvement with the secretory pathway for exosome release. This identifies a cellular pathway that LRRK2 interacts with and a potential biomarker for PD, if levels or activity status of LRRK2 could be correlated with disease phenotype. In fact recently, urine exosome-associated LRRK2 phosphorylation at Ser1292 was positively correlated with genotype and disease risk [77]. If 14-3-3 interaction is blocked using the small peptide difopein, LRRK2 becomes dephosphorylated at Ser910/Ser935 [70, 76]. Difopein binds and prevents phosphopeptide-dependent binding of 14-3-3 proteins [78] and when co-expressed with LRRK2 seems to "uncap" the sites by binding 14-3-3, allowing access to a phosphatase that dephosphorylates Ser910/Ser935 [70, 76]. This indicates that 14-3-3 proteins can protect phosphorylation of LRRK2, but also that there are active phosphatases that regulate LRRK2 regardless of its kinase activity status. LRRK2 Ser1292 phosphorylation is dependent on direct LRRK2 kinase activity and is also dephosphorylated in cells after LRRK2 kinase inhibitor treatment [41, 42]. The dephosphorylation of LRRK2 after inhibition has been utilized as a measure of LRRK2 inhibition by small molecules in cells and tissues by multiple groups. This has enabled the determination of LRRK2 kinase inhibitors that are engaging the target in cells and tissues as a pharmacodynamic marker to correlate the inhibition of LRRK2 in dosing studies [79].

LRRK2 Kinases

The cluster of serines, including S910, S935, S955, and S973, found preceding the namesake LRR domain, appear to be constitutively phosphorylated. It has been proposed that these sites are phosphorylated by kinases other than LRRK2 itself [38, 52, 58, 60, 68]. LRRK2 is highly expressed in immune cells, and there is ample evidence that LRRK2 responds to stimulation of the immune system. Interferon-γ treatment of immune cells increases LRRK2 expression [80]. Stimulation of Toll-like receptor (TLR) 2, TLR4, TLR6, and TLR9 has been shown to increase the phosphorylation of LRRK2 at Ser910/Ser935 [54]. Interestingly, TLR stimulation can overcome the induced dephosphorylation of LRRK2 caused by acute inhibition [54]. This could be through activation of a kinase signaling cascade that tilts the balance away from dephosphorylation or by causing inhibition of phosphatase, with both scenarios assuming there is an active ongoing phosphatase activity on LRRK2. Though this issue is unresolved, genetic and pharmacological interrogation implicated the inhibitor of IκB kinases (IKKα/β and IKKε/TBK1) as activated kinases that could mediate the phosphorylation of LRRK2. This has broad implications in the potential role of LRRK2 in immune signaling. LRRK2 knockout rats or long-term kinase inhibition with PF475 reduces the CD63+ cell response to LPS or α-synuclein-mediated nigral degeneration after stereotactic injection to the area [81, 82]. Extracellular α-synuclein could activate an LRRK2-dependent TLR signaling cascade that, when stimulated, contributes to nigral cell death. The absence of LRRK2 kinase activity may relieve toxic stress responses that diminish nigral death, which is a positive indicator that inhibition of LRRK2 is a viable neuroprotective strategy.

The regulation of LRRK2 phosphorylation is likely mediated by several kinases that respond to different stimuli. To screen recombinant kinases that could modify

LRRK2 would be resource intensive. Alternatively, Chia et al. screened for LRRK2 upstream kinases using a reverse genetic screen with siRNAs targeting kinases within the kinome with a pSer935 immunoblot readout and found casein kinase CK1α1 as an upstream kinase for LRRK2 [83]. Repression of expression or inhibition of CK1α1, but not LRRK2 inhibition, resulted in loss of Ser910/Ser935 phosphorylation and an increase in ARHGEF7 association with LRRK2, which in turn decreases GTP binding. CK1α1 siRNA treatment also reduces the Rab7L1-dependent Golgi fragmentation caused by LRRK2, indicating that constitutive phosphorylation of LRRK2 modulates the recruitment of LRRK2 to the trans-Golgi network [83]. This could mean that the cytoplasmic accumulations of LRRK2 caused by its kinase inhibition are vesicular in nature.

Protein kinase A has been proposed as a candidate upstream kinase of LRRK2 Ser910/Ser935 [68] and also for Ser1444 site near the R1441C/G/H PD-related mutations [84]. The in vivo role for PKA in phosphorylating LRRK2 at Ser910/Ser935 has yet to be fully resolved. Forskolin treatment, an activator of PKA through activation of adenylate cyclase to increase cAMP levels, was indicated to increase LRRK2 phosphorylation at Ser910/Ser935 and Ser1444. This was proposed to provide an alternate 14-3-3 dimer phosphosite interaction between pSer910 and pSer1444. However, forskolin treatment was shown to decrease LRRK2 phosphorylation at Ser935/955/973 and reduce 14-3-3 interaction on LRRK2 expressed in HEK293 cells and endogenous LRRK2 [41]. Further, mutations of LRRK2 Ser910 or Ser935 to Ala ablate 14-3-3 interaction with LRRK2 (Section "LRRK2 Kinases") [58, 68].

LRRK2 Phosphatases

Phosphorylation is a reversible process where dephosphorylation of Ser/Thr/Tyr residues is mediated by protein phosphatases. The induction of LRRK2 dephosphorylation after inhibition and the loss of phosphorylation in PD-related mutations of LRRK2 predict that the phosphatases that regulate LRRK2 are novel targets for PD investigations. Protein phosphatase 1 (PP1) was found to be a regulator of LRRK2 phosphorylation after inhibition. Co-treatment of cells with LRRK2 inhibitor with calyculin A, a PP1 inhibitor, blocked the induced dephosphorylation of LRRK2. Further, LRRK2 inhibition and in the N1437H, R1441G, and Y1699C PD-related mutations increased the association of PP1 with LRRK2 in expression studies and on endogenous LRRK2. PP1 target specificity is driven by the association of regulatory subunits. We do not yet know which subunits form the active PP1 holoenzyme that acts on LRRK2. To fully understand how LRRK2 dephosphorylation is regulated, it is necessary to know which subunits mediate the interaction with LRRK2 and if they are different in pathogenic conditions or if there is tissue specificity to complex assembly. PP1 dephosphorylation of LRRK2 Ser910/Ser935/Ser955/Ser973 provides a novel target for PD therapeutic development.

Ser1292 autophosphorylation not only reveals intrinsic LRRK2 kinase activity but also mediates some pathogenic phenotypes (described above). Since dephosphorylation of this site reveals kinase inhibition, it must also be subject to phosphatase activity. In a pharmacological approach, Reynolds et al. found that calyculin A and okadaic acid, inhibitors of PP1 and PP2, respectively, could increase the phosphorylation status of LRRK2 at Ser1292, indicating that both families of phosphatases act on LRRK2. Although PP1 exhibits increased association with LRRK2 in conditions in which it is dephosphorylated, the pharmacological evidence that both PP1 and PP2 type enzymes act on LRRK2 opens the repertoire of dephosphorylating enzymes to both major Ser/Thr phosphatase families and thus provides even more molecular targets that modulate LRRK2 function

A LRRK2 Dephosphorylation and Ubiquitination Cycle

The R1441C, A1442P, and I2020T mutations of LRRK2 exhibit intracellular instability with a shortened half-life compared to wild-type LRRK2 [85–87]. One difference between these mutations and wild-type LRRK2 is that they are dephosphorylated at the Ser910/Ser935/Ser955/Ser973 sites, providing insight into a potential mechanism for degradation. When 42 different LRRK2 mutations were analyzed simultaneously, several mutants that proved to be dephosphorylated at Ser910/Ser935 exhibited lower steady-state accumulation of the protein [58]. Zhao et al. reported that inhibition of LRRK2 kinase activity reduces the stability of the LRRK2 enzyme [70]. This instability was found on expressed LRRK2 and endogenous LRRK2 in cell culture and in animals dosed with the LRRK2 inhibitor GNE1023. Further, inhibition of LRRK2 kinase activity induced the ubiquitination of the enzyme, which could be blocked by inhibiting phosphatases with calyculin A, therefore linking LRRK2 inhibition-induced dephosphorylation and ubiquitination of the protein. Difopein co-expression with LRRK2 further linked ubiquitination of LRRK2 to a cycle of phosphorylation/dephosphorylation by inducing dephosphorylation of Ser935, ubiquitination, and a decrease in stability of LRRK2. Interestingly, PD-related mutations of LRRK2 that are dephosphorylated at the upstream kinase phosphorylation sites also exhibit increased ubiquitination of the protein in the basal state, which could be reversed with calyculin A treatment. The clear inverse relationship between the phosphorylation and ubiquitination status of LRRK2 reveals novel targets and regulatory pathways associated with PD [70].

Ubiquitin is covalently linked to proteins through Lys residues, and additional ubiquitin molecules can be conjugated in a successive manner to the previously added ubiquitin to generate extended chains of ubiquitin molecules attached to the target protein [88]. In these chains, ubiquitin can be linked to adjoining ubiquitins via different Lys residues in the protein to generate a variety of architectures of ubiquitin chains with specific functions. LRRK2 is subject to several different types of ubiquitin linkages including K48 and K63, which result in altered protein stability and likely other signaling mechanisms for LRRK2. The degradation of LRRK2 is complex, with both the ubiquitin proteasome system (generally driven by K48) and the autophagy lysosome system (driven by K63 for aggregated proteins) being implicated in the degradation of LRRK2. LRRK2 protein accumulates after proteasome and autophagy disruption [70, 89–92]. LRRK2 is also subject to chaperone-mediated autophagy, a process that is also dysregulated by PD mutant LRRK2 forming a complex interrelationship [89]. If indeed LRRK2 is degraded by these three mechanisms, it will be important to determine the phosphorylation and ubiquitination status in each of these cases. The ubiquitin ligases and deubiquitinases that act on LRRK2 in response to altered phosphorylation will provide insight into how LRRK2 is regulated in normal and pathogenic conditions. The ligases that act on LRRK2 specifically after dephosphorylation have yet to be elucidated; however, carboxy terminal Hsp70 interacting protein (CHIP) has been identified as a ubiquitin ligase that acts on LRRK2 [90, 91], but the mechanism to stimulate this function remains unresolved.

LRRK2 ubiquitination has a major impact on the potential for LRRK2 inhibitor-based therapeutic strategies. Although we still need to understand what the ubiquitin modifiers of LRRK2 are and how their activities toward LRRK2 are stimulated, their modification of LRRK2 in response to dephosphorylation induced by inhibition will need to be taken into account. Complete ablation of LRRK2 activity with LRRK2 inhibitors will lead to dephosphorylation and ubiquitination of LRRK2 and be an on-target molecular effect of inhibition [93].

Conclusions

Since the discovery of LRRK2 mutations that cause Parkinson's disease, there has been a large investment of public and private resources to understanding how they precipitate disease onset. One way to gain insight is to understand how LRRK2 is regulated and how that regulation is perturbed in the mutated state. Phosphorylation of LRRK2 appears to be a major regulatory mechanism of intrinsic (autophosphorylation) and extrinsic action (upstream kinase phosphorylation). This chapter has described the identification of LRRK2 phosphorylation sites, the impact that PD-related mutations have on phosphorylation, the potential enzymes that regulate the phosphorylation, and the interplay between phosphorylation and ubiquitination. The phosphoregulation of LRRK2 is a complex and dynamic system that integrates the biology of several PD-related mutations with inhibitors of LRRK2 kinase activity. Dephosphorylation of the upstream kinase phosphosites links PD mutations and inhibition with similar molecular outcomes such as relocalization to cytoplasmic accumulations and filamentous skein-like structures [36, 58, 71, 94], loss of 14-3-3 binding [58, 68], and increased binding of PP1 [59]. PD mutations in LRRK2 such as R1441C/G and Y1699C exhibit increased kinase activity but decreased LRRK2 phosphorylation with concomitant calyculin A-sensitive dependent ubiquitination. These results are in contradistinction to the most common mutation G2019S, which displays increased kinase activity and no change in phosphorylation or ubiquitination. LRRK2 inhibition results in decreased kinase activity and dephosphorylation with calyculin A-sensitive ubiquitination [70]. These similar but opposite scenarios imply that the dynamic dephosphorylation of LRRK2 likely precipitates the ubiquitination activity. These biochemically measurable differences among LRRK2 pathogenic mutations likely reveal different aspects of LRRK2 biology that are disrupted in mutations and also provide potential understanding on how to target PD based on these mutations.

There is an extensive literature that links LRRK2 to vesicular trafficking, neurotransmitter signaling, and neurite outgrowth that could also reflect similar underlying mechanisms, reviewed in [95]. Several reports link LRRK2 to dysregulation of autophagy. However, there are no verified reports of the direct regulation of known autophagy proteins by LRRK2 to explain the observed global phenotypic alterations of autophagy flux in loss of function or mutant protein expression. If pathogenic LRRK2 alters membrane organelle biology, changing the rate of trafficking and/or amounts of particular vesicle pools, this would likely induce changes in the flux of autophagy in cells. One potential mechanism is that LRRK2 phosphorylates vesicular proteins to regulate vesicular dynamics. Indeed, LRRK2 phosphorylates a subset of Rab GTPase proteins, regulating their interaction with the guanine nucleotide dissociation inhibitor Rabin8 and membrane association [16]. The dysregulation of these Rabs could change the cycling of vesicles in cells, which, in cell type or assay-dependent measurements, show lysosomal/autophagy, neurotransmitter, or neurite outgrowth defects. It will be important to determine how the LRRK2 phosphorylation and ubiquitination status affect the regulation of these GTPases since these have been shown to affect the localization of LRRK2.

The prevailing model of LRRK2 phosphoregulation is a feedback mechanism where LRRK2 either stimulates a kinase activity or represses a phosphatase activity, of which one or both aspects impact the ubiquitination status of LRRK2. Figure 3.2

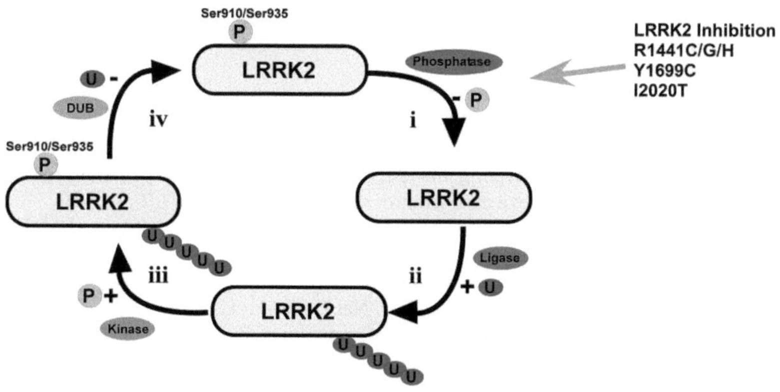

Fig. 3.2 LRRK2 dephosphorylation/ubiquitination cycle. LRRK2 likely exists in a phosphorylated and un-ubiquitinated state that is subject to dephosphorylation by a phosphatase (*i*), which results in changes in the LRRK2 macromolecular complex. Dephosphorylation of LRRK2 promotes its ubiquitination by a ubiquitin ligase (*ii*). This could lead to degradation or altered signaling based on the types of ubiquitin chains conjugated to LRRK2. This leads to degradation or signaling of LRRK2 via alternate linkages. Kinases phosphorylate LRRK2 (*iii*), which leads to deubiquitination by a deubiquitinase (*iv*), restoring LRRK2 to its typical state

presents a potential model that incorporates these aspects of LRRK2 biology. In a basal state, LRRK2 is phosphorylated at Ser910/Ser935/Ser955/Ser973. In some pathogenic mutations such as N1437H, R1441C/G/H, Y1699C, and I2020T or after LRRK2 inhibition, a phosphatase dephosphorylates LRRK2 (**i**). PP1 is a candidate enzyme for this activity. This is followed by a ubiquitin ligase conjugating ubiquitin molecules to LRRK2 (**ii**). Ubiquitination is a reversible process, and though we have yet to identify the order of deubiquitination and rephosphorylation of LRRK2, this model shows that a kinase, potentially IKKs or CK1, acts to phosphorylate LRRK2 (**iii**) with a subsequent deubiquitinase enzyme removing the ubiquitin chains (**iv**). This model identifies four states of LRRK2 in the phosphorylation/ubiquitination cycle, which may or may not exist exclusive of each other. It is likely that LRRK2 activity or localization is driven by these states to change the signaling from LRRK2 with disruptive effects of inhibition and mutations.

Conflict of Interest The author declares no conflicts of interest.

References

1. Dorsey ER et al (2007) Projected number of people with Parkinson disease in the most populous nations, 2005 through 2030. Neurology 68(5):384–6
2. Dorsey ER et al (2013) The coming crisis: obtaining care for the growing burden of neurodegenerative conditions. Neurology 80(21):1989–96
3. Di Monte DA (2003) The environment and Parkinson's disease: is the nigrostriatal system preferentially targeted by neurotoxins? Lancet Neurol 2(9):531–8
4. Warner TT, Schapira AH (2003) Genetic and environmental factors in the cause of Parkinson's disease. Ann Neurol 53(3):S16–23, discussion S23-5
5. Gorrell JM, DiMonte D, Graham D (1996) The role of the environment in Parkinson's disease. Environ Health Perspect 104(6):652–4
6. Zimprich A et al (2004) Mutations in LRRK2 cause autosomal-dominant parkinsonism with pleomorphic pathology. Neuron 44(4):601–7
7. Di Fonzo A et al (2005) A frequent LRRK2 gene mutation associated with autosomal dominant Parkinson's disease. Lancet 365(9457):412–5
8. Farrer M et al (2005) LRRK2 mutations in Parkinson disease. Neurology 65(5):738–40
9. Healy DG et al (2008) Phenotype, genotype, and worldwide genetic penetrance of LRRK2-associated Parkinson's disease: a case–control study. Lancet Neurol 7(7):583–90
10. Paisan-Ruiz C et al (2004) Cloning of the gene containing mutations that cause PARK8-linked Parkinson's disease. Neuron 44(4):595–600
11. Satake W et al (2009) Genome-wide association study identifies common variants at four loci as genetic risk factors for Parkinson's disease. Nat Genet 41(12):1303–7
12. Simon-Sanchez J et al (2009) Genome-wide association study reveals genetic risk underlying Parkinson's disease. Nat Genet 41(12):1308–12
13. Langston JW et al (2015) Multisystem Lewy body disease and the other parkinsonian disorders. Nat Genet 47(12):1378–84
14. Hanks SK, Quinn AM, Hunter T (1988) The protein kinase family: conserved features and deduced phylogeny of the catalytic domains. Science 241(4861):42–52
15. Hanks SK, Hunter T (1995) Protein kinases 6. The eukaryotic protein kinase superfamily: kinase (catalytic) domain structure and classification. Faseb J 9(8):576–96
16. Steger M et al (2016) Phosphoproteomics reveals that Parkinson's disease kinase LRRK2 regulates a subset of Rab GTPases. Elife 5:e12813.

17. Jaleel M et al (2007) LRRK2 phosphorylates moesin at threonine-558: characterization of how Parkinson's disease mutants affect kinase activity. Biochem J 405(2):307–17
18. Marin I, van Egmond WN, van Haastert PJ (2008) The Roco protein family: a functional perspective. FASEB J 22(9):3103–10
19. Lewis PA (2009) The function of ROCO proteins in health and disease. Biol Cell 101(3):183–91
20. Jorgensen ND et al (2009) The WD40 domain is required for LRRK2 neurotoxicity. PLoS One 4(12), e8463
21. Greggio E et al (2008) The Parkinson disease-associated leucine-rich repeat kinase 2 (LRRK2) is a dimer that undergoes intramolecular autophosphorylation. J Biol Chem 283(24):16906–14
22. Liao J et al (2014) Parkinson disease-associated mutation R1441H in LRRK2 prolongs the "active state" of its GTPase domain. Proc Natl Acad Sci U S A 111:4055–60
23. Deng J et al (2008) Structure of the ROC domain from the Parkinson's disease-associated leucine-rich repeat kinase 2 reveals a dimeric GTPase. Proc Natl Acad Sci U S A 105(5):1499–504
24. Rudi K et al (2015) Conformational heterogeneity of the Roc domains in C. tepidum Roc-COR and implications for human LRRK2 Parkinson mutations. Biosci Rep 35(5), e00254
25. Gotthardt K et al (2008) Structure of the Roc-COR domain tandem of C. tepidum, a prokaryotic homologue of the human LRRK2 Parkinson kinase. EMBO J 27(17):2352
26. Gilsbach BK, Kortholt A (2014) Structural biology of the LRRK2 GTPase and kinase domains: implications for regulation. Front Mol Neurosci 7:32
27. Stafa K et al (2012) GTPase activity and neuronal toxicity of Parkinson's disease-associated LRRK2 is regulated by ArfGAP1. PLoS Genet 8(2), e1002526
28. Habig K et al (2013) LRRK2 guides the actin cytoskeleton at growth cones together with ARHGEF7 and Tropomyosin 4. Biochim Biophys Acta 1832(12):2352–67
29. Haebig K et al (2010) ARHGEF7 (Beta-PIX) acts as guanine nucleotide exchange factor for leucine-rich repeat kinase 2. PLoS One 5(10), e13762
30. Lin MK, Farrer MJ (2014) Genetics and genomics of Parkinson's disease. Genome Med 6(6):48
31. Mata IF et al (2006) LRRK2 in Parkinson's disease: protein domains and functional insights. Trends Neurosci 29(5):286–93
32. Hanks SK, Quinn AM (1991) Protein kinase catalytic domain sequence database: identification of conserved features of primary structure and classification of family members. Methods Enzymol 200:38–62
33. Greggio E, Cookson MR (2009) Leucine rich repeat kinase 2 mutations and Parkinson's disease: three questions. ASN Neuro 1:13–24
34. Gilsbach BK et al (2012) Roco kinase structures give insights into the mechanism of Parkinson disease-related leucine-rich-repeat kinase 2 mutations. Proc Natl Acad Sci U S A 109(26):10322–7
35. Ray S et al (2014) The Parkinson disease-linked LRRK2 protein mutation I2020T stabilizes an active state conformation leading to increased kinase activity. J Biol Chem 289(19):13042–53
36. Greggio E et al (2006) Kinase activity is required for the toxic effects of mutant LRRK2/dardarin. Neurobiol Dis 23(2):329–41
37. Tsika E, Moore DJ (2013) Contribution of GTPase activity to LRRK2-associated Parkinson disease. Small GTPases 4(3):164–70
38. West AB et al (2007) Parkinson's disease-associated mutations in LRRK2 link enhanced GTP-binding and kinase activities to neuronal toxicity. Hum Mol Genet 16(2):223–32
39. Kamikawaji S, Ito G, Iwatsubo T (2009) Identification of the autophosphorylation sites of LRRK2. Biochemistry 48(46):10963–75
40. Li X et al (2007) Leucine-rich repeat kinase 2 (LRRK2)/PARK8 possesses GTPase activity that is altered in familial Parkinson's disease R1441C/G mutants. J Neurochem 103(1):238–47

41. Reynolds A et al (2014) LRRK2 kinase activity and biology are not uniformly predicted by its autophosphorylation and cellular phosphorylation site status. Front Mol Neurosci 7:54
42. Sheng Z et al (2012) Ser1292 autophosphorylation is an indicator of LRRK2 kinase activity and contributes to the cellular effects of PD mutations. Sci Transl Med 4(164), 164ra161
43. Rudenko IN et al (2012) The G2385R variant of leucine-rich repeat kinase 2 associated with Parkinson's disease is a partial loss-of-function mutation. Biochem J 446(1):99–111
44. Gardoni F et al (2001) Hippocampal synaptic plasticity involves competition between Ca2+/calmodulin-dependent protein kinase II and postsynaptic density 95 for binding to the NR2A subunit of the NMDA receptor. J Neurosci 21(5):1501–9
45. Strack S et al (1997) Translocation of autophosphorylated calcium/calmodulin-dependent protein kinase II to the postsynaptic density. J Biol Chem 272(21):13467–70
46. Giese KP et al (1998) Autophosphorylation at Thr286 of the alpha calcium-calmodulin kinase II in LTP and learning. Science 279(5352):870–3
47. Bronstein JM, Farber DB, Wasterlain CG (1986) Autophosphorylation of calmodulin kinase II: functional aspects. FEBS Lett 196(1):135–8
48. Greggio E et al (2009) The Parkinson's disease kinase LRRK2 autophosphorylates its GTPase domain at multiple sites. Biochem Biophys Res Commun 389(3):449–54
49. Kamikawaji S et al (2013) Differential effects of familial parkinson mutations in LRRK2 revealed by a systematic analysis of autophosphorylation. Biochemistry 52(35):6052–62
50. Pungaliya PP et al (2010) Identification and characterization of a leucine-rich repeat kinase 2 (LRRK2) consensus phosphorylation motif. PLoS One 5(10), e13672
51. Nichols RJ et al (2009) Substrate specificity and inhibitors of LRRK2, a protein kinase mutated in Parkinson's disease. Biochem J 424(1):47–60
52. Gloeckner CJ et al (2010) Phosphopeptide analysis reveals two discrete clusters of phosphorylation in the N-terminus and the Roc domain of the Parkinson-disease associated protein kinase LRRK2. J Proteome Res 9(4):1738–45
53. Li X et al (2010) Reevaluation of phosphorylation sites in the Parkinson disease-associated leucine-rich repeat kinase 2. J Biol Chem 285(38):29569–76
54. Dzamko N et al (2012) The IkappaB kinase family phosphorylates the Parkinson's disease kinase LRRK2 at Ser935 and Ser910 during Toll-like receptor signaling. PLoS One 7(6), e39132
55. Webber PJ et al (2011) Autophosphorylation in the leucine-rich repeat kinase 2 (LRRK2) GTPase domain modifies kinase and GTP-binding activities. J Mol Biol 412(1):94–110
56. Liu Z et al (2015) LRRK2 autophosphorylation enhances its GTPase activity. FASEB J 30:336–47
57. Plowey ED et al (2008) Role of autophagy in G2019S-LRRK2-associated neurite shortening in differentiated SH-SY5Y cells. J Neurochem 105(3):1048–56
58. Nichols RJ et al (2010) 14-3-3 binding to LRRK2 is disrupted by multiple Parkinson's disease-associated mutations and regulates cytoplasmic localization. Biochem J 430(3):393–404
59. Lobbestael E et al (2013) Identification of protein phosphatase 1 as a regulator of the LRRK2 phosphorylation cycle. Biochem J 456:119–28
60. Doggett EA et al (2012) Phosphorylation of LRRK2 serines 955 and 973 is disrupted by Parkinson's disease mutations and LRRK2 pharmacological inhibition. J Neurochem 120(1):37–45
61. Dzamko N et al (2010) Inhibition of LRRK2 kinase activity leads to dephosphorylation of Ser(910)/Ser(935), disruption of 14-3-3 binding and altered cytoplasmic localization. Biochem J 430(3):405–13
62. Miller ML et al (2008) Linear motif atlas for phosphorylation-dependent signaling. Sci Signal 1(35):ra2
63. Yaffe MB et al (1997) The structural basis for 14-3-3:phosphopeptide binding specificity. Cell 91(7):961–71
64. Mackintosh C (2004) Dynamic interactions between 14-3-3 proteins and phosphoproteins regulate diverse cellular processes. Biochem J 381(Pt 2):329–42

65. Johnson C et al (2010) Bioinformatic and experimental survey of 14-3-3-binding sites. Biochem J 427(1):69–78
66. Coblitz B et al (2006) C-terminal binding: an expanded repertoire and function of 14-3-3 proteins. FEBS Lett 580(6):1531–5
67. Masters SC et al (1999) Interaction of 14-3-3 with a nonphosphorylated protein ligand, exoenzyme S of Pseudomonas aeruginosa. Biochemistry 38(16):5216–21
68. Li X et al (2011) Phosphorylation-dependent 14-3-3 binding to LRRK2 is impaired by common mutations of familial Parkinson's disease. PLoS One 6(3), e17153
69. Wiebe MS et al (2010) Mice deficient in the serine/threonine protein kinase VRK1 are infertile due to a progressive loss of spermatogonia. Biol Reprod 82(1):182–93
70. Zhao J et al (2015) LRRK2 Dephosphorylation increases its ubiquitination. Biochem J 469:107–120
71. Kett LR et al (2012) LRRK2 Parkinson disease mutations enhance its microtubule association. Hum Mol Genet 21(4):890–9
72. Beilina A et al (2014) Unbiased screen for interactors of leucine-rich repeat kinase 2 supports a common pathway for sporadic and familial Parkinson disease. Proc Natl Acad Sci U S A 111(7):2626–31
73. Lavalley NJ et al (2016) 14-3-3 Proteins regulate mutant LRRK2 kinase activity and neurite shortening. Hum Mol Genet 25(1):109–22
74. Deng X et al (2011) Characterization of a selective inhibitor of the Parkinson's disease kinase LRRK2. Nat Chem Biol 7(4):203–5
75. Hermanson SB et al (2012) Screening for novel LRRK2 inhibitors using a high-throughput TR-FRET cellular assay for LRRK2 Ser935 phosphorylation. PLoS One 7(8), e43580
76. Fraser KB et al (2013) LRRK2 secretion in exosomes is regulated by 14-3-3. Hum Mol Genet 22(24):4988–5000
77. Fraser KB et al (2016) Urinary LRRK2 phosphorylation predicts parkinsonian phenotypes in G2019S LRRK2 carriers. Neurology 86:994–9
78. Masters SC, Fu H (2001) 14-3-3 proteins mediate an essential anti-apoptotic signal. J Biol Chem 276(48):45193–200
79. Fuji RN et al (2015) Effect of selective LRRK2 kinase inhibition on nonhuman primate lung. Sci Transl Med 7(273), 273ra15
80. Hakimi M et al (2011) Parkinson's disease-linked LRRK2 is expressed in circulating and tissue immune cells and upregulated following recognition of microbial structures. J Neural Transm 118(5):795–808
81. Daher JP et al (2015) LRRK2 Pharmacological Inhibition Abates alpha-Synuclein Induced Neurodegeneration., J Biol Chem
82. Daher JP et al (2014) Abrogation of alpha-synuclein-mediated dopaminergic neurodegeneration in LRRK2-deficient rats. Proc Natl Acad Sci U S A 111(25):9289–94
83. Chia R et al (2014) Phosphorylation of LRRK2 by casein kinase 1alpha regulates trans-Golgi clustering via differential interaction with ARHGEF7. Nat Commun 5:5827
84. Muda K et al (2014) Parkinson-related LRRK2 mutation R1441C/G/H impairs PKA phosphorylation of LRRK2 and disrupts its interaction with 14-3-3. Proc Natl Acad Sci U S A 111(1):E34–43
85. Greene ID et al (2014) Evidence that the LRRK2 ROC domain Parkinson's disease-associated mutants A1442P and R1441C exhibit increased intracellular degradation. J Neurosci Res 92(4):506–16
86. Ohta E et al (2013) Dominant-negative effects of LRRK2 heterodimers: a possible mechanism of neurodegeneration in Parkinson's disease caused by LRRK2 I2020T mutation. Biochem Biophys Res Commun 430(2):560–6
87. Ohta E, Kubo M, Obata F (2010) Prevention of intracellular degradation of I2020T mutant LRRK2 restores its protectivity against apoptosis. Biochem Biophys Res Commun 391(1):242–7
88. Komander D, Rape M (2012) The ubiquitin code. Annu Rev Biochem 81:203–29

89. Orenstein SJ et al (2013) Interplay of LRRK2 with chaperone-mediated autophagy. Nat Neurosci 16(4):394–406
90. Ko HS et al (2009) CHIP regulates leucine-rich repeat kinase-2 ubiquitination, degradation, and toxicity. Proc Natl Acad Sci U S A 106(8):2897–902
91. Ding X, Goldberg MS (2009) Regulation of LRRK2 stability by the E3 ubiquitin ligase CHIP. PLoS One 4(6), e5949
92. Friedman LG et al (2012) Disrupted autophagy leads to dopaminergic axon and dendrite degeneration and promotes presynaptic accumulation of alpha-synuclein and LRRK2 in the brain. J Neurosci 32(22):7585–93
93. Melrose HL (2015) LRRK2 and ubiquitination: implications for kinase inhibitor therapy. Biochem J 470(3):e21–4
94. Alegre-Abarrategui J et al (2009) LRRK2 regulates autophagic activity and localizes to specific membrane microdomains in a novel human genomic reporter cellular model. Hum Mol Genet 18(21):4022–34
95. Wallings R, Manzoni C, Bandopadhyay R (2015) Cellular processes associated with LRRK2 function and dysfunction. FEBS J 282(15):2806–26

Chapter 4
Understanding the GTPase Activity of LRRK2: Regulation, Function, and Neurotoxicity

An Phu Tran Nguyen and Darren J. Moore

Abstract Mutations in the *leucine-rich repeat kinase 2* (*LRRK2*) gene are the most frequent cause of Parkinson's disease (PD) with late-onset and autosomal-dominant inheritance. LRRK2 belongs to the ROCO superfamily of proteins, characterized by a Ras-of-complex (Roc) GTPase domain in tandem with a C-terminal-of-Roc (COR) domain. LRRK2 also contains a protein kinase domain adjacent to the Roc-COR tandem domain in addition to multiple repeat domains. Disease-causing familial mutations cluster within the Roc-COR tandem and kinase domains of LRRK2, where they act to either impair GTPase activity or enhance kinase activity. Familial LRRK2 mutations share in common the capacity to induce neuronal toxicity in cultured cells. While the contribution of the frequent G2019S mutation, located within the kinase domain, to kinase activity and neurotoxicity has been extensively investigated, the contribution of GTPase activity has received less attention. The GTPase domain has been shown to play an important role in regulating kinase activity, in dimerization, and in mediating the neurotoxic effects of LRRK2. Accordingly, the GTPase domain has emerged as a potential therapeutic target for inhibiting the pathogenic effects of LRRK2 mutations. Many important mechanisms remain to be elucidated, including how the GTPase cycle of LRRK2 is regulated, whether GTPase effectors exist for LRRK2, and how GTPase activity contributes to the overall functional output of LRRK2. In this review, we discuss the importance of the GTPase domain for LRRK2-linked PD focusing in particular on its regulation, function, and contribution to neurotoxic mechanisms.

Keywords LRRK2 • GTPase • ROCO protein • Roc • COR • Roc-COR • Dimerization • Parkinson's disease • PARK8

A.P.T. Nguyen • D.J. Moore (✉)
Center for Neurodegenerative Science, Van Andel Research Institute, 333 Bostwick Ave NE, Grand Rapids, MI 49503, USA
e-mail: AnPhu.TranNguyen@vai.org; darren.moore@vai.org

H.J. Rideout (ed.), *Leucine-Rich Repeat Kinase 2 (LRRK2)*,
Advances in Neurobiology 14, DOI 10.1007/978-3-319-49969-7_4

Introduction

Missense mutations in the *leucine-rich repeat kinase 2* (LRRK2) gene cause late-onset, autosomal-dominant Parkinson's disease (PD) and represent the most common cause of familial PD accounting for 5–15 % of dominant PD [1–4]. Moreover, genome-wide association studies have identified common variants in the LRRK2 genomic locus that are associated with risk for idiopathic PD [5–7]. Several putative variants in the LRRK2 gene have been identified with at least seven missense mutations (i.e., N1437H, R1441C, R1441G, R1441H, Y1699C, G2019S, and I2020T) considered to be truly pathogenic based upon segregation with disease in PD families (Fig. 4.1, [8, 9]). The most frequent mutation is G2019S which is responsible for up to 40 % of familial PD depending on ethnicity and 1–2 % of idiopathic PD [2, 8]. The presence of the G2019S variant in idiopathic cases has been attributed to age-dependent but incomplete penetrance [2, 10]. Interestingly, familial mutations tend to cluster within the catalytic triad of the LRRK2 protein composed of a Ras-of-complex (Roc) GTPase domain (i.e., N1437H, R1441C, R1441G, R1441H) and a protein kinase domain (i.e., G2019S, I2020T) separated by a C-terminal-of-Roc domain (COR; i.e., Y1699C) (Fig. 4.1, [11]). In the past decade since the first discovery of LRRK2 mutations, the kinase activity of LRRK2 has been investigated

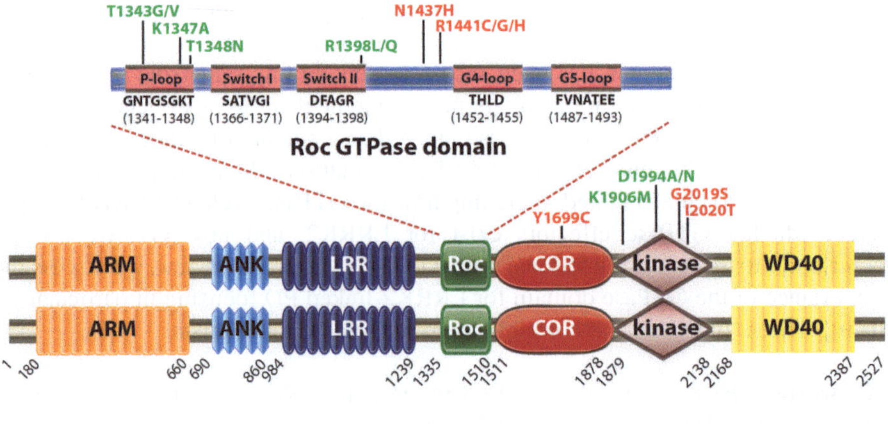

Fig. 4.1 Domain architecture, familial mutations, and functional residues of human LRRK2. LRRK2 is shown as a homodimeric multi-domain protein. Individual LRRK2 domains are depicted in the full-length protein with their respective amino acid positions: ARM, armadillo repeat region; ANK, ankyrin repeat region; LRR, leucine-rich repeats; Roc, Ras-of-complex GTPase domain; COR, C-terminal-of-Roc domain; kinase, protein tyrosine kinase-like domain; WD40, WD40 repeat region. Familial pathogenic mutations that segregate with PD are indicated in red that cluster within the Roc-COR-kinase catalytic region, whereas key functional residues that alter enzymatic activity within the Roc GTPase and kinase domains are indicated in green. The Roc GTPase domain (shown at the top) contains five G-box motifs that are conserved in members of the small GTPase superfamily: guanine nucleotide phosphate-binding loop (P-loop) that binds to GDP or GTP, switch I and II motifs that change conformation upon GTP binding and regulate GTP hydrolysis, and G4 and G5 motifs. K1347A and T1348N mutations impair GDP/GTP binding, whereas R1398L or R1398Q/T1343G increases GTP hydrolysis and R1398L/T1343V impairs GTP hydrolysis

intensively largely in the context of the most common G2019S mutation which has been shown to produce a hyperactive kinase. Although a large number of familial mutations are located within the Roc-COR tandem domain, the role of GTPase activity in regulating the normal function of LRRK2 has received considerably less attention and remains incompletely understood. However, as will be discussed, the GTPase domain is likely key for understanding LRRK2 function and neurotoxic mechanisms, and therefore the GTPase domain represents an important therapeutic target for the treatment of PD.

LRRK2 Domain Architecture

LRRK2 encodes a multi-domain protein of 2527 amino acids that exists predominantly in a dimeric form in cells and tissues [11–15]. At present, the structure of LRRK2 has not been determined related largely to technical issues of purifying sufficient quantities of soluble full-length recombinant human LRRK2 protein. X-ray crystallography has so far been applied to isolated LRRK2 domains, including the kinase domain and the Roc domain revealing a controversial domain-swapped dimeric conformation [16, 17]. LRRK2 is predicted to contain multiple domains including a central Roc-COR tandem domain, a tyrosine kinase-like protein kinase domain, and at least four repeat domains located within N-terminal (armadillo, ankyrin, and leucine-rich repeats) and C-terminal (WD40 repeats) regions (Fig. 4.1, [18]). The Roc-COR tandem domain classifies LRRK2 as a member of the ROCO protein superfamily which represents a unique multi-domain family of Ras-like G proteins [19–21]. Important structural and functional understanding of LRRK2 has been inferred from other ROCO protein members in bacteria and *Dictyostelium discoideum* where the ROCO family was first described [17, 22, 23]. ROCO proteins are characterized by the presence of a Roc-COR tandem domain often (but not always) in association with a kinase domain [20, 24]. The Roc domain contains five G-box motifs that are required for guanine nucleotide binding and hydrolysis (Fig. 4.1, [11, 19, 25]). In mammals, four ROCO proteins have been identified including LRRK1, LRRK2, MASL1 (malignant fibrous histiocytoma-amplified sequence with leucine-rich tandem repeats 1), and DAPK1 (death-associated protein kinase 1) [24]. LRRK1 and LRRK2 share the closest sequence homology and differ only in the N-terminal region which is ~650 amino acids longer in LRRK2. LRRK1, LRRK2, and DAPK1 possess a kinase domain, whereas MASL1 does not. The evolutionary conservation of the Roc-COR tandem domain, independent of a kinase domain, implies that GTPase activity is most likely the primary functional output of ROCO proteins with the kinase domain potentially serving to regulate the GTPase domain. However, it is not yet known whether LRRK2 conforms to the classic model of a ROCO protein since there is evidence that the kinase domain may serve to regulate the intrinsic GTPase domain as well as extrinsic protein substrate phosphorylation. Interestingly, only LRRK2 and DAPK1 have so far been linked to human disease (PD and cancer, respectively) [24, 25].

GTPase Domain and Activity

Genetic Mutations Located in Roc and COR Domains Cause Familial PD

Several familial mutations of LRRK2 are located within the Roc-COR tandem domain suggesting that GTPase activity plays an important role in the development of PD [11]. Although the Roc GTPase domain comprises only a small fraction (residues 1335–1510) of full-length LRRK2 protein (~7 % of total), this domain represents something of a hotspot for mutations perhaps best exemplified by the identification of three PD mutations at a single R1441 residue (i.e., R1441C, R1441G, and R1441H) [8, 11]. The R1441G variant is frequent in PD families from the Basque region of Spain (~20 % of familial PD) but rare elsewhere, whereas the R1441C and R1441H variants are found in many populations but are not frequent [8]. Another variant nearby in the Roc domain, N1437H, has been identified in a large Norwegian family with autosomal-dominant PD [9]. Additional variants in the Roc domain such as I1371V are found in individual PD cases but have not yet been confirmed by segregation analyses in families [8, 26]. The protective R1398H variant is associated with PD in certain populations suggesting that variation within the GTPase domain may also be beneficial [27–29]. The familial Y1699C mutation, identified in German-Canadian and UK families, is the only known disease-causing variant located within the COR domain (residues 1510–1850) [4, 30]. Human genetic studies highlight the importance of the Roc-COR tandem domain to the development of PD with multiple independent mutations located within this region. In contrast, only two mutations, G2019S and I2020T, located in adjacent residues of the kinase activation loop, are known to unambiguously cause familial PD [8]. Pathogenic mutations located outside of the catalytic triad are either rare or their pathogenicity has not been confirmed. The major challenge now becomes how can we best reconcile the functional effects of disease-causing mutations in the Roc-COR tandem and kinase domains for understanding their effects on overall LRRK2 function.

LRRK2 Is a Functional GTPase

LRRK2 is known to possess GTPase activity, at least when measured in in vitro assays using recombinant protein. LRRK2 can selectively bind to guanine nucleotides (GDP and GTP) with similar affinity via a phosphate-binding “P-loop” motif (^{1341}GNTGSGKT^{1348}) within its GTPase domain (Fig. 4.1, [31–35]). This has been measured by binding of LRRK2 or the isolated Roc domain to immobilized or radiolabeled GTP and its non-hydrolyzable analogs (i.e., GTPγS, GppCp), either using purified recombinant LRRK2 or LRRK2 expressed in cell extracts. Competition with an excess of free GTP or GDP can reduce this binding as can synthetic mutations that disrupt key P-loop residues such as K1347A or T1348N, thereby confirming binding specificity. The impact of familial mutations on GTP-binding activity of LRRK2 is somewhat inconsistent but with some evidence that mutations located in the Roc-COR tandem domain tend to increase GTP binding [9, 35]. LRRK2 also exhibits a low rate of GTP hydrolysis activity in vitro. Interestingly, familial mutations in the Roc-COR tandem domain have been shown to reduce GTPase activity to varying degrees. R1441C, R1441G, R1441H, and Y1699C variants all exhibit decreased GTP hydrolysis compared to wild-type LRRK2 [31–33, 36–38]. Familial mutations located in the kinase domain such as G2019S have no discernible effects on GTP binding or hydrolysis suggesting that their pathogenic effects are solely mediated through altered kinase activity [15, 35, 37]. Oppositely, mutations located in the Roc-COR domain have inconsistent effects on kinase activity with the general consensus across most laboratories being that they do not alter kinase activity [18, 35, 39, 40]. Therefore, a clearer picture has emerged that familial mutations in the Roc-COR domain consistently impair GTP hydrolysis, and some may also correspondingly increase the affinity for GTP binding, whereas mutations located in the kinase domain influence only kinase activity. One interesting exception is the familial N1437H mutation, located in the Roc domain, which is reported to simultaneously increase both GTP-binding and kinase activity [9].

At a functional level, only a handful of key residues within the LRRK2 Roc domain have so far been identified to influence GTPase activity based upon highly conserved residues in related small G proteins. The P-loop null mutants, K1347A and T1348N, are useful tools that effectively disrupt GTP binding, but they also have the undesirable effect of impairing LRRK2 dimerization and compromising protein stability [15, 31, 34, 35]. Therefore, these mutants should be used with some caution when attempting to attribute the contribution of GTP binding to LRRK2 activity or cellular properties. Nonetheless, the effects of P-loop mutants on LRRK2 suggest that either guanine nucleotide binding or the conformation of the P-loop (or Roc domain in general) is important for dimerization and that dimeric LRRK2 is generally more stable than monomeric LRRK2 in cells [15]. Notably, familial PD mutations have not been identified in the P-loop motif. The switch II motif (^{1394}DFAGR^{1398}) has been shown to be critical for the GTP hydrolysis activity of LRRK2 (Fig. 4.1, [31]). The R1398 residue in LRRK2 is typically a glutamine (Gln, Q) in the vast majority of small GTPases, whereas likewise the T1343 residue in the P-loop is typically a glycine (Gly, G). Replacing both R1398 with a Gln (R1398Q) and T1343 with a Gly (T1343G) to create a Ras-like form of LRRK2 (R1398Q/T1343G) increases GTPase activity [31]. A Gln → Leu (Q → L) substitution in the switch II region of Ras GTPases oppositely impairs GTP hydrolysis and creates a "GTP-locked" protein that is constitutively bound to GTP. However, introduction of a Leu at 1398 (R1398L) unexpectedly increases the GTP hydrolysis activity (by two- to threefold) of LRRK2, similar to the Ras-like R1398Q/T1343G mutant, creating a functional mutant equivalent to a predominant "GDP bound" protein [15, 37]. Combining the R1398L mutant with a T1343V mutation in the P-loop (R1398L/T1343V) is now sufficient to create a "GTP-bound" form of LRRK2 with impaired GTPase activity [15], similar to corresponding mutations in some related Ras proteins. The Ras-like R1398Q/T1343G, GDP-bound R1398L, and GTP-bound R1398L/T1343V mutant forms of LRRK2 exhibit normal GTP binding, are relatively stable in cells, and do not influence LRRK2 dimerization, in contrast to mutants disrupting GTP binding (i.e., K1347A, T1348N) [15]. These hypothesis-testing mutations provide important tools for exploring the contribution of GTPase activity to other functions of LRRK2, such as kinase activity, dimerization, or cellular phenotypes. The availability of GTPase mutations that can create GDP-bound and GTP-bound forms of LRRK2 may also prove useful for identifying GTPase effector proteins that bind in the GTP-bound "on" state as well as guanine nucleotide exchange factors (GEFs) and GTPase-activating proteins (GAPs) that regulate the GTPase cycle, if indeed they exist for LRRK2.

Regulation of GTPase Activity

Members of the Ras GTPase superfamily conventionally function as binary molecular switches cycling between a GDP-bound "off" and GTP-bound "on" state. When small GTPases are bound to GDP, they adopt an inactive conformation, but GTP binding induces conformational changes which allow the enzyme to bind and activate effector proteins and initiate signal transduction cascades. Most small GTPases are regulated by GAPs that promote the hydrolysis of GTP to GDP and render the GTPase inactive (GDP bound) and GEFs that promote the exchange of GDP with GTP leading to an active GTPase (GTP bound). GTPase activation is often coupled to the activation of a kinase effector similar to a classical Ras GTPase/Raf kinase mechanism, and thus parallels have been drawn with LRRK2 which contains an intrinsic kinase domain adjacent to the Roc-COR tandem domain. This has led to the notion that the GTPase and kinase domains of LRRK2 could be linked together via an intramolecular mechanism with GTPase activity serving to regulate kinase activity. However, current evidence is not consistent with such a simple mechanism. For example, GTP binding to LRRK2 does indeed increase kinase activity, whereas disrupting GTP binding by mutating key P-loop residues impairs kinase activity (albeit with the caveat that such mutations also disrupt dimerization) [15, 31, 34, 35]. However, GTP binding to LRRK2 only increases kinase activity in the context of a cell extract, whereas GTP binding directly to purified recombinant LRRK2 has no impact on kinase activity [15, 41]. This observation has led to the suggestion that GTP-binding capacity rather than direct GTP binding per se drives kinase activation and may hint at the requirement for a yet to be identified GTP-dependent accessory protein (presumably only present in cell extracts) [41]. A second inconsistency with a "simple" Ras-/Raf-like mechanism is the observation that GTP hydrolysis unexpectedly appears to contribute to kinase activation rather than serving to terminate kinase activity as might be predicted from such a model. For example, the GTP-locked R1398L/T1343V mutant exhibits markedly reduced kinase activity, whereas ArfGAP1, a GAP-like protein for LRRK2 which promotes its GTP hydrolysis activity, also enhances its kinase activity [15, 42]. Furthermore, while GTP and

GppCp binding have been shown to promote LRRK2 kinase activity [15, 41], this effect is attenuated when GTP hydrolysis activity is impaired (via the R1398L/T1343V mutation) [15], suggesting that GTPase activity might be required in part for kinase activation. It is notable that GTP increases kinase activity to a greater extent than non-hydrolyzable GppCp again suggesting a requirement for an actual GTP hydrolysis event (presuming that GTP and GppCp have similar affinity for binding to LRRK2 in this assay) [15]. These observations would be consistent with the effects of ArfGAP1 on promoting kinase activity [42]. While GTP binding can promote LRRK2 kinase activity in an unconventional manner, the GTP hydrolysis event itself may also contribute to kinase activation. Therefore, the intramolecular regulation of GTPase and kinase activities is rather complex, and LRRK2 does not appear to function as a canonical GTPase.

Although LRRK2 is an unconventional GTPase, there is some evidence for the regulation of LRRK2 GTPase activity by GAPs and GEFs. ArhGEF7 was first nominated as a GEF and was shown to interact with LRRK2 in cells and mouse brain and increases its GTPase activity [43]. The familial R1441C mutation causes reduced binding of LRRK2 to ArhGEF7 consistent with the known reduction of GTPase activity in this mutant. ArhGEF7 binding to LRRK2 is regulated by the phosphorylation of LRRK2 by CK1α, and loss of this constitutive phosphorylation increases ArhGEF7 binding and alters ArhGEF7-mediated LRRK2 GTP binding [44]. LRRK2 also phosphorylates ArhGEF7 in vitro at two threonine residues within its N-terminus [43]. It is unclear whether ArhGEF7 activity imparts any effect on LRRK2 kinase activity and also what impact phosphorylation has on ArhGEF7 activity.

GAP-like proteins for LRRK2 have also been identified. Deletion of *GCS1* (an ortholog of mammalian ArfGAP1) in yeast was originally identified as a suppressor of LRRK2-induced toxicity [37]. LRRK2 interacts with ArfGAP1 in human cells and in rodent brain, and mutations within the Roc-COR domain that alter GTPase activity modulate the interaction with ArfGAP1 [42, 45]. GTP hydrolysis activity of LRRK2 is markedly enhanced by ArfGAP1 in vitro, in a manner dependent on the GAP domain, consistent with a role for ArfGAP1 as a GAP-like protein for LRRK2 [42, 45]. The impact of ArfGAP1 on LRRK2 kinase activity is unclear with reports of either increased or reduced activity [42, 45]. Unexpectedly, ArfGAP1 also serves as a robust substrate of LRRK2-mediated phosphorylation, with multiple putative sites of phosphorylation identified, although the impact of phosphorylation on ArfGAP1 activity and function is not yet clear [42, 45]. In rodent primary neurons, silencing of ArfGAP1 expression consistently rescues G2019S LRRK2-induced toxicity [42, 45], similar to findings in yeast [37], supporting a critical role for ArfGAP1 in mediating toxicity downstream of LRRK2. Whether this toxic pathway involves the phosphorylation and/or GAP activity of ArfGAP1, or ArfGAP1-dependent effects on Golgi vesicle sorting, is not yet clear. In a *Drosophila* model, the overexpression of mutant LRRK2 or ArfGAP1 alone induces dopaminergic neuronal loss and motor deficits, yet surprisingly their co-expression ameliorates these neurotoxic effects through an unknown mechanism [45], a finding that is difficult to reconcile with yeast and neuronal culture data. Whether ArfGAP1 is required for mutant

LRRK2-dependent phenotypes in rodent models remains to be determined. A second putative GAP-like protein, RGS2, has also been identified for LRRK2. RGS2 was originally identified in a *C.elegans* screen as a genetic modifier of the susceptibility of LRRK2 transgenic worms to rotenone-induced dopaminergic neuronal loss [46]. RGS2 was subsequently shown to interact with LRRK2, enhance its GTP hydrolysis activity in vitro, and also serve as a modest substrate of LRRK2 kinase activity [46]. However, opposite to the effects of ArfGAP1, RGS2 reduces LRRK2 kinase activity in vitro, and RGS2 overexpression rescues LRRK2-induced neuronal toxicity suggesting a neuroprotective capacity [46]. How these distinct GAPs both acting upon the GTPase domain can have opposing effects on LRRK2 kinase activity and neuronal toxicity remains to be clarified. However, it is not yet clear whether ArfGAP1 or RGS2 serves as authentic physiological GAPs for LRRK2 in vivo or whether they may serve to modulate GTPase activity via an alternative mechanism perhaps, for example, by acting as GTPase effector proteins or by stabilizing LRRK2 dimers. At this juncture, while LRRK2 can interact with known GEFs and GAPs that can modify its GTPase activity at least in vitro, whether these proteins modulate the GTPase cycle of LRRK2 in a conventional manner requires additional investigation.

An alternative mechanism for the regulation of the LRRK2 GTPase cycle has been proposed based upon the structure of simpler ROCO proteins. LRRK2 is suggested to function as a G protein activated by nucleotide-dependent dimerization (GAD) [22, 23]. GADs do not require GEFs or GAPs but instead rely upon nucleotide binding and dimerization to regulate the GTPase cycle. For the related ROCO proteins from *Chlorobium tepidum* and *Methanosarcina barkeri*, constitutive dimerization is mediated through the COR domains where upon GTP binding a conformational shift induces the juxtaposition of the adjacent Roc domains of the dimer so that they complement each other to form an active GTPase [23, 47]. Subsequent GTP hydrolysis, presumably coupled to the activation of an effector protein, restores the Roc domains to the inactive GDP-bound conformation [22, 23, 47]. Critical for this noncanonical mechanism is COR domain-mediated dimerization with the COR domain dimer interface being the most highly conserved region among ROCO proteins [23, 47]. LRRK1 and LRRK2 share a great deal of sequence conservation with simpler ROCO proteins within the COR domain suggesting that a potential dimer interface may also be present in LRRK2 [23, 47]. While LRRK2 has been shown to predominantly exist as a homodimeric protein in cells [12–15], in the absence of high-resolution structural data, domain mapping interaction studies have highlighted potential roles for the Roc or WD40 repeat domains as the putative dimer interface [13, 16, 36, 38, 48]. A recent study has provided evidence for an interaction between the isolated COR domains of LRRK2, but the relative strength of this interaction compared to those between other domains (i.e., Roc-Roc or Roc-COR) is not yet clear [47]. An intermolecular interaction between the Roc domains, albeit potentially weaker, is to be expected based upon the proposed GAD mechanism. Recent studies have shown that the isolated Roc domain of LRRK2 can form stable monomeric or dimeric conformations in solution that are catalytically active [38, 49], suggesting that dimerization is not essential for activity although whether this applies in the context of the full-length LRRK2 protein is not yet clear. One of the most robust interactions within LRRK2 appears to be between the Roc and COR domains, being more robust than the interaction between Roc domains [13, 36]. The interaction between isolated Roc and COR domains is strengthened by the familial Y1699C mutation that is suggested to occupy an intramolecular interaction interface and limit the conformational flexibility of these domains within a monomer [36], thereby potentially explaining how this mutation impairs GTPase activity [36, 37]. Such an effect of this mutation would be consistent with a GAD mechanism for LRRK2 based on similar studies with *C. tepidum* ROCO protein using mutations analogous to R1441C and Y1699C [23]. At present, it is not known which domains and residues of LRRK2 are important for dimerization in the full-length protein since most studies have relied upon isolated domains, therefore highlighting the need for structural data. One inconsistency with the GAD model is that guanine nucleotide binding does not appear to regulate LRRK2 dimerization [36]. Instead, functional mutations that disrupt the P-loop (i.e., K1347A or T1348N), pharmacological kinase inhibition, and association with cellular membranes have been reported to modulate LRRK2 dimerization [12–15, 47]. It is possible that dimerization alone is sufficient to regulate the GTPase cycle of LRRK2. While a GAD mechanism for LRRK2 is an attractive hypothesis, there is currently limited evidence to support such a mechanism, and additional biochemical and structural studies will be required to further understand how the GTPase cycle is regulated.

While there is some evidence that GTPase activity can regulate LRRK2 kinase activity in an unconventional manner, there is also emerging evidence that kinase activity may reciprocally serve to regulate GTPase activity. The mapping of in vitro autophosphorylation sites within LRRK2 by mass spectrometry reveals that many of these sites tend to cluster within the Roc domain at multiple serine and threonine residues including key P-loop residues (T1343, T1348), S1403, T1404, T1410, T1491, and T1503 [50–53]. How phosphorylation at each of these sites regulates GTP binding and hydrolysis activity is not yet clear. However, in full-length LRRK2, disrupting kinase activity has no appreciable effect on GTPase activity suggesting that phosphorylation is likely to have rather subtle or dynamic effects on GTPase activity depending upon the combination of sites modified in a single dimer [15]. Phosphorylation at individual sites is not particularly abundant and varies between sites and may modify only a small proportion of LRRK2 at any given time [51, 53]. For this reason it has so far been difficult to confirm the phosphorylation at individual sites occurring in cells and tissues using phospho-specific antibodies [53], with the exception of phosphorylation at S1292 located between the leucine-rich repeat and Roc domains [54]. Studies of the functional impact of individual phospho-sites within the Roc domain are limited, but one study suggests that the T1503 residue may regulate the GTP binding and kinase activity of LRRK2 [53]. A recent study using the isolated Roc domain from LRRK2 suggests that autophosphorylation enhances the rate of GTP hydrolysis and promotes the formation of Roc dimers, potentially by altering the conformation of the P-loop structure [49]. P-loop phosphorylation appears to be common to many GTPases suggesting a novel mechanism for the control of GTPase activity by kinases [49]. Although these studies are insightful, it remains to be determined how autophosphorylation regulates GTPase activity in the context of full-length LRRK2 protein. GTPase activity may also be regulated by extrinsic kinases. PKA has been shown to phosphorylate LRRK2 at S1444, and this modification is reduced by the familial mutations R1441C/R1441G/R1441H which occupy a consensus PKA recognition site [55]. Phosphorylation at S1444 by PKA serves as a 14-3-3 docking site, and binding leads to decreased kinase activity in vitro, whereas inhibition of S1444 phosphorylation impairs 14-3-3 binding and increases kinase activity [55]. The mechanism by which S1444 phosphorylation influences kinase activity is not yet clear, and the impact of phosphorylation and 14-3-3 binding on the GTPase cycle of LRRK2 has not been determined. Intrinsic and extrinsic phosphorylation may therefore provide an additional level of regulation of the LRRK2 GTPase cycle potentially by altering P-loop structure or by the recruitment of accessory proteins.

Contribution of **GTPase** *Activity to LRRK2-Induced Toxicity*

While familial mutations in the kinase domain, such as G2019S, have been extensively shown to induce cellular toxicity in culture models in a kinase-dependent manner [34, 56–59], there is still uncertainty about the contribution of GTPase activity to cellular toxicity. Studies in a yeast model expressing the catalytic core of human LRRK2 have highlighted a critical requirement of the GTPase domain and GTPase activity for cellular toxicity. While mutations that disrupt GTP binding and hydrolysis (i.e., K1347A or T1348N) cause a dramatic increase in yeast toxicity, enhancing GTP hydrolysis (i.e., R1398L or R1398Q/T1343G) improves viability [37]. Similar effects of these LRRK2 synthetic mutants on the viability of primary neuronal models have been confirmed. LRRK2-induced toxicity in yeast correlates with severe defects in endosomal trafficking to the vacuole and the accumulation of autophagosomes [37]. Genetic suppressors of LRRK2 toxicity also restore endosomal trafficking suggesting a causal role. Similar cellular phenotypes have been observed with full-length mutant LRRK2 in mammalian cells or neurons [58, 60–64]. Consistent with the importance of GTPase activity for LRRK2 toxicity, the GTP-locked R1398L/T1343V mutation enhances neuronal toxicity induced by LRRK2 in primary cultures comparable to the effects of the pathogenic G2019S mutation [15]. Interestingly, however, GTPase-hyperactive R1398L or GTPase-impaired R1398L/T1343V mutations are not able to modify the elevated kinase activity or neurotoxic effects of G2019S LRRK2 [15], suggesting that the G2019S variant may act independent of GTPase activity. A prior study suggested that disruption of GTP binding (via K1347A) can rescue neuronal toxicity induced by G2019S LRRK2 [34], although it is likely that neuroprotection may result instead from the impaired dimerization and destabilization of LRRK2 known to be caused by the K1347A variant in neurons [15]. As mentioned above, ArfGAP1 expression is required for G2019S LRRK2-induced neuronal toxicity, whereas RGS2 is neuroprotective [42, 46]. It is not known whether these GAPs act upon LRRK2 GTPase or kinase activity to regulate neurotoxicity or whether they act in pathways downstream of LRRK2 potentially as GTPase effectors or kinase substrates. Aside from the G2019S mutation, it is not yet known whether GTPase activity contributes to neuronal toxicity induced by familial mutations in the Roc-COR domain such as R1441C and Y1699C, although some evidence suggests that kinase activity may be required for the toxicity of the R1441C mutant [34, 56, 59]. A number of model organisms with interesting phenotypes have been developed based upon familial mutations in the Roc-COR tandem domain of LRRK2 although mechanistic insight into the contribution of GTPase activity is so far lacking [65]. For example, R1441C or Y1699C LRRK2 selectively inhibit axonal transport and cause locomotor deficits in neuronal and *Drosophila* models that may result from their preferential association with deacetylated microtubules [66]. Increasing microtubule acetylation prevents the association of mutant LRRK2 with microtubules and restores axonal transport [66]. Transgenic or knockin mouse models expressing R1441G or R1441C LRRK2 exhibit a combination of motor deficits, impaired dopaminergic neurotransmission, axonopathy, tau pathology, altered autophagy, or abnormal nuclear envelope architecture [63, 67–70]. Understanding how GTPase activity contributes to these LRRK2-dependent phenotypes will be challenging and may rely in the future upon genetic or pharmacological manipulation of the GTPase domain.

Conclusion and Future Perspectives

LRRK2 is a central player in PD and an attractive target for therapeutic development. However, LRRK2-related mechanisms leading to neurotoxicity remain incompletely understood. So far, most studies have highlighted the kinase activity of LRRK2 as a key therapeutic target since the most common G2019S mutation elevates kinase activity and induces neuronal toxicity in a kinase-dependent manner [18, 34, 40, 57]. Genetic or pharmacological inhibition of kinase activity has been proven to be protective in viral-based G2019S LRRK2 rodent models [57, 71] and also protects against neurodegeneration induced by human α-synuclein or LPS-induced neuroinflammation in rat models [72]. A robust in vivo substrate of LRRK2 kinase activity is still lacking to be able to fully explain the neuroprotective effects of kinase inhibition, although one such substrate could be LRRK2 itself via autophosphorylation [54].

LRRK2 contains an evolutionarily conserved Roc-COR tandem domain, and many familial mutations are clustered within the Roc and COR domains and impair GTPase activity. Therefore, GTPase activity is clearly important for LRRK2 function, for regulating kinase activity, and for the development of PD. Potential therapeutic strategies for targeting the Roc-COR tandem domain might include (1) inhibition of GTP binding, (2) modulation of GTP hydrolysis, (3) disrupting dimerization, (4) kinase inhibition, or (5) targeting GAPs, GEFs, or GTPase effectors. The validation of each of these strategies is now required in disease-relevant models and may rely upon the future development of small-molecule compounds. There have been a paucity of such studies and compounds, although recent studies have reported the development of novel compounds that simultaneously inhibit LRRK2 GTP binding and kinase activity and attenuate LRRK2 toxicity, although the mechanism of action is not yet clear [73, 74]. Yeast LRRK2 models could prove to be a useful tool for screening and identifying small-molecule GTPase modulators since cellular toxicity in this model is fully dependent on GTPase activity [37]. The incorporation of hypothesis-testing mutations that create "GTP-locked" and "GDP-locked" forms of LRRK2 into adenoviral-mediated rodent models expressing mutant LRRK2 may prove informative for understanding how best to modulate GTPase activity to attenuate neurodegenerative phenotypes [15]. Such models produce rapid and robust phenotypes and it is relatively simple to produce viruses containing new LRRK2 variants [75]. A similar approach has been employed to demonstrate that kinase activity is required for neuropathology induced by G2019S LRRK2 [71]. This approach could also be used to evaluate key residues that are important for LRRK2 dimerization or to modulate the expression of GAPs such as ArfGAP1 or RGS2.

The Roc-COR tandem domain and GTPase activity of LRRK2 represent attractive and potentially tractable targets for the development of new therapeutics to treat PD. A deeper mechanistic understanding of how the GTPase cycle is regulated and how GTPase activity modulates kinase activity and neurotoxicity will be critical to fully understand LRRK2 function and its role in the development of PD.

Acknowledgments The authors are grateful for funding support from the National Institutes of Health (R01 NS091719), the Swiss National Science Foundation (grant no. 31003A_144063), and the Van Andel Research Institute.

Conflict of Interest The author declares no conflicts of interest.

References

1. Gasser T (2009) Mendelian forms of Parkinson's disease. Biochim Biophys Acta 1792:587–596
2. Healy DG, Falchi M, O'Sullivan SS, Bonifati V, Durr A, Bressman S, Brice A, Aasly J, Zabetian CP, Goldwurm S, Ferreira JJ, Tolosa E, Kay DM, Klein C, Williams DR, Marras C, Lang AE, Wszolek ZK, Berciano J, Schapira AHV, Lynch T, Bhatia KP, Gasser T, Lees AJ, Wood NW (2008) Phenotype, genotype, and worldwide genetic penetrance of LRRK2-associated Parkinson's disease: a case–control study. Lancet Neurol 7(7):583–590. doi:10.1016/S1474-4422(08)70117-0
3. Paisán-Ruiz C, Jain S, Evans EW, Gilks WP, Simón J, van der Brug M, de Munain AL, Aparicio S, Gil AM, Khan N, Johnson J, Martinez JR, Nicholl D, Carrera IM, Pena AS, de Silva R, Lees A, MartI-Massó JF, Pérez-Tur J, Wood NW, Singleton AB (2004) Cloning of the gene containing mutations that cause PARK8-linked Parkinson's disease. Neuron 44(4):595–600. doi:10.1016/j.neuron.2004.10.023
4. Zimprich A, Biskup S, Leitner P, Lichtner P, Farrer M, Lincoln S, Kachergus J, Hulihan M, Uitti RJ, Calne DB, Stoessl AJ, Pfeiffer RF, Patenge N, Carbajal IC, Vieregge P, Asmus F, Müller-Myhsok B, Dickson DW, Meitinger T, Strom TM, Wszolek ZK, Gasser T (2004)

Mutations in LRRK2 cause autosomal-dominant Parkinsonism with pleomorphic pathology. Neuron 44(4):601–607. doi:10.1016/j.neuron.2004.11.005
5. Satake W, Nakabayashi Y, Mizuta I, Hirota Y, Ito C, Kubo M, Kawaguchi T, Tsunoda T, Watanabe M, Takeda A, Tomiyama H, Nakashima K, Hasegawa K, Obata F, Yoshikawa T, Kawakami H, Sakoda S, Yamamoto M, Hattori N, Murata M, Nakamura Y, Toda T (2009) Genome-wide association study identifies common variants at four loci as genetic risk factors for Parkinson's disease. Nat Genet 41(12):1303–1307
6. Simon-Sanchez J, Schulte C, Bras JM, Sharma M, Gibbs JR, Berg D, Paisan-Ruiz C, Lichtner P, Scholz SW, Hernandez DG, Kruger R, Federoff M, Klein C, Goate A, Perlmutter J, Bonin M, Nalls MA, Illig T, Gieger C, Houlden H, Steffens M, Okun MS, Racette BA, Cookson MR, Foote KD, Fernandez HH, Traynor BJ, Schreiber S, Arepalli S, Zonozi R, Gwinn K, van der Brug M, Lopez G, Chanock SJ, Schatzkin A, Park Y, Hollenbeck A, Gao J, Huang X, Wood NW, Lorenz D, Deuschl G, Chen H, Riess O, Hardy JA, Singleton AB, Gasser T (2009) Genome-wide association study reveals genetic risk underlying Parkinson's disease. Nat Genet 41(12):1308–1312
7. Nalls MA, Pankratz N, Lill CM, Do CB, Hernandez DG, Saad M, DeStefano AL, Kara E, Bras J, Sharma M, Schulte C, Keller MF, Arepalli S, Letson C, Edsall C, Stefansson H, Liu X, Pliner H, Lee JH, Cheng R, International Parkinson's Disease Genomics Consortium (IPDGC), Parkinson's Study Group (PSG) Parkinson's Research: The Organized GENetics Initiative (PROGENI), 23andMe, GenePD, NeuroGenetics Research Consortium (NGRC), Hussman Institute of Human Genomics (HIHG), Ashkenazi Jewish Dataset Investigator, Cohorts for Health and Aging Research in Genetic Epidemiology (CHARGE), North American Brain Expression Consortium (NABEC), United Kingdom Brain Expression Consortium (UKBEC), Greek Parkinson's Disease Consortium, Alzheimer Genetic Analysis Group, Ikram MA, Ioannidis JP, Hadjigeorgiou GM, Bis JC, Martinez M, Perlmutter JS, Goate A, Marder K, Fiske B, Sutherland M, Xiromerisiou G, Myers RH, Clark LN, Stefansson K, Hardy JA, Heutink P, Chen H, Wood NW, Houlden H, Payami H, Brice A, Scott WK, Gasser T, Bertram L, Eriksson N, Foroud T, Singleton AB (2014) Large-scale meta-analysis of genome-wide association data identifies six new risk loci for Parkinson's disease. Nat Genet 46(9):989–993. doi:10.1038/ng.3043
8. Biskup S, West AB (2009) Zeroing in on LRRK2-linked pathogenic mechanisms in Parkinson's disease. Biochim Biophys Acta 1792(7):625–633. doi:10.1016/j.bbadis.2008.09.015
9. Aasly JO, Vilariño-Güell C, Dachsel JC, Webber PJ, West AB, Haugarvoll K, Johansen KK, Toft M, Nutt JG, Payami H, Kachergus JM, Lincoln SJ, Felic A, Wider C, Soto-Ortolaza AI, Cobb SA, White LR, Ross OA, Farrer MJ (2010) Novel pathogenic LRRK2 p.Asn1437His substitution in familial Parkinson's disease. Mov Disord 25(13):2156–2163. doi:10.1002/mds.23265
10. Hulihan MM, Ishihara-Paul L, Kachergus J, Warren L, Amouri R, Elango R, Prinjha RK, Upmanyu R, Kefi M, Zouari M, Sassi SB, Yahmed SB, El Euch-Fayeche G, Matthews PM, Middleton LT, Gibson RA, Hentati F, Farrer MJ (2008) LRRK2 Gly2019Ser penetrance in Arab-Berber patients from Tunisia: a case–control genetic study. Lancet Neurol 7(7):591–594. doi:10.1016/S1474-4422(08)70116-9
11. Tsika E, Moore DJ (2013) Contribution of GTPase activity to LRRK2-associated Parkinson disease. Small GTPases 4(3):164–170. doi:10.4161/sgtp.25130
12. Berger Z, Smith KA, LaVoie MJ (2010) Membrane localization of LRRK2 is associated with increased formation of the highly active LRRK2 dimer and changes in its phosphorylation. Biochemistry 49(26):5511–5523. doi:10.1021/bi100157u
13. Greggio E, Zambrano I, Kaganovich A, Beilina A, Taymans J-M, Daniëls V, Lewis P, Jain S, Ding J, Syed A, Thomas KJ, Baekelandt V, Cookson MR (2008) The Parkinson disease-associated leucine-rich repeat kinase 2 (LRRK2) is a dimer that undergoes intramolecular autophosphorylation. J Biol Chem 283(24):16906–16914. doi:10.1074/jbc.M708718200
14. Sen S, Webber PJ, West AB (2009) Dependence of leucine-rich repeat kinase 2 (LRRK2) kinase activity on dimerization. J Biol Chem 284(52):36346–36356. doi:10.1074/jbc.M109.025437

15. Biosa A, Trancikova A, Civiero L, Glauser L, Bubacco L, Greggio E, Moore DJ (2013) GTPase activity regulates kinase activity and cellular phenotypes of Parkinson's disease-associated LRRK2. Hum Mol Genet 22(6):1140–1156. doi:10.1093/hmg/dds522
16. Deng J, Lewis PA, Greggio E, Sluch E, Beilina A, Cookson MR (2008) Structure of the ROC domain from the Parkinson's disease-associated leucine-rich repeat kinase 2 reveals a dimeric GTPase. Proc Natl Acad Sci U S A 105(5):1499–1504. doi:10.1073/pnas.0709098105
17. Gilsbach BK, Ho FY, Vetter IR, van Haastert PJ, Wittinghofer A, Kortholt A (2012) Roco kinase structures give insights into the mechanism of Parkinson disease-related leucine-rich-repeat kinase 2 mutations. Proc Natl Acad Sci U S A 109(26):10322–10327. doi:10.1073/pnas.1203223109
18. Greggio E, Cookson M (2009) Leucine-rich repeat kinase 2 mutations and Parkinson's disease: three questions. ASN Neuro 1(1), e00002
19. Marín I (2006) The parkinson disease gene LRRK2: evolutionary and structural insights. Mol Biol Evol 23(12):2423–2433. doi:10.1093/molbev/msl114
20. Marín I, van Egmond WN, van Haastert PJM (2008) The Roco protein family: a functional perspective. FASEB J 22(9):3103–3110. doi:10.1096/fj.08-111310
21. Bosgraaf L, Van Haastert PJM (2003) Roc, a Ras/GTPase domain in complex proteins. Biochim Biophys Acta 1643(1–3):5–10. doi:10.1016/j.bbamcr.2003.08.008
22. Gasper R, Meyer S, Gotthardt K, Sirajuddin M, Wittinghofer A (2009) It takes two to tango: regulation of G proteins by dimerization. Nat Rev Mol Cell Biol 10(6):423–429
23. Gotthardt K, Weyand M, Kortholt A, Van Haastert PJM, Wittinghofer A (2008) Structure of the Roc-COR domain tandem of *C. tepidum*, a prokaryotic homologue of the human LRRK2 Parkinson kinase. EMBO J 27(16):2239–2249
24. Civiero L, Dihanich S, Lewis PA, Greggio E (2014) Genetic, structural, and molecular insights into the function of ras of complex proteins domains. Chem Biol 21(7):809–818. doi:10.1016/j.chembiol.2014.05.010
25. Gilsbach BK, Kortholt A (2014) Structural biology of the LRRK2 GTPase and kinase domains: implications for regulation. Front Mol Neurosci 7:32. doi:10.3389/fnmol.2014.00032
26. Giordana MT, D'Agostino C, Albani G, Mauro A, Di Fonzo A, Antonini A, Bonifati V (2007) Neuropathology of Parkinson's disease associated with the LRRK2 Ile1371Val mutation. Mov Disord 22(2):275–278. doi:10.1002/mds.21281
27. Tan EK, Peng R, Teo YY, Tan LC, Angeles D, Ho P, Chen ML, Lin CH, Mao XY, Chang XL, Prakash KM, Liu JJ, Au WL, Le WD, Jankovic J, Burgunder JM, Zhao Y, Wu RM (2010) Multiple LRRK2 variants modulate risk of Parkinson disease: a Chinese multicenter study. Hum Mutat 31(5):561–568. doi:10.1002/humu.21225
28. Ross OA, Soto-Ortolaza AI, Heckman MG, Aasly JO, Abahuni N, Annesi G, Bacon JA, Bardien S, Bozi M, Brice A, Brighina L, Van Broeckhoven C, Carr J, Chartier-Harlin MC, Dardiotis E, Dickson DW, Diehl NN, Elbaz A, Ferrarese C, Ferraris A, Fiske B, Gibson JM, Gibson R, Hadjigeorgiou GM, Hattori N, Ioannidis JP, Jasinska-Myga B, Jeon BS, Kim YJ, Klein C, Kruger R, Kyratzi E, Lesage S, Lin CH, Lynch T, Maraganore DM, Mellick GD, Mutez E, Nilsson C, Opala G, Park SS, Puschmann A, Quattrone A, Sharma M, Silburn PA, Sohn YH, Stefanis L, Tadic V, Theuns J, Tomiyama H, Uitti RJ, Valente EM, van de Loo S, Vassilatis DK, Vilarino-Guell C, White LR, Wirdefeldt K, Wszolek ZK, Wu RM, Farrer MJ, Genetic Epidemiology of Parkinson's Disease (GEO-PD) Consortium. (2011) Association of LRRK2 exonic variants with susceptibility to Parkinson's disease: a case–control study. Lancet Neurol 10(10):898–908. doi:10.1016/S1474-4422(11)70175-2
29. Heckman MG, Soto-Ortolaza AI, Aasly JO, Abahuni N, Annesi G, Bacon JA, Bardien S, Bozi M, Brice A, Brighina L, Carr J, Chartier-Harlin MC, Dardiotis E, Dickson DW, Diehl NN, Elbaz A, Ferrarese C, Fiske B, Gibson JM, Gibson R, Hadjigeorgiou GM, Hattori N, Ioannidis JP, Boczarska-Jedynak M, Jasinska-Myga B, Jeon BS, Kim YJ, Klein C, Kruger R, Kyratzi E, Lesage S, Lin CH, Lynch T, Maraganore DM, Mellick GD, Mutez E, Nilsson C, Opala G, Park SS, Petrucci S, Puschmann A, Quattrone A, Sharma M, Silburn PA, Sohn YH, Stefanis L, Tadic V, Theuns J, Tomiyama H, Uitti RJ, Valente EM, Van Broeckhoven C, van de Loo S, Vassilatis

DK, Vilarino-Guell C, White LR, Wirdefeldt K, Wszolek ZK, Wu RM, Hentati F, Farrer MJ, Ross OA, Genetic Epidemiology of Parkinson's Disease (GEO-PD) Consortium (2013) Population-specific frequencies for LRRK2 susceptibility variants in the Genetic Epidemiology of Parkinson's Disease (GEO-PD) Consortium. Mov Disord 28(12):1740–1744. doi:10.1002/mds.25600

30. Khan NL, Jain S, Lynch JM, Pavese N, Abou-Sleiman P, Holton JL, Healy DG, Gilks WP, Sweeney MG, Ganguly M, Gibbons V, Gandhi S, Vaughan J, Eunson LH, Katzenschlager R, Gayton J, Lennox G, Revesz T, Nicholl D, Bhatia KP, Quinn N, Brooks D, Lees AJ, Davis MB, Piccini P, Singleton AB, Wood NW (2005) Mutations in the gene LRRK2 encoding dardarin (PARK8) cause familial Parkinson's disease: clinical, pathological, olfactory and functional imaging and genetic data. Brain J Neurol 128(Pt 12):2786–2796. doi:10.1093/brain/awh667
31. Ito G, Okai T, Fujino G, Takeda K, Ichijo H, Katada T, Iwatsubo T (2007) GTP binding is essential to the protein kinase activity of LRRK2, a causative gene product for familial Parkinson's disease. Biochemistry 46(5):1380–1388. doi:10.1021/bi061960m
32. Lewis PA, Greggio E, Beilina A, Jain S, Baker A, Cookson MR (2007) The R1441C mutation of LRRK2 disrupts GTP hydrolysis. Biochem Biophys Res Commun 357(3):668–671. doi:10.1016/j.bbrc.2007.04.006
33. Li X, Tan Y-C, Poulose S, Olanow CW, Huang X-Y, Yue Z (2007) Leucine-rich repeat kinase 2 (LRRK2)/PARK8 possesses GTPase activity that is altered in familial Parkinson's disease R1441C/G mutants. J Neurochem 103(1):238–247. doi:10.1111/j.1471-4159.2007.04743.x
34. Smith WW, Pei Z, Jiang H, Dawson VL, Dawson TM, Ross CA (2006) Kinase activity of mutant LRRK2 mediates neuronal toxicity. Nat Neurosci 9(10):1231–1233
35. West AB, Moore DJ, Choi C, Andrabi SA, Li X, Dikeman D, Biskup S, Zhang Z, Lim K-L, Dawson VL, Dawson TM (2007) Parkinson's disease-associated mutations in LRRK2 link enhanced GTP-binding and kinase activities to neuronal toxicity. Hum Mol Genet 16(2):223–232. doi:10.1093/hmg/ddl471
36. Daniëls V, Vancraenenbroeck R, Law BMH, Greggio E, Lobbestael E, Gao F, De Maeyer M, Cookson MR, Harvey K, Baekelandt V, Taymans J-M (2011) Insight into the mode of action of the LRRK2 Y1699C pathogenic mutant. J Neurochem 116(2):304–315. doi:10.1111/j.1471-4159.2010.07105.x
37. Xiong Y, Coombes CE, Kilaru A, Li X, Gitler AD, Bowers WJ, Dawson VL, Dawson TM, Moore DJ (2010) GTPase activity plays a key role in the pathobiology of LRRK2. PLoS Genet 6(4), e1000902
38. Liao J, Wu CX, Burlak C, Zhang S, Sahm H, Wang M, Zhang ZY, Vogel KW, Federici M, Riddle SM, Nichols RJ, Liu D, Cookson MR, Stone TA, Hoang QQ (2014) Parkinson disease-associated mutation R1441H in LRRK2 prolongs the "active state" of its GTPase domain. Proc Natl Acad Sci U S A 111(11):4055–4060. doi:10.1073/pnas.1323285111
39. Jaleel M, Nichols RJ, Deak M, Campbell DG, Gillardon F, Knebel A, Alessi DR (2007) LRRK2 phosphorylates moesin at threonine-558: characterization of how Parkinson's disease mutants affect kinase activity. Biochem J 405(2):307–317. doi:10.1042/bj20070209
40. West AB, Moore DJ, Biskup S, Bugayenko A, Smith WW, Ross CA, Dawson VL, Dawson TM (2005) Parkinson's disease-associated mutations in leucine-rich repeat kinase 2 augment kinase activity. Proc Natl Acad Sci U S A 102(46):16842–16847
41. Taymans J-M, Vancraenenbroeck R, Ollikainen P, Beilina A, Lobbestael E, De Maeyer M, Baekelandt V, Cookson MR (2011) LRRK2 kinase activity is dependent on LRRK2 GTP binding capacity but independent of LRRK2 GTP binding. PLoS One 6(8), e23207. doi:10.1371/journal.pone.0023207
42. Stafa K, Trancikova A, Webber PJ, Glauser L, West AB, Moore DJ (2012) GTPase activity and neuronal toxicity of Parkinson's disease–associated LRRK2 is regulated by ArfGAP1. PLoS Genet 8(2), e1002526. doi:10.1371/journal.pgen.1002526
43. Haebig K, Gloeckner CJ, Miralles MG, Gillardon F, Schulte C, Riess O, Ueffing M, Biskup S, Bonin M (2010) ARHGEF7 (beta-PIX) acts as guanine nucleotide exchange factor for leucine-rich repeat kinase 2. PLoS One 5(10), e13762. doi:10.1371/journal.pone.0013762

44. Chia R, Haddock S, Beilina A, Rudenko IN, Mamais A, Kaganovich A, Li Y, Kumaran R, Nalls MA, Cookson MR (2014) Phosphorylation of LRRK2 by casein kinase 1alpha regulates trans-Golgi clustering via differential interaction with ARHGEF7. Nat Commun 5:5827. doi:10.1038/ncomms6827
45. Xiong Y, Yuan C, Chen R, Dawson TM, Dawson VL (2012) ArfGAP1 is a GTPase activating protein for LRRK2: reciprocal regulation of ArfGAP1 by LRRK2. J Neurosci 32(11):3877–3886. doi:10.1523/jneurosci.4566-11.2012
46. Dusonchet J, Li H, Guillily M, Liu M, Stafa K, Derada Troletti C, Boon JY, Saha S, Glauser L, Mamais A, Citro A, Youmans KL, Liu L, Schneider BL, Aebischer P, Yue Z, Bandopadhyay R, Glicksman MA, Moore DJ, Collins JJ, Wolozin B (2014) A Parkinson's disease gene regulatory network identifies the signaling protein RGS2 as a modulator of LRRK2 activity and neuronal toxicity. Hum Mol Genet 23(18):4887–4905. doi:10.1093/hmg/ddu202
47. Terheyden S, Ho FY, Gilsbach BK, Wittinghofer A, Kortholt A (2015) Revisiting the Roco G-protein cycle. Biochem J 465(1):139–147. doi:10.1042/BJ20141095
48. Jorgensen ND, Peng Y, Ho CC, Rideout HJ, Petrey D, Liu P, Dauer WT (2009) The WD40 domain is required for LRRK2 neurotoxicity. PLoS One 4(12), e8463. doi:10.1371/journal.pone.0008463
49. Liu Z, Mobley JA, DeLucas LJ, Kahn RA, West AB (2016) LRRK2 autophosphorylation enhances its GTPase activity. FASEB J 30(1):336–347. doi:10.1096/fj.15-277095
50. Gloeckner CJ, Boldt K, von Zweydorf F, Helm S, Wiesent L, Sarioglu H, Ueffing M (2010) Phosphopeptide analysis reveals two discrete clusters of phosphorylation in the N-terminus and the Roc domain of the Parkinson-disease associated protein kinase LRRK2. J Proteome Res 9(4):1738–1745. doi:10.1021/pr9008578
51. Greggio E, Taymans JM, Zhen EY, Ryder J, Vancraenenbroeck R, Beilina A, Sun P, Deng J, Jaffe H, Baekelandt V, Merchant K, Cookson MR (2009) The Parkinson's disease kinase LRRK2 autophosphorylates its GTPase domain at multiple sites. Biochem Biophys Res Commun 389(3):449–454. doi:10.1016/j.bbrc.2009.08.163
52. Kamikawaji S, Ito G, Iwatsubo T (2009) Identification of the autophosphorylation sites of LRRK2. Biochemistry 48(46):10963–10975. doi:10.1021/bi9011379
53. Webber PJ, Smith AD, Sen S, Renfrow MB, Mobley JA, West AB (2011) Autophosphorylation in the leucine-rich repeat kinase 2 (LRRK2) GTPase domain modifies kinase and GTP-binding activities. J Mol Biol 412(1):94–110. doi:10.1016/j.jmb.2011.07.033
54. Sheng Z, Zhang S, Bustos D, Kleinheinz T, Le Pichon CE, Dominguez SL, Solanoy HO, Drummond J, Zhang X, Ding X, Cai F, Song Q, Li X, Yue Z, van der Brug MP, Burdick DJ, Gunzner-Toste J, Chen H, Liu X, Estrada AA, Sweeney ZK, Scearce-Levie K, Moffat JG, Kirkpatrick DS, Zhu H (2012) Ser1292 autophosphorylation is an indicator of LRRK2 kinase activity and contributes to the cellular effects of PD mutations. Sci Trans Med 4(164), 164ra161. doi:10.1126/scitranslmed.3004485
55. Muda K, Bertinetti D, Gesellchen F, Hermann JS, von Zweydorf F, Geerlof A, Jacob A, Ueffing M, Gloeckner CJ, Herberg FW (2014) Parkinson-related LRRK2 mutation R1441C/G/H impairs PKA phosphorylation of LRRK2 and disrupts its interaction with 14-3-3. Proc Natl Acad Sci U S A 111(1):E34–E43. doi:10.1073/pnas.1312701111
56. Greggio E, Jain S, Kingsbury A, Bandopadhyay R, Lewis P, Kaganovich A, van der Brug MP, Beilina A, Blackinton J, Thomas KJ, Ahmad R, Miller DW, Kesavapany S, Singleton A, Lees A, Harvey RJ, Harvey K, Cookson MR (2006) Kinase activity is required for the toxic effects of mutant LRRK2/dardarin. Neurobiol Dis 23(2):329–341
57. Lee BD, Shin J-H, VanKampen J, Petrucelli L, West AB, Ko HS, Lee Y-I, Maguire-Zeiss KA, Bowers WJ, Federoff HJ, Dawson VL, Dawson TM (2010) Inhibitors of leucine-rich repeat kinase-2 protect against models of Parkinson's disease. Nat Med 16(9):998–1000
58. MacLeod D, Dowman J, Hammond R, Leete T, Inoue K, Abeliovich A (2006) The familial Parkinsonism gene LRRK2 regulates neurite process morphology. Neuron 52(4):587–593
59. Ramsden N, Perrin J, Ren Z, Lee BD, Zinn N, Dawson VL, Tam D, Bova M, Lang M, Drewes G, Bantscheff M, Bard F, Dawson TM, Hopf C (2011) Chemoproteomics-based design of

potent LRRK2-selective lead compounds that attenuate Parkinson's disease-related toxicity in human neurons. ACS Chem Biol 6(10):1021–1028. doi:10.1021/cb2002413
60. Manzoni C, Mamais A, Dihanich S, Abeti R, Soutar MP, Plun-Favreau H, Giunti P, Tooze SA, Bandopadhyay R, Lewis PA (2013) Inhibition of LRRK2 kinase activity stimulates macroautophagy. Biochim Biophys Acta 1833(12):2900–2910. doi:10.1016/j.bbamcr.2013.07.020
61. Manzoni C, Mamais A, Dihanich S, McGoldrick P, Devine MJ, Zerle J, Kara E, Taanman JW, Healy DG, Marti-Masso JF, Schapira AH, Plun-Favreau H, Tooze S, Hardy J, Bandopadhyay R, Lewis PA (2013) Pathogenic Parkinson's disease mutations across the functional domains of LRRK2 alter the autophagic/lysosomal response to starvation. Biochem Biophys Res Commun 441(4):862–866. doi:10.1016/j.bbrc.2013.10.159
62. Plowey ED, Cherra SJ 3rd, Liu YJ, Chu CT (2008) Role of autophagy in G2019S-LRRK2-associated neurite shortening in differentiated SH-SY5Y cells. J Neurochem 105(3):1048–1056. doi:10.1111/j.1471-4159.2008.05217.x
63. Ramonet D, Daher JP, Lin BM, Stafa K, Kim J, Banerjee R, Westerlund M, Pletnikova O, Glauser L, Yang L, Liu Y, Swing DA, Beal MF, Troncoso JC, McCaffery JM, Jenkins NA, Copeland NG, Galter D, Thomas B, Lee MK, Dawson TM, Dawson VL, Moore DJ (2011) Dopaminergic neuronal loss, reduced neurite complexity and autophagic abnormalities in transgenic mice expressing G2019S mutant LRRK2. PLoS One 6(4), e18568. doi:10.1371/journal.pone.0018568
64. Alegre-Abarrategui J, Christian H, Lufino MM, Mutihac R, Venda LL, Ansorge O, Wade-Martins R (2009) LRRK2 regulates autophagic activity and localizes to specific membrane microdomains in a novel human genomic reporter cellular model. Hum Mol Genet 18(21):4022–4034. doi:10.1093/hmg/ddp346
65. Daniel G, Moore DJ (2015) Modeling LRRK2 pathobiology in Parkinson's disease: from yeast to rodents. Curr Top Behav Neurosci 22:331–368. doi:10.1007/7854_2014_311
66. Godena VK, Brookes-Hocking N, Moller A, Shaw G, Oswald M, Sancho RM, Miller CC, Whitworth AJ, De Vos KJ (2014) Increasing microtubule acetylation rescues axonal transport and locomotor deficits caused by LRRK2 Roc-COR domain mutations. Nat Commun 5:5245. doi:10.1038/ncomms6245
67. Li Y, Liu W, Oo TF, Wang L, Tang Y, Jackson-Lewis V, Zhou C, Geghman K, Bogdanov M, Przedborski S, Beal MF, Burke RE, Li C (2009) Mutant LRRK2(R1441G) BAC transgenic mice recapitulate cardinal features of Parkinson's disease. Nat Neurosci 12(7):826–828
68. Tagliaferro P, Kareva T, Oo TF, Yarygina O, Kholodilov N, Burke RE (2015) An early axonopathy in a hLRRK2(R1441G) transgenic model of Parkinson disease. Neurobiol Dis 82:359–371. doi:10.1016/j.nbd.2015.07.009
69. Tong Y, Pisani A, Martella G, Karouani M, Yamaguchi H, Pothos EN, Shen J (2009) R1441C mutation in LRRK2 impairs dopaminergic neurotransmission in mice. Proc Natl Acad Sci U S A 106(34):14622–14627. doi:10.1073/pnas.0906334106
70. Tsika E, Kannan M, Foo CS, Dikeman D, Glauser L, Gellhaar S, Galter D, Knott GW, Dawson TM, Dawson VL, Moore DJ (2014) Conditional expression of Parkinson's disease-related R1441C LRRK2 in midbrain dopaminergic neurons of mice causes nuclear abnormalities without neurodegeneration. Neurobiol Dis 71:345–358. doi:10.1016/j.nbd.2014.08.027
71. Tsika E, Nguyen AP, Dusonchet J, Colin P, Schneider BL, Moore DJ (2015) Adenoviral-mediated expression of G2019S LRRK2 induces striatal pathology in a kinase-dependent manner in a rat model of Parkinson's disease. Neurobiol Dis 77:49–61. doi:10.1016/j.nbd.2015.02.019
72. Daher JP, Abdelmotilib HA, Hu X, Volpicelli-Daley LA, Moehle MS, Fraser KB, Needle E, Chen Y, Steyn SJ, Galatsis P, Hirst WD, West AB (2015) Leucine-rich repeat kinase 2 (LRRK2) pharmacological inhibition abates alpha-synuclein gene-induced neurodegeneration. J Biol Chem 290(32):19433–19444. doi:10.1074/jbc.M115.660001
73. Li T, Yang D, Zhong S, Thomas JM, Xue F, Liu J, Kong L, Voulalas P, Hassan HE, Park JS, MacKerell AD Jr, Smith WW (2014) Novel LRRK2 GTP-binding inhibitors reduced degeneration in Parkinson's disease cell and mouse models. Hum Mol Genet 23(23):6212–6222. doi:10.1093/hmg/ddu341

74. Li T, He X, Thomas JM, Yang D, Zhong S, Xue F, Smith WW (2015) A novel GTP-binding inhibitor, FX2149, attenuates LRRK2 toxicity in Parkinson's disease models. PLoS One 10(3), e0122461. doi:10.1371/journal.pone.0122461
75. Dusonchet J, Kochubey O, Stafa K, Young SM Jr, Zufferey R, Moore DJ, Schneider BL, Aebischer P (2011) A rat model of progressive nigral neurodegeneration induced by the Parkinson's disease-associated G2019S mutation in LRRK2. J Neurosci 31(3):907–912. doi:10.1523/jneurosci.5092-10.2011

Chapter 5
LRRK2 and Autophagy

Claudia Manzoni and Patrick A. Lewis

Abstract Leucine-rich repeat kinase 2 (LRRK2) has been implicated in a wide range of cellular processes, including the catabolic pathways collectively described as autophagy. In this chapter, the evidence linking LRRK2 to autophagy will be examined, along with how regulation of autophagy and lysosomal pathways may provide a nexus between the physiological function of this protein and the different diseases with which it has been associated. Data from cellular and animal models for LRRK2 function and dysfunction support a role in the regulation and control of autophagic pathways in the cell, although the extant results do not provide a clear indication as to whether LRRK2 is a positive or negative regulator of these pathways, and there are conflicting data as to the impact of mutations in LRRK2 causative for Parkinson's disease. Given that LRRK2 is a priority drug target for Parkinson's, the evidence suggesting that knockout or inhibition of LRRK2 can result in deregulation of autophagy may have important implications and is discussed in the context of our wider understanding of LRRK2.

Keywords LRRK2 • Macroautophagy • Catabolism • Parkinson's disease • Lysosomes • Autophagosomes

Introduction

LRRK2 has been linked to the cellular processes collectively called autophagy by a wide range of studies and in a number of model systems. While the precise mechanisms governing how LRRK2 interacts with or regulates autophagy remain obscure as well as the impact this may have in disease, it is indisputable that the manipulation of LRRK2 significantly alters the major pathways involved in autophagy. In this chapter, the evidence linking LRRK2 with autophagy is assessed, and the various potential scenarios as to how this might relate to disease are played out.

C. Manzoni • P.A. Lewis (✉)
School of Pharmacy, University of Reading, Whiteknights, Reading, UK

Department of Molecular Neuroscience, UCL Institute of Neurology,
Queen Square, London, UK
e-mail: c.manzoni@reading.ac.uk; p.a.lewis@reading.ac.uk

H.J. Rideout (ed.), *Leucine-Rich Repeat Kinase 2 (LRRK2)*,
Advances in Neurobiology 14, DOI 10.1007/978-3-319-49969-7_5

Autophagy

To understand the role of LRRK2 in autophagy, it is first essential to briefly summarize what is meant by this term. Autophagy (from the Greek auto—self—and phagein/phagos, eat) is an essential catabolic process for the degradation of exhausted, damaged, and potentially dangerous materials within eukaryotic cells [1]. Since materials to be degraded range from misfolded proteins to damaged organelles, autophagy can be seen as a routine process of cleaning and waste removal. Specialized cells, such as immune cells able to perform phagocytosis, specifically use autophagy to destroy the potentially hazardous content of the phagocytic vesicle; therefore, in this respect, autophagy can also be seen as a first-line defense for keeping a clean and safe cellular environment [2]. Another key connection is between autophagy and cell death; a type of programmed cell death (PCD) called PCD type II occurs by hyperactivation of autophagy, where lysosomes and vacuoles take over and degrade the whole cellular body [3]. Finally, autophagy can be initiated under a cocktail of stress and nutrient depletion; under these circumstances, the self-digestion of portions of cytoplasm becomes a fundamental buffer system for recycling amino acids and fuelling cell energy with substrates for oxidation [4]. Modified cells in which autophagy has been ablated do not survive starvation periods, emphasizing the essential nature of this function for the evolution of life.

It is quite clear from the above that autophagy is a fundamental, and complex, process—and the molecular mechanisms behind it are equally complex. A wide variety of signals, checkpoints, molecular players, and feedback loops are involved in its regulation, and our understanding of these is far from complete [5]. Moreover, even if the end point of autophagy is the delivery and degradation of materials into the lysosomes, several routes exist to achieving that end point: macroautophagy, chaperone-mediated autophagy (CMA), and microautophagy (Fig. 5.1).

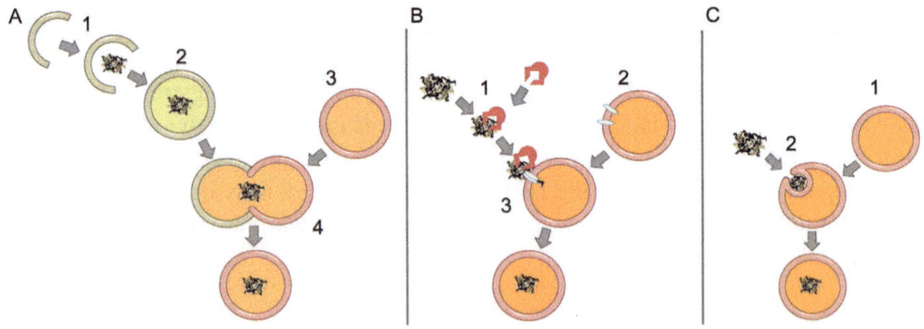

Fig. 5.1 The different forms of autophagy. (**a**) Simplified scheme for macroautophagy: (*1*) formation of omegasome, (*2*) encapsulation of waste for degradation, (*3*) lysosome, (*4*) fusion of autophagosome with lysosome. (**b**) Simplified scheme for chaperone-mediated autophagy: (*1*) HSC70 recognizes and binds to waste for targeting, (*2*) Lamp2A forms complex in membrane of lysosomal vesicle, (*3*) protein for degradation is passed into the lumen of the lysosome via Lamp2A complex. (**c**) Simplified scheme for microautophagy: (*1*) lysosome, (*2*) invagination of lysosomal membrane allows direct sequestration of waste for degradation into the lumen of the lysosome

Macroautophagy is probably the best-known, and certainly most studied, form of autophagy; it is characterized by the production of a specific double-membraned vesicle called an autophagosome. The vesicle production starts in discrete parts of the endoplasmic reticulum called omegasomes. The cup-shaped membrane that emerges from the omegasome is referred to as isolation membrane or phagophore; it will eventually detach and close around the material to be degraded thus sequestering portions of the cytosol and, in some cases, organelles. Non-specific, bulk macroautophagy occurs when parts of the cytosols are non-specifically engulfed during the closure of the autophagosome. Specific macroautophagy occurs when the autophagosome chooses the material to be degraded; in this case cargo proteins (such as p62, NBR1, MDP52, optineurin, and NIX) recognize the material to be degraded, tag this material, and simultaneously interact with microtubule-associated proteins 1A/1B light chain 3A (more commonly referred to as LC3), a protein which is specifically inserted into the nascent vesicle. The generation of this bridge is what eventually guides the autophagosome to close around the cargo to be degraded. The fully loaded autophagosome is then dispatched for degradation following a final fusion with the lysosome. In the process referred to as *canonical macroautophagy*, the phosphorylation state of the UNC-like kinase 1 (ULK1) complex drives the creation of the isolation membrane [6]. The ULK1 complex is directly phosphorylated by the mammalian target of rapamycin complex 1 (mTORc1) and AMP kinase (AMPK). mTOR is sensitive to nitrogen availability, and amino acid depletion is able to activate autophagy through mTOR, while AMPK is sensitive to cellular energy levels and is activated by depletion of ATP, along with the concomitant increase in ADP and AMP. The second key player in canonical macroautophagy is the BECLIN-1 complex, able to activate the class III phosphatidylinositol 3-kinase (PI3K) VPS34 necessary for the production of inositol-3 phosphate (I3P) that is enriched at the omegasome, remains in the isolation membrane, and promotes the recruitment of effector proteins, such as the WIPI proteins, to the autophagosome [7]. Finally, the third crucial player in canonical macroautophagy is the aforementioned LC3, a protein that undergoes proteolytic cleavage and conjugation with phosphatidylethanolamine prior insertion into the autophagosome membrane [8]. LC3 is important for specific cargo recognition and for the final closure of the autophagosome. Moreover, since its modifications are detectable as a shift in electrophoretic mobility, LC3 has become the marker of choice for biochemically monitoring autophagosome formation [9].

The description of canonical macroautophagy suggests that *noncanonical* pathways also exist, and indeed this is the case. Examples of noncanonical pathways are those that do not require ULK1 (e.g., macroautophagy induced by ammonia), BECLIN-1 (macroautophagy induced by MK801 and arsenic trioxide), or even LC3 -since small monomeric GTPases like RAB9 may sometimes take its place [10]. Noncanonical macroautophagy is not well understood, and we still do not have a detailed comprehension of the stimuli able to trigger these alternative pathways nor of the protein complexes involved in their regulation.

In contrast to macroautophagy, *CMA* does not require the generation of autophagic vesicles: the materials to be degraded, for the most part proteins, are delivered directly to the lysosomes. Proteins containing the pentapeptide motif KFERQ are recognized by the heat shock cognate protein 70 (HSC-70) and transported to the lysosomal surface where this chaperone is recognized by the lysosomal protein LAMP2A and the cargo internalized to be degraded [11]. The process of CMA has been investigated in a great deal of detail with regard to disorders such as Alzheimer's, the prion disorders, and Huntington's chorea, where deposition of amyloid material is seen as a consequence of an imbalance between production and catabolism of amyloidogenic proteins that are known substrates of CMA. This is likewise the case for proteins involved in Parkinson's, for example, α-synuclein and LRRK2 (of which more later).

Finally, *microautophagy* occurs through direct invagination of the lysosomal membrane, with target proteins/waste imported straight into the lysosome. Little is known about microautophagy [12]. It is likely that it represents a constitutive form of autophagy; however, starvation and inhibition of mTORc1 by rapamycin are able to potentiate it. One explanation is that microautophagy acts to balance the dimensions of the lysosome, which would otherwise continuously increase its surface by fusion with autophagosomes and late endosomes.

To add another layer of complexity, autophagy can also be classified by the cargo targeted for degradation, for example, *mitophagy* for mitochondria, *pexophagy* for the peroxisomes, *xenophagy* for the degradation of engulfed, exogenous foreign bodies in phagocytic cells, and so on [13–15].

Evidence for a Role for LRRK2 in Autophagy

Introductory Note

There are a wide range of data supporting a role for LRRK2 in the regulation and function of autophagy. At this juncture, and before proceeding any further, it is worth considering several aspects of exactly how LRRK2, a large and complex protein, interacts with autophagy, a likewise complex and dynamic process [16].

First, the distinction between the physiological function of LRRK2 and its role in disease needs to be considered. The evidence suggesting the involvement of LRRK2 in autophagy, while not conclusive, is extensive; however, how this involvement relates with disease is still a matter of great discussion. We face three possible scenarios:

1. The role of LRRK2 in autophagy may be relevant in physiology, but it may not be related with the molecular mechanism of disease.
2. Alterations of autophagy due to changes in LRRK2 may influence the disease, but this may not be the primary event leading to disease.
3. Finally the alteration of autophagy by LRRK2 may truly be the primary, causal, molecular mechanism of disease.

It is unclear which scenario holds true, and the situation is complicated by the involvement of LRRK2 in diseases other than Parkinson's (cancer, inflammatory bowel disorder, and leprosy), where the contribution of autophagy to pathogenesis may be distinct from that in PD [17].

A second remark relates to the experimental approaches used to assess LRRK2 and autophagy. Experiments are conducted in many different model systems; data in the literature are mixed—comparing results from endogenous LRRK2, overexpression models, and ablation of kinase, GTPase, or both activities alongside the analysis of Mendelian mutations that are relevant to Parkinson's disease. The heterogeneous experimental system, in combination with the nuances of the autophagic process itself, accounts for the disturbing amount of conflicting results in the literature. This does not, per se, invalidate the extant literature; however, it represents a major challenge in terms of identifying a unifying role for LRRK2 in autophagy and acts as a cautionary note with regard to extrapolating conclusions from single model systems and unique experimental settings.

The Case of Parkinson's Disease

Alterations of Autophagy in Parkinson's Disease Brains

Parkinson's disease (PD) is pathologically defined by the deposition of misfolded protein, predominantly α-synuclein, in the brain as intracellular inclusions known as *Lewy bodies* [18]. The accumulation of undegraded proteins and their deposition in amyloid bodies are common to a number of different misfolding disorders frequently referred to as amyloidoses. This phenomenon has been seen as consequence of an intrinsic imbalance between protein production and turnover. Since degradation of misfolded proteins is carried out both by the proteosome and the autophagic systems, the presence of amyloid deposits like Lewy bodies in PD can be taken as indirect evidence of alterations in proteostasis and, by implication, autophagy. The majority of PD cases are idiopathic, also called sporadic, since they happen with no proof of causality for Mendelian transmission. Accumulation of autophagic vesicle reminiscent of a possible impairment of macroautophagy, as well as reduction in the LAMP2A and HSC-70 protein levels important for CMA, has indeed been described in postmortem human specimens from idiopathic PD cases [19]. Moreover, autophagy-deficient mouse models display brain abnormalities like formation of ubiquitin-positive inclusions and presynaptic accumulation of α-synuclein and LRRK2, further emphasizing autophagy as one of the cellular mechanisms contributing to the general pathogenesis of PD [20]. A small percentage (between 5 and 10 %) of PD cases are familial, caused by a mutation transmitted through generations in a Mendelian fashion. Of the Mendelian causes of Parkinson's, LRRK2 is the most frequently mutated gene [21]. The case of familial PD associated with LRRK2 mutations is quite unusual due to the variability in protein deposition observed in postmortem brains [22]. The majority of LRRK2 (PD) cases present with deposition of α-synuclein as observed in idiopathic PD cases; however, a small minority of patients show inclusions of proteins other than α-synuclein—such as the microtubule-associated protein tau or the RNA-binding protein TDP-43—or no identifiable inclusions at all [23]. Whether this pleomorphic pathology is indicative of a fundamental disruption of aggregate disposal mechanisms, including autophagy, due to LRRK2 mutations is unclear.

Alterations of Autophagy in LRRK2 Mutant PD Cell Models

A number of different cellular approaches have been used to examine autophagy in the context of LRRK2 (PD) mutations using multiple cellular systems. The first studies investigating LRRK2 and autophagy did so using immortalized cell lines expressing either exogenous wild type or mutant LRRK2. Plowey and co-workers, using SHSY-5Y neuroblastoma cells as a model, discovered alterations in autophagic vesicles upon transfection with LRRK2 containing the G2019S pathogenic mutation [24]. Examining a previously reported LRRK2 phenotype, neurite retraction, the authors were able to present alterations in this by knocking down key components of the autophagy machinery such as LC3 or ATG7—providing the first direct evidence of a link between LRRK2 and macroautophagy. Soon after, a group based at the University of Oxford demonstrated that LRRK2 regulates autophagic vesicle formation and associates with specific vesicle microdomains [25]. Following these early reports, data has accumulated implicating LRRK2 in the regulation of macroautophagy; however, there is as yet no consensus on the molecular mechanism underlying this association with some papers suggesting that this could be through regulation of calcium levels and the NAADP pathway [26], the MEK/ERK pathway [27], and via the Beclin-associated protein BCL-2 [28]. As tool compounds targeting LRRK2 became available, enabling researchers to effectively inhibit LRRK2 kinase activity, it also became clear that reducing LRRK2 activity in this manner acted to induce autophagy [29, 30]. Likewise, knockdown of LRRK2 has been reported to cause an induction of autophagy [25]—although in both cases (knockdown and inhibition) the precise mechanism remains obscure. There appears to be a close relationship between LRRK2 kinase activity and cellular localization, and a study utilizing BV2 and RAW cells (microglial and macrophage in origin, respectively) highlighted an intimate connection between membrane localization of LRRK2, kinase activity, and regulation of macroautophagy [31]. Intriguingly, given the complex enzymatic complex formed by LRRK2, a yeast-based study revealed a role for the GTPase domain of LRRK2 in regulating autophagy [32].

One approach to directly test how mutations in LRRK2 impact on autophagy in disease contest is to use cells derived from PD patients, compared with appropriate nonmutant controls. *Fibroblasts* have been used as a primary experimental model system as well as a source for *induced pluripotent stem (iPS)* cells to be then differentiated into human neuronal populations. Data generated through the use of these models are highly variable and discordant and (similar to the models described above) fail to identify a clear molecular mechanism for LRRK2 in disease. Taken in toto, however, they present again an undeniable association between LRRK2 and autophagy. Fibroblasts carrying the G2019S mutation in the kinase domain of LRRK2 have been shown to have an increased basal rate of macroautophagy. This alteration of the basal flux was not related with the canonical control of mTOR over macroautophagy, but was dependent upon an abnormal activity of the MEK1/3 pathway [27]. Another mTOR-independent alteration of macroautophagy was reported in human fibroblasts with PD mutations encompassing the entire catalytic core of the LRRK2 protein—suggesting that at an endogenous level, diverse LRRK2 mutations could alter autophagic response [33]. Although this latter study did not observe a basal increase in macroautophagy as previously seen in G2019S carriers, impaired macroautophagic response to starvation was detected for all mutant lines. G2019S mutant fibroblasts have also been documented to have a raised sensitivity to inhibition of mitochondrial fission, a sensitivity that is linked to autophagy [34], and also to the mitochondrial toxin 1-methyl-4-phenylpyridinium (MPTP)—likewise dependent upon autophagy [35].

An early study presenting data from neurons derived from LRRK2 mutant patient iPS cells suggested that the G2019S mutation could modify autophagic vesicle formation in dopaminergic neuronal cells [36], although this was matched by data from idiopathic PD patient-specific cells (i.e., not carrying a genetic defect), making interpretation of these results complicated.

Using patient-derived neuronal cells, LRRK2 has also been implicated in CMA [37]. In this model system, LRRK2 was degraded via CMA, with proof of mutant LRRK2 acting to inhibit this pathway. Given the evidence linking α-synuclein to CMA for its degradation, this provides a potential mechanism for protein aggregate buildup in disease—although it should be noted that LRRK2 has been reported to be degraded by the ubiquitin proteasome system as well [38], and how these pathways intersect and compensate each other in the context of LRRK2 is still not clear.

Alterations of Autophagy in LRRK2 PD Animal Models

In vivo models for genetic forms of PD have, almost without exception, proved disappointing in terms of reproducing disease-relevant phenotypes—most notably clear nigral cell death and protein aggregation. Despite this, there are large numbers of animal models for LRRK2 disease ranging from *Caenorhabditis elegans* to rodent transgenics. Data from these models have proved an important source of evidence for a link between LRRK2 and autophagy, although again the precise molecular events linking the two are ambiguous. *Caenorhabditis elegans,* worms expressing mutant LRRK2 (both the R1441C and G2019S mutations), displayed reduced autophagic activity, while wild-type LRRK2 improved autophagic function throughout aging [39]. In *Drosophila melanogaster*, knockout of the *Drosophila* LRRK2 paralog (*dLRRK*) caused severe deficits in the autophagy/lysosomal system—with accumulation of enlarged, dysfunctional lysosomes [40]. A caveat to both worm and fly studies is that neither *C. elegans* nor *D. melanogaster* possesses clear

LRRK2 paralogs—each possessing only one LRRK gene. Whether these species provide a suitable model for LRRK2, or LRRK1 for that matter, is not yet clear [41].

The generation of *LRRK2 rodent knockout* models appeared at first to be an underwhelming avenue of investigation: the absence of neurodegeneration and the complete lack of pathological hallmarks typical of PD in the brain of LRRK2 knockout animals led to these models being dismissed as something of a footnote to the LRRK2 field. Once investigators looked beyond the brain, however, intriguing data began to emerge. Two independently generated LRRK2 knockout mice from the Shen laboratory displayed peculiar alterations in their kidneys, with signs of enhanced autophagy at young ages and reduced autophagy at 20 months accompanied by progressive accumulation of secondary lysosomes, α-synuclein, and ubiquitinated proteins [42]. A third LRRK2 knockout mouse was generated at the Novartis laboratories [43]. A very similar alteration in the kidneys was reported, with accumulation of secondary lysosomes even though the absence of LC3 alterations, the presence of mTOR changes, and the end point of the study at only 14 months make it difficult to fully understand the implication for macroautophagy. Interestingly, changes were detected in the lungs where lamellar bodies accumulated abnormally in type II cells. A fourth LRRK2 knockout mouse reproduced a similar profile of progressive kidney alterations. Interestingly, accumulation of the autophagy marker p62 was similar to that observed in the Shen models, while LC3-II was shown to accumulate rather than decrease as previously reported [44]. Similar to the data from mice, age-dependent results with accumulation of lysosomes in the kidneys and lamellar bodies in the lungs have been subsequently reproduced by at least two independent groups in a LRRK2 knockout rat generated by the Sigma Advanced Genetic Engineering (SAGE) laboratories [45]. Although the comparison between knockout of LRRK2 and inhibition of its kinase activity is not a simple one (noting that LRRK2 has a number of other activities that are lost if the gene is knocked out), it is striking that treatment of nonhuman primates with a specific inhibitor of LRRK2 kinase activity recapitulates many of the characteristics of the knockout phenotype [46]. In particular, the acutely treated primates developed vacuolar pathology in their lungs, with potential consequences for lung function and striking a note of caution with regard to the use of LRRK2 kinase inhibitors in a clinical setting.

There exist a range of rodent LRRK2 mutation models, using a variety of strategies (knock in and overexpression) to assess the impact of LRRK2 mutations. A number of these have reported alterations in autophagic markers such as LC3 and p62, although these appear not to be consistent between mutations studied. For example, G2019S mice exhibit alterations in these markers [47], whereas mice expressing R1441C do not [48]. As for the cellular studies of LRRK2 and autophagy, the conclusion to be drawn from these animal models is that there is proof of an impact of LRRK2 on autophagy, but that it is not obvious from the extant experimental data as to the precise mechanism or direction of this alteration.

Finally, and providing a valuable illustration of the reciprocal nature of the pathways that govern autophagy, selective knockout of *Atg7*, a key component of the macroautophagy machinery, in dopaminergic neurons results in the accumulation of LRRK2 and alpha-synuclein [20]. This study demonstrates that, just as LRRK2 can alter autophagy, autophagy can have an impact on LRRK2.

Parkinson's Disease: A Possible Scenario

As with much of the science relating to LRRK2, it is important to examine the biological insights in the context of the preeminent driver of research into this protein: Parkinson's disease. The vast majority of PD cases are idiopathic in nature, with no positive familial history of disease and with no clear association with genetic risk factors. The presence of Lewy bodies, predominantly composed of undigested, misfolded, and ubiquitinated α-synuclein, paired with the finding of vacuolation and alterations in CMA in idiopathic brains, provided a clear indication that autophagy could be implicated in the etiology of PD. However, the absence of specific genes to investigate built a nearly insurmountable wall between the hypothesis and the actual possibility of proving it with further investigations.

A small percentage of PD cases, however, are familial in origin and are therefore associated with mutations in known genes. LRRK2 is the gene product most frequently mutated in familial PD but other mutated proteins have been identified: α-synuclein, GBA, VPS35, PINK1, Parkin, DJ-1, and FBXO-7 to name just a selection [21]. Unlike the idiopathic cases, familial cases are easier to study in model systems, and the autophagy theory can be dissected in greater depth.

In this chapter we have focused on how LRRK2 has been linked to autophagy; it would therefore be interesting to consider how LRRK2 and autophagy relates to other PD proteins. The most prominent of these is α-synuclein, with mutant α-synuclein having been demonstrated to impair CMA in a process that highly resembles the one proposed as justification for LRRK2 inhibition of CMA, and equally aggregated α-synuclein can inhibit autophagy [49, 50]. GBA is a lysosomal enzyme, important for lysosomal integrity and functionality. When altered, the generation of a primary lysosomal defect leads to impaired autophagy and mitophagy [51]. The same can be said for ATP13A2, an ATPase involved in cation transport in acidic vesicles, associated with Kufor Rakeb syndrome, a form of dementia that presents with parkinsonism [52]. Loss of function mutations in ATP13A2 result in a severe lysosomal deficiency and a reduction in autophagosomes clearance. VPS35 is a well-known component of the retromer complex that is implicated in the trafficking of proteins and enzymes in the trans-Golgi network thus supporting the correct loading of lysosomes with hydrolytic enzymes [53]. PINK1, Parkin, and FBXO-7 have all been implicated in mitophagy, and it is now well established they have a fundamental role in mitochondria quality control [14]. Finally WDR45 is a protein associated with a form of neurodegeneration with brain iron accumulation that presents with parkinsonism [54]. WDR45 has not been extensively studied, but it is often referred to as WIPI-4, and we know at least other two proteins, WIPI-1 and WIPI-2, that are important for the initial stages of macroautophagy suggesting a strong link to autophagy.

By looking at a broad spectrum of familial PD and parkinsonism, it seems like there is a common theme associated with the disease. All familial genes are for some extent associated with catabolism [55, 56]. Interestingly this association is not sharp or precise; it is not a single degradation pathway or a single form of autophagy that is being implicated. The leitmotif behind different forms of familial PD is much more broad, but it points in the direction of impairment of catabolic processes related with autophagy, lysosomes, and possibly vesicle recycling. This may account for the clinical and phenotypic differences observed with different mutations; it may also suggest that rather of being a strong molecular mechanism of disease, alterations in catabolism may act by creating a susceptible environment in which a combination of other genetic and environmental factors may push the system above the threshold eventually culminating in the propagation of neurodegeneration.

Further Diseases, Proteins, and Circumstantial Evidence

One of the most fascinating aspects of the LRRK2 gene and human pathobiology is the implication of LRRK2 in multiple human disorders. While the bulk of research into LRRK2 has focused on its role in PD, it has also been linked to inflammatory bowel disease, cancer, and leprosy [17]. It is notable that autophagy is a unifying feature of all of these disorders. Unfortunately there are very few papers investigating the molecular connection of LRRK2 with the pathogenesis of these diseases leaving only genetic evidence, most certainly circumstantial in nature, to suggest that LRRK2 may play a role in the regulation of autophagy and that this may consequently impact on the pathogenesis of disease. An isolated publication reported a possible molecular association of LRRK2 and the receptor tyrosine kinase MET signaling in cancer, thus implicating LRRK2 with alterations in apoptosis and autophagy [57].

It is also noteworthy that at least one other protein in the ROCO family, to which LRRK2 belongs, has been strongly implicated in the regulation of autophagy. Death-associated protein kinase 1 (DAPK1) has been closely linked to type II PCD and has been reported to directly regulate the BECLIN-1 pathway by phosphorylation of the BECLIN-1 protein complex [58]. How this relates to the changes observed in LRRK2 models, however, has not been investigated.

Conclusions

Mechanisms Linking LRRK2 to Autophagy

How do we bring together these discordant data to gain insight into the role of LRRK2 as an orchestrator, or at least conductor, of autophagic activity in the cell? The extant data relating to LRRK2 and autophagy are summarized in Table 5.1. This highlights the volume of evidence linking LRRK2 to autophagy while also emphasizing the challenge presented by attempting to dissect out the precise mechanisms and pathways involved.

Table 5.1 Alterations in autophagy pathways linked to LRRK2

Autophagy type	Model	Impact	Citation
Macroautophagy	SHSY-5Y cells, mutations (G2019S, K1906M)	Increased autophagosomes in presence	Plowey et al. [24]
Macroautophagy	HEK293T cells, VERO cells, mutations (R1441C/G2019S), knockdown	Impairment of autophagosome formation due to R1441C mutation, increase in autophagic activity due to LRRK2 knockdown	Alegre-Abarrategui [25]
Macroautophagy	HEK293T cells, PC12 cells, mutations (G2019S, K1906M)	Increased autophagosome formation	Gomez-Suaga et al. [26]
CMA	SHSY-5Y cells, HEK293 cells, human-derived iPSC neuronal cells, mutations (G2019S)	Inhibition of CMA by mutated LRRK2	Orenstein et al. [37]
Macroautophagy	Human fibroblasts, mutations (G2019S)	Increase in basal autophagy	Bravo-San Pedro et al. [27]
Macroautophagy	HeLa cells, human fibroblast, mutation (G2019S)	Increased autophagy due to mutation, Bcl-2 mediated	Su et al. [28]
Macroautophagy	H4 neuroglioma cells, primary rat astrocytes, LRRK2 kinase inhibition	Increase in autophagic flux	Manzoni et al. [29]
Macroautophagy	RAW264.7 cells, BV2 cells, LRRK2 knockdown, LRRK2 kinase inhibition	Decrease in rapamycin-induced autophagic flux	Schapansky et al. [31]
Macroautophagy	Human-derived iPSC neuronal cells, mutations (G2019S)	Accumulation of autophagic vesicles, decreased lysosomal function	Sanchez-Danes et al. [36]
Macroautophagy	*C. elegans*, mutations (G2019S)	Accelerated loss of autophagic function due to G2019S mutation	Saha et al. [39]
Macroautophagy	*Mus musculus*, mutations (G2019S transgene)	Enlarged autophagic vesicle-like structures	Ramonet et al. [47]
Macroautophagy	*Saccharomyces cerevisiae,* mutations (R1441C, artificial GTPase mutations)	Increased autophagic vesicles due to loss of GTPase activity	Xiong et al. [32]

(continued)

Table 5.1 (continued)

Autophagy type	Model	Impact	Citation
Macroautophagy	*Mus musculus*, knockout, knock in kinase dead, mutation (G2019S)	Autophagic/lysosomal abnormalities due to loss of LRRK2	Herzig et al. [43]
Macroautophagy	*Drosophila melanogaster*, knockout	Accumulation of autophagosomes due to loss of *dLRRK*	Dodson et al. [40]
Macroautophagy	SHSY-5Y cells, LRRK2 kinase inhibition	Induction of autophagy following inhibition of LRRK2 kinase activity	Saez-Atienzar et al. [30]
Macroautophagy	Human fibroblasts, LRRK2 mutations (R1441G, Y1699C, G2019S)	Reduced autophagic response to starvation due to mutations.	Manzoni et al. [33]
Macroautophagy	*Mus musculus*, knockout	Altered autophagic markers, vesicle accumulation	Tong et al. [42]
Macroautophagy	*Mus musculus*, knockout	Altered autophagic markers, vesicle accumulation	Hinkle et al. [44]
Macroautophagy	*Macaca fascicularis*, LRRK2 kinase inhibitors	Altered lysosomal function	Fuji et al. [46]
Macroautophagy	*Rattus norvegicus* (Long-Evans rat), knockout	Altered lysosomal function	Baptista et al. [45]

As a foundation to our understanding of LRRK2 in this context, it is worth revisiting the presumed role of LRRK2 within the cell. Based upon the functional domains that make up LRRK2, i.e., protein-protein interaction/scaffolding domains, a GTPase domain, and a kinase domain, the working hypothesis for LRRK2 function is that it operates as a signaling node within the cell, controlling cellular processes via phosphorylation of substrates in signaling pathways or by acting as a molecular switch for other kinases via the ROC/GTPase domain. As noted above, the complex domain organization of LRRK2 makes it difficult to be categorical about which of these activities predominates or whether there is a sequential series of events within LRRK2 that leads to stimulation/suppression of downstream pathways.

It is clear that there are alterations in autophagic function that can be measured as a consequence of changes in LRRK2; however, no specific event has been identified close to the known regulators of autophagy. This, in turn, raises an important question as to whether the changes that are observed in autophagy are proximal, immediate events to LRRK2 or whether LRRK2 is acting distal to these changes, perhaps a number of signaling events removed. The second key question is what aspects of LRRK2 biology are critical for these cellular phenotypes—is kinase activity the central player? Or GTPase switch activity? With the available published data, it is impossible to rule out any of these mechanisms. Indeed, there is potential for LRRK2 to act simultaneously as a regulator of the initiator of autophagic processes, perhaps by a kinase-mediated signaling cascade, while also influencing CMA and autophagosome recycling through a Rab-like GTPase function. Finally the heterogeneity of data obtained in different experimental systems may stimulate the suspicion that LRRK2 may have different roles in different cell types; thus, the relation of LRRK2 and autophagy could be cell/tissue and maybe time dependent rather than a simple general feature.

It is certainly the case that further research is justified and perhaps even required to clarify which of these molecular scenarios is correct.

Acknowledgments The authors wish to acknowledge generous funding from the Medical Research Council (grants MR/L010933/1 and MR/N026004/1), Parkinson's UK (Fellowship F1002), the Rosetrees Trust, and the Michael J. Fox Foundation.

Conflict of Interest The author declares no conflicts of interest.

References

1. Yang Z, Klionsky DJ (2010) Mammalian autophagy: core molecular machinery and signaling regulation. Curr Opin Cell Biol 22(2):124–131. doi:10.1016/j.ceb.2009.11.014
2. Oczypok EA, Oury TD, Chu CT (2013) It's a cell-eat-cell world: autophagy and phagocytosis. Am J Pathol 182(3):612–622. doi:10.1016/j.ajpath.2012.12.017
3. Fuchs Y, Steller H (2015) Live to die another way: modes of programmed cell death and the signals emanating from dying cells. Nat Rev Mol Cell Biol 16(6):329–344. doi:10.1038/nrm3999
4. Kaur J, Debnath J (2015) Autophagy at the crossroads of catabolism and anabolism. Nat Rev Mol Cell Biol 16(8):461–472. doi:10.1038/nrm4024
5. Parzych KR, Klionsky DJ (2014) An overview of autophagy: morphology, mechanism, and regulation. Antioxid Redox Signal 20(3):460–473. doi:10.1089/ars.2013.5371
6. Alers S, Loffler AS, Wesselborg S, Stork B (2012) Role of AMPK-mTOR-Ulk1/2 in the regulation of autophagy: cross talk, shortcuts, and feedbacks. Mol Cell Biol 32(1):2–11. doi:10.1128/MCB.06159-11
7. Kang R, Zeh HJ, Lotze MT, Tang D (2011) The Beclin 1 network regulates autophagy and apoptosis. Cell Death Differ 18(4):571–580. doi:10.1038/cdd.2010.191
8. Kabeya Y, Mizushima N, Ueno T, Yamamoto A, Kirisako T, Noda T, Kominami E, Ohsumi Y, Yoshimori T (2000) LC3, a mammalian homologue of yeast Apg8p, is localized in autophagosome membranes after processing. EMBO J 19(21):5720–5728. doi:10.1093/emboj/19.21.5720
9. Mizushima N, Yoshimori T (2007) How to interpret LC3 immunoblotting. Autophagy 3(6):542–545
10. Codogno P, Mehrpour M, Proikas-Cezanne T (2012) Canonical and non-canonical autophagy: variations on a common theme of self-eating? Nat Rev Mol Cell Biol 13(1):7–12. doi:10.1038/nrm3249
11. Kaushik S, Cuervo AM (2012) Chaperone-mediated autophagy: a unique way to enter the lysosome world. Trends Cell Biol 22(8):407–417. doi:10.1016/j.tcb.2012.05.006

12. Li WW, Li J, Bao JK (2012) Microautophagy: lesser-known self-eating. Cell Mol Life Sci 69(7):1125–1136. doi:10.1007/s00018-011-0865-5
13. Sorbara MT, Girardin SE (2015) Emerging themes in bacterial autophagy. Curr Opin Microbiol 23:163–170. doi:10.1016/j.mib.2014.11.020
14. Deas E, Wood NW, Plun-Favreau H (2011) Mitophagy and Parkinson's disease: the PINK1-parkin link. Biochim Biophys Acta 1813(4):623–633. doi:10.1016/j.bbamcr.2010.08.007
15. Manjithaya R, Nazarko TY, Farre JC, Subramani S (2010) Molecular mechanism and physiological role of pexophagy. FEBS Lett 584(7):1367–1373. doi:10.1016/j.febslet.2010.01.019
16. Wallings R, Manzoni C, Bandopadhyay R (2015) Cellular processes associated with LRRK2 function and dysfunction. FEBS J 282(15):2806–2826. doi:10.1111/febs.13305
17. Lewis PA, Manzoni C (2012) LRRK2 and human disease: a complicated question or a question of complexes? Sci Signal 5(207):pe2. doi:10.1126/scisignal.2002680
18. Lashuel HA, Overk CR, Oueslati A, Masliah E (2013) The many faces of alpha-synuclein: from structure and toxicity to therapeutic target. Nat Rev Neurosci 14(1):38–48. doi:10.1038/nrn3406
19. Alvarez-Erviti L, Rodriguez-Oroz MC, Cooper JM, Caballero C, Ferrer I, Obeso JA, Schapira AH (2010) Chaperone-mediated autophagy markers in Parkinson disease brains. Arch Neurol 67(12):1464–1472. doi:10.1001/archneurol.2010.198
20. Friedman LG, Lachenmayer ML, Wang J, He L, Poulose SM, Komatsu M, Holstein GR, Yue Z (2012) Disrupted autophagy leads to dopaminergic axon and dendrite degeneration and promotes presynaptic accumulation of alpha-synuclein and LRRK2 in the brain. J Neurosci 32(22):7585–7593. doi:10.1523/JNEUROSCI.5809-11.2012
21. Hardy J, Lewis P, Revesz T, Lees A, Paisan-Ruiz C (2009) The genetics of Parkinson's syndromes: a critical review. Curr Opin Genet Dev 19:254–265. doi:S0959-437X(09)00055-0 [pii]10.1016/j.gde.2009.03.008
22. Zimprich A, Biskup S, Leitner P, Lichtner P, Farrer M, Lincoln S, Kachergus J, Hulihan M, Uitti RJ, Calne DB, Stoessl AJ, Pfeiffer RF, Patenge N, Carbajal IC, Vieregge P, Asmus F, Muller-Myhsok B, Dickson DW, Meitinger T, Strom TM, Wszolek ZK, Gasser T (2004) Mutations in LRRK2 cause autosomal-dominant parkinsonism with pleomorphic pathology. Neuron 44(4):601–607. doi:S0896627304007202 [pii]10.1016/j.neuron.2004.11.005
23. Wider C, Dickson DW, Wszolek ZK (2010) Leucine-rich repeat kinase 2 gene-associated disease: redefining genotype-phenotype correlation. Neurodegener Dis 7(1–3):175–179. doi:10.1159/000289232
24. Plowey ED, Cherra SJ, III, Liu YJ, Chu CT (2008) Role of autophagy in G2019S-LRRK2-associated neurite shortening in differentiated SH-SY5Y cells. J Neurochem 105(3):1048–1056
25. Alegre-Abarrategui J, Christian H, Lufino MM, Mutihac R, Venda LL, Ansorge O, Wade-Martins R (2009) LRRK2 regulates autophagic activity and localizes to specific membrane microdomains in a novel human genomic reporter cellular model. Hum Mol Genet 18(21):4022–4034. doi:ddp346 [pii]10.1093/hmg/ddp346
26. Gomez-Suaga P, Luzon-Toro B, Churamani D, Zhang L, Bloor-Young D, Patel S, Woodman PG, Churchill GC, Hilfiker S (2011) Leucine-rich repeat kinase 2 regulates autophagy through a calcium-dependent pathway involving NAADP. Hum Mol Genet 21:511–525. doi:10.1093/hmg/ddr481
27. Bravo-San Pedro JM, Niso-Santano M, Gomez-Sanchez R, Pizarro-Estrella E, Aiastui-Pujana A, Gorostidi A, Climent V, Lopez de Maturana R, Sanchez-Pernaute R, Lopez de Munain A, Fuentes JM, Gonzalez-Polo RA (2012) The LRRK2 G2019S mutant exacerbates basal autophagy through activation of the MEK/ERK pathway. Cell Mol Life Sci 8:1537–1539. doi:10.1007/s00018-012-1061-y
28. Su YC, Guo X, Qi X (2015) Threonine 56 phosphorylation of Bcl-2 is required for LRRK2 G2019S-induced mitochondrial depolarization and autophagy. Biochim Biophys Acta 1852(1):12–21. doi:10.1016/j.bbadis.2014.11.009

29. Manzoni C, Mamais A, Dihanich S, Abeti R, Soutar MP, Plun-Favreau H, Giunti P, Tooze SA, Bandopadhyay R, Lewis PA (2013) Inhibition of LRRK2 kinase activity stimulates macroautophagy. Biochim Biophys Acta 1833(12):2900–2910. doi:10.1016/j.bbamcr.2013.07.020
30. Saez-Atienzar S, Bonet-Ponce L, Blesa JR, Romero FJ, Murphy MP, Jordan J, Galindo MF (2014) The LRRK2 inhibitor GSK2578215A induces protective autophagy in SH-SY5Y cells: involvement of Drp-1-mediated mitochondrial fission and mitochondrial-derived ROS signaling. Cell Death Dis 5, e1368. doi:10.1038/cddis.2014.320
31. Schapansky J, Nardozzi JD, Felizia F, LaVoie MJ (2014) Membrane recruitment of endogenous LRRK2 precedes its potent regulation of autophagy. Hum Mol Genet 23(16):4201–4214. doi:10.1093/hmg/ddu138
32. Xiong Y, Coombes CE, Kilaru A, Li X, Gitler AD, Bowers WJ, Dawson VL, Dawson TM, Moore DJ (2010) GTPase activity plays a key role in the pathobiology of LRRK2. PLoS Genet 6(4), e1000902. doi:10.1371/journal.pgen.1000902
33. Manzoni C, Mamais A, Dihanich S, McGoldrick P, Devine MJ, Zerle J, Kara E, Taanman JW, Healy DG, Marti-Masso JF, Schapira AH, Plun-Favreau H, Tooze S, Hardy J, Bandopadhyay R, Lewis PA (2013) Pathogenic Parkinson's disease mutations across the functional domains of LRRK2 alter the autophagic/lysosomal response to starvation. Biochem Biophys Res Commun 441(4):862–866. doi:10.1016/j.bbrc.2013.10.159
34. Su YC, Qi X (2013) Inhibition of excessive mitochondrial fission reduced aberrant autophagy and neuronal damage caused by LRRK2 G2019S mutation. Hum Mol Genet 22(22):4545–4561. doi:10.1093/hmg/ddt301
35. Yakhine-Diop SM, Bravo-San Pedro JM, Gomez-Sanchez R, Pizarro-Estrella E, Rodriguez-Arribas M, Climent V, Aiastui A, Lopez de Munain A, Fuentes JM, Gonzalez-Polo RA (2014) G2019S LRRK2 mutant fibroblasts from Parkinson's disease patients show increased sensitivity to neurotoxin 1-methyl-4-phenylpyridinium dependent of autophagy. Toxicology 324:1–9. doi:10.1016/j.tox.2014.07.001
36. Sanchez-Danes A, Richaud-Patin Y, Carballo-Carbajal I, Jimenez-Delgado S, Caig C, Mora S, Di Guglielmo C, Ezquerra M, Patel B, Giralt A, Canals JM, Memo M, Alberch J, Lopez-Barneo J, Vila M, Cuervo AM, Tolosa E, Consiglio A, Raya A (2012) Disease-specific phenotypes in dopamine neurons from human iPS-based models of genetic and sporadic Parkinson's disease. EMBO Mol Med 4:380–395. doi:10.1002/emmm.201200215
37. Orenstein SJ, Kuo SH, Tasset I, Arias E, Koga H, Fernandez-Carasa I, Cortes E, Honig LS, Dauer W, Consiglio A, Raya A, Sulzer D, Cuervo AM (2013) Interplay of LRRK2 with chaperone-mediated autophagy. Nat Neurosci 16:394–406. doi:10.1038/nn.3350
38. Lichtenberg M, Mansilla A, Zecchini VR, Fleming A, Rubinsztein DC (2011) The Parkinson's disease protein LRRK2 impairs proteasome substrate clearance without affecting proteasome catalytic activity. Cell Death Dis 2, e196. doi:10.1038/cddis.2011.81
39. Saha S, Ash PE, Gowda V, Liu L, Shirihai O, Wolozin B (2015) Mutations in LRRK2 potentiate age-related impairment of autophagic flux. Mol Neurodegener 10:26. doi:10.1186/s13024-015-0022-y
40. Dodson MW, Leung LK, Lone M, Lizzio MA, Guo M (2014) Novel ethyl methanesulfonate (EMS)-induced null alleles of the Drosophila homolog of LRRK2 reveal a crucial role in endolysosomal functions and autophagy in vivo. Dis Model Mech 7(12):1351–1363. doi:10.1242/dmm.017020
41. Marin I (2008) Ancient origin of the Parkinson disease gene LRRK2. J Mol Evol 67(1):41–50. doi:10.1007/s00239-008-9122-4
42. Tong Y, Yamaguchi H, Giaime E, Boyle S, Kopan R, Kelleher RJ, 3rd, Shen J (2010) Loss of leucine-rich repeat kinase 2 causes impairment of protein degradation pathways, accumulation of alpha-synuclein, and apoptotic cell death in aged mice. Proc Natl Acad Sci U S A 107(21):9879–9884. doi:1004676107 [pii]10.1073/pnas.1004676107
43. Herzig MC, Kolly C, Persohn E, Theil D, Schweizer T, Hafner T, Stemmelen C, Troxler TJ, Schmid P, Danner S, Schnell CR, Mueller M, Kinzel B, Grevot A, Bolognani F, Stirn M, Kuhn RR, Kaupmann K, van der Putten PH, Rovelli G, Shimshek DR (2011) LRRK2 protein levels

are determined by kinase function and are crucial for kidney and lung homeostasis in mice. Hum Mol Genet 20(21):4209–4223. doi:10.1093/hmg/ddr348
44. Hinkle KM, Yue M, Behrouz B, Dachsel JC, Lincoln SJ, Bowles EE, Beevers JE, Dugger B, Winner B, Prots I, Kent CB, Nishioka K, Lin WL, Dickson DW, Janus CJ, Farrer MJ, Melrose HL (2012) LRRK2 knockout mice have an intact dopaminergic system but display alterations in exploratory and motor co-ordination behaviors. Mol Neurodegener 7:25. doi:10.1186/1750-1326-7-25
45. Baptista MA, Dave KD, Frasier MA, Sherer TB, Greeley M, Beck MJ, Varsho JS, Parker GA, Moore C, Churchill MJ, Meshul CK, Fiske BK (2013) Loss of leucine-rich repeat kinase 2 (LRRK2) in rats leads to progressive abnormal phenotypes in peripheral organs. PLoS One 8(11), e80705. doi:10.1371/journal.pone.0080705
46. Fuji RN, Flagella M, Baca M, Baptista MA, Brodbeck J, Chan BK, Fiske BK, Honigberg L, Jubb AM, Katavolos P, Lee DW, Lewin-Koh SC, Lin T, Liu X, Liu S, Lyssikatos JP, O'Mahony J, Reichelt M, Roose-Girma M, Sheng Z, Sherer T, Smith A, Solon M, Sweeney ZK, Tarrant J, Urkowitz A, Warming S, Yaylaoglu M, Zhang S, Zhu H, Estrada AA, Watts RJ (2015) Effect of selective LRRK2 kinase inhibition on nonhuman primate lung. Sci Transl Med 7(273):273ra215. doi:10.1126/scitranslmed.aaa3634
47. Ramonet D, Daher JP, Lin BM, Stafa K, Kim J, Banerjee R, Westerlund M, Pletnikova O, Glauser L, Yang L, Liu Y, Swing DA, Beal MF, Troncoso JC, McCaffery JM, Jenkins NA, Copeland NG, Galter D, Thomas B, Lee MK, Dawson TM, Dawson VL, Moore DJ (2011) Dopaminergic neuronal loss, reduced neurite complexity and autophagic abnormalities in transgenic mice expressing G2019S mutant LRRK2. PLoS One 6(4), e18568. doi:10.1371/journal.pone.0018568
48. Tsika E, Kannan M, Foo CS, Dikeman D, Glauser L, Gellhaar S, Galter D, Knott GW, Dawson TM, Dawson VL, Moore DJ (2014) Conditional expression of Parkinson's disease-related R1441C LRRK2 in midbrain dopaminergic neurons of mice causes nuclear abnormalities without neurodegeneration. Neurobiol Dis 71:345–358. doi:10.1016/j.nbd.2014.08.027
49. Winslow AR, Chen CW, Corrochano S, Acevedo-Arozena A, Gordon DE, Peden AA, Lichtenberg M, Menzies FM, Ravikumar B, Imarisio S, Brown S, O'Kane CJ, Rubinsztein DC (2010) alpha-Synuclein impairs macroautophagy: implications for Parkinson's disease. J Cell Biol 190(6):1023–1037. doi:10.1083/jcb.201003122
50. Vogiatzi T, Xilouri M, Vekrellis K, Stefanis L (2008) Wild type alpha-synuclein is degraded by chaperone-mediated autophagy and macroautophagy in neuronal cells. J Biol Chem 283(35):23542–23556. doi:10.1074/jbc.M801992200
51. Sidransky E, Lopez G (2012) The link between the GBA gene and parkinsonism. Lancet Neurol 11(11):986–998. doi:10.1016/S1474-4422(12)70190-4
52. Ramirez A, Heimbach A, Grundemann J, Stiller B, Hampshire D, Cid LP, Goebel I, Mubaidin AF, Wriekat AL, Roeper J, Al-Din A, Hillmer AM, Karsak M, Liss B, Woods CG, Behrens MI, Kubisch C (2006) Hereditary parkinsonism with dementia is caused by mutations in ATP13A2, encoding a lysosomal type 5 P-type ATPase. Nat Genet 38(10):1184–1191. doi:10.1038/ng1884
53. Vilarino-Guell C, Wider C, Ross OA, Dachsel JC, Kachergus JM, Lincoln SJ, Soto-Ortolaza AI, Cobb SA, Wilhoite GJ, Bacon JA, Behrouz B, Melrose HL, Hentati E, Puschmann A, Evans DM, Conibear E, Wasserman WW, Aasly JO, Burkhard PR, Djaldetti R, Ghika J, Hentati F, Krygowska-Wajs A, Lynch T, Melamed E, Rajput A, Rajput AH, Solida A, Wu RM, Uitti RJ, Wszolek ZK, Vingerhoets F, Farrer MJ (2011) VPS35 mutations in Parkinson disease. Am J Hum Genet 89(1):162–167. doi:10.1016/j.ajhg.2011.06.001
54. Hayflick SJ, Kruer MC, Gregory A, Haack TB, Kurian MA, Houlden HH, Anderson J, Boddaert N, Sanford L, Harik SI, Dandu VH, Nardocci N, Zorzi G, Dunaway T, Tarnopolsky M, Skinner S, Holden KR, Frucht S, Hanspal E, Schrander-Stumpel C, Mignot C, Heron D, Saunders DE, Kaminska M, Lin JP, Lascelles K, Cuno SM, Meyer E, Garavaglia B, Bhatia K, de Silva R, Crisp S, Lunt P, Carey M, Hardy J, Meitinger T, Prokisch H, Hogarth P (2013)

beta-Propeller protein-associated neurodegeneration: a new X-linked dominant disorder with brain iron accumulation. Brain J Neurol 136(Pt 6):1708–1717. doi:10.1093/brain/awt095

55. Manzoni C, Lewis PA (2013) Dysfunction of the autophagy/lysosomal degradation pathway is a shared feature of the genetic synucleinopathies. FASEB J 27:3424–3429. doi:10.1096/fj.12-223842
56. Tofaris GK (2012) Lysosome-dependent pathways as a unifying theme in Parkinson's disease. Mov Disord 27(11):1364–1369. doi:10.1002/mds.25136
57. Looyenga BD, Furge KA, Dykema KJ, Koeman J, Swiatek PJ, Giordano TJ, West AB, Resau JH, Teh BT, MacKeigan JP (2011) Chromosomal amplification of leucine-rich repeat kinase-2 (LRRK2) is required for oncogenic MET signaling in papillary renal and thyroid carcinomas. Proc Natl Acad Sci U S A 108(4):1439–1444. doi:10.1073/pnas.1012500108
58. Civiero L, Dihanich S, Lewis PA, Greggio E (2014) Genetic, structural, and molecular insights into the function of ras of complex proteins domains. Chem Biol 21(7):809–818. doi:10.1016/j.chembiol.2014.05.010

Chapter 6
Molecular Insights and Functional Implication of LRRK2 Dimerization

Laura Civiero, Isabella Russo, Luigi Bubacco, and Elisa Greggio

Abstract The Parkinson's disease protein leucine-rich repeat kinase 2 (LRRK2) is a multidomain protein with an enzymatic core comprising serine-threonine kinase and GTPase activities and a number of protein-protein interaction domains. While the complex domain architecture of LRRK2 has hampered its structural investigation, there is convincing evidence that LRRK2 can form dimers in solution and in the cell and that the GTPase/ROC domain plays a central role in this process. This chapter focuses on recent studies addressing the molecular nature, the functional significance, and the pathological implication of LRRK2 dimerization.

Keywords LRRK2 • Dimerization • Kinase • GTPase • Inhibitor • Phosphorylation

Introduction

Dimerization is a widespread process that in protein kinases governs the switch between active and inactive states.

Receptor tyrosine kinases, such as epidermal growth factor (EGF) receptors, are activated upon extracellular ligand binding. The latter induces conformational changes resulting in lateral diffusion, dimerization of monomers, and kinase activation by trans-autophosphorylation. Autophosphorylation of tyrosine residues promotes intracellular substrate binding and initiation of downstream signaling (for a detailed review, see [1]).

Serine/threonine kinases can be also activated by dimerization. For example, the mitogen-activated protein kinase kinase kinase (MAPKKK) RAF exists as an autoinhibited monomer in the quiescent state. Upon GTP binding, the small GTPase RAS recruits RAF to the plasma membrane inducing a conformational change that results in dephosphorylation, loss of 14-3-3 binding at the RAF N-terminal region, and stabilization of a side-to-side catalytically active dimer (for a detailed review, see

L. Civiero • I. Russo • L. Bubacco • E. Greggio (✉)
Department of Biology, University of Padova, Via Ugo Bassi 58/B, Padova 35131, Italy
e-mail: elisa.greggio@unipd.it

H.J. Rideout (ed.), *Leucine-Rich Repeat Kinase 2 (LRRK2)*,
Advances in Neurobiology 14, DOI 10.1007/978-3-319-49969-7_6

[2]). On the contrary, p21-activated kinase 1 (PAK1), a master regulator of actin cytoskeleton dynamics, exists as a self-inhibited dimer and is activated upon binding of Cdc42 or Rac1. Interaction with small GTPases results in dimer dissociation with consequent release of PAK1 trans-autoinhibition. Monomeric PAK1 undergoes autophosphorylation at multiple sites, a process required for full kinase activation [3].

As for many serine/threonine kinases, multiple lines of evidence suggest that also LRRK2 can form dimers in solution. However, in contrast to other serine/threonine kinases, the kinase domain of LRRK2 is embedded in a complex protein architecture, which has hampered or limited the structural investigation of LRRK2 dimerization properties, as it will be discussed in the next paragraphs. There is also good evidence suggesting that LRRK2 dimer represents the active conformation of the kinase, although the molecular mechanism governing monomer/dimer transition and the upstream events required for dimer formation remain poorly understood. In this chapter we will attempt to provide a comprehensive overview of the relevant structural features, functional significance, and pathological implication of LRRK2 dimerization and critically discuss the possibility of modulating dimerization as a putative therapeutic avenue to target LRRK2-linked Parkinson's disease (PD).

LRRK2 Dimerization: Structural Evidence

LRRK2 belongs to the family of ROCO (Ras Of COmplex) proteins [4]. ROCOs possess a Ras-like GTPase domain (ROC) invariably followed by a COR (C-terminal of ROC) domain of unclear function [5]. ROC-COR is never found isolated in nature: the simplest ROCO proteins, such as human MASL1 and bacterial ROCO, present with N-terminal leucine-rich repeats (LRRs). In addition to a LRR-ROC-COR module, LRRK2, LRRK1, and other ROCO proteins from metazoans and *Dictyostelium discoideum* also possess a serine/threonine kinase domain C-terminal of ROC-COR, in addition to one or more extra N- or C-terminal domains (Fig. 6.1) [6]. Recent evidence from structural investigations of prokaryotic ROCO proteins suggests that COR operates as a dimerization device to stimulate ROC GTPase activity. Such a mechanism would place ROCO proteins among the GAD class of molecular switches (G proteins activated by nucleotide dependent dimerization) [7]. Structural information is also available for the kinase domain of Roco4, a ROCO protein from *D. discoideum* which shares similar domain architecture of LRRK2 [8] and has been recently shown to phosphorylate human ROC with similar kinetics as human LRRK2 [9]. In the crystal used to obtain the structure of the isolated Roco4 kinase domain, the protein is monomeric [8], but whether it engages in homophilic *cis*- or *trans*-interactions with other domains in the context of the full-length protein remains unresolved at this time (Fig. 6.1, green structure).

The kinase domain of LRRK2 is predicted to belong to the class of mitogen-activated protein kinase kinase kinase (MAPKKK), such as RAF [10], and one feature of these kinases is the formation of dimers [2]. A major challenge in the

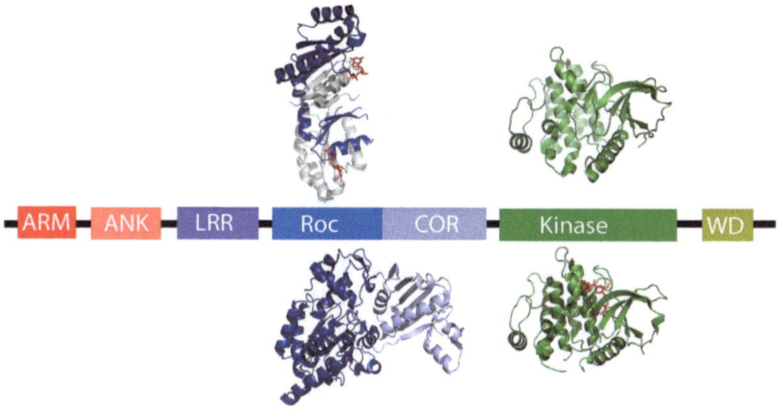

Fig. 6.1 Domain topology of LRRK2. LRRK2 domains are armadillo repeats (ARM), ankyrin repeats (ANK), leucine-rich repeats (LRR), Ras of complex proteins (Roc), C-terminal of Roc (COR), and kinase (kinase) and WD40 repeats (WD). Crystal structures of the human swapped Roc dimer (PDB 2ZEJ) (*blue* and *gray*, *upper panel*) and a cartoon representation of the *C. tepidum* RocCOR structure (PDB 3DPU) (*blue* and *violet*, *bottom panel*). In green, *D. discoideum* Roco4 kinase domain 3D model is represented in the apo conformation (PDB 4F0G) (*upper panel*). In the bottom panel, humanized Roco4 kinase domain bound to LRRK2 IN-1 inhibitor is visualized (PDB 4YZM)

investigation of LRRK2 tertiary and quaternary structure is due to the technical hurdle of isolating sufficient high-quality/high-quantity full-length LRRK2 and/or domains. As a consequence, structural information on full-length LRRK2 is still missing, and clues on LRRK2 quaternary structure come from indirect evidence. A few years ago, we found that highly purified full-length LRRK2 proteins are capable of forming dimers in solution as revealed by immune-gold, looking at the distribution of distances among metal particles imaged by transmission electron microscopy [11]. Interestingly, the pathogenic LRRK2 G2019S with a kinase activity two- to threefold higher than the wild type and the kinase and GTPase-inactive LRRK2 functional mutants do not lose the ability of forming dimers in solution, suggesting that dimer formation of isolated LRRK2 proteins does not obviously correlate with LRRK2 enzymatic activity [11].

Published in 2008, the first structure of the isolated human ROC domain (amino acids 1333–1516) in complex with GDP revealed a dimeric GTPase [12]. Although the proposed model describes a canonical GTPase fold, the catalytic core of LRRK2-ROC adopts an unusual topology due to domain swapping, in which the N-terminal of one monomer interacts with the C-terminal of the second monomer, thereby forming a constitutive dimer (Fig. 6.1, blue and gray structure). Two follow-up studies investigated the quaternary structure of isolated ROC in solution using size-exclusion chromatography. Liu et al. showed that ROC construct, which crystalizes as a swapped dimer [12], purifies as a mixture of dimers and monomers and ROC phosphorylation by LRRK2 or Roco4 enhances the proportion of monomers [9]. Of interest, the GTPase activity of monomers and dimers are nearly identical. A second study by Liao and collaborators [13] purified an extended ROC construct (amino acids 1329–1520) that resulted in highly stable protein, and exhibited two chromatographic peaks corresponding to the apparent size of a dimer and a monomer. The authors further demonstrated that the dimeric fraction can be converted into monomers by addition of guanine nucleotides [13]. Therefore, the existence of both dimers and monomers in solution suggests that isolated ROC may not form a constitutive dimer and that the swapped 3D model might be a crystallographic artifact.

One important consideration to make is that the ROC domain always occurs in tandem with the C-terminal of ROC (COR) domain, and the ROC-COR module is conserved across species. Thus, it is likely that ROC-COR represents a single functional unit, and structural and biophysical studies should be focused on this fragment rather than on isolated ROC. However, the difficulty of purifying sufficient amounts of stable recombinant human ROC-COR has hampered its structural investigation. Important structural information on the role of COR in ROC dimerization comes from studies with related ROCO proteins from bacteria. In 2008, a crystallographic study of the ROC-COR unit from the thermophilic green bacteria *Chlorobium tepidum* disclosed a dimeric organization of ROCO proteins. However, in contrast to the previously determined structure of human ROC domain, the structural analysis of bacterial ROC-COR revealed a canonical G protein domain where dimerization is mediated by the C-terminal half of the COR domain and by highly conserved residues on the ROC-COR interface (Fig. 6.1, blue and violet structures) [14]. That COR is a stable dimerization device serving as a scaffold for the ROC domain was also confirmed by a more recent study in which site-direct spin labeling was used to evaluate by pulse EPR the distances defined by the constrains that govern dimerization [15]. The ability of ROC-COR to hydrolyze GTP depends on its dimeric conformation and mutations analogous to the Parkinson's disease mutations (R1441C, Y1699C, I1371V) located in the ROC-COR interface lead to a reduction of GTPase activity [14]. Supporting this notion, LRRK2 carrying R1441C or Y1699C mutations, which lacks GTPase activity [16, 17], binds the ROC domain with lower affinity in vitro compared to the wild-type protein [18]. In 2015, a more detailed study on the structure and function of the ROC-COR from *Methanosarcina barkeri* convincingly demonstrated that dimerization is independent from nucleotide binding, but dimerization is essential for GTPase activity [19]. The structure of the GDP-bound fragment ROC-CORΔC from *M. barkeri* that does not contain the C-terminal subdomain of ROC responsible for dimer formation in *C. tepidum* is monomeric, and ROC-CORΔC displays an approximately tenfold decreased GTPase activity compared to ROC-COR dimer from *M. barkeri* [19].

As shown for the bacterial orthologues, co-immunoprecipitation experiments using human LRRK2 fragments suggest that COR is essential for stable dimer formation [19]. Furthermore, the hydrolysis rate of recombinant ROC-COR-kinase fragment is around 40-fold faster than that of the recombinant human ROC domain [19], suggesting that dimerization regulates GTPase activity of LRRK2, similar to what was observed for bacterial Roco proteins.

Based on the low affinity (in the range of μM) for nucleotides [11, 13, 14] and the capability of dimerization/oligomerization [11, 20], LRRK2 and other ROCO proteins were recently suggested to act as GADs [7]. GADs are a group of multidomain proteins that do not require guanine exchange factors (GEFs) to exchange GDP for GTP [7]. The GTP-bound dimer is the active form responsible for the biological process, which is inactivated by hydrolysis of GTP. However, in contrast to other GADs where the constitutive dimer adopts an open conformation in the nucleotide free/GDP-bound and an open/closed two-state equilibrium in the GTP-bound state [21], no differences were observed for the GTPase domain of the bacterial ROC-COR dimer with the ROC domain remaining associated during the GTP hydrolysis [15]. Furthermore, while the isolated G domains of GADs such as MnME and human GBP1 are sufficient to form a functional dimer [21], the COR dimer scaffold is necessary to keep the ROC domain in close proximity, at least for the bacterial Roco proteins. Although LRRK2 shows strong analogies with other Roco proteins and, more in general, with GADs, a detailed understanding of the G protein cycle of LRRK2 supported by structural information of full-length proteins is missing and represents a high-priority challenge in the field.

LRRK2 Dimerization in Cells

Although not directly supported by a crystallographic structure, there is robust evidence that isolated full-length LRRK2 forms dimers in solution. But what is the evidence supporting LRRK2 dimerization also in the cellular environment? The first indication for potential LRRK2 cellular dimerization was obtained using tandem affinity purification [22]. The authors showed that HA-tagged LRRK2 is efficiently co-precipitated by Strep/Flag-tagged LRRK2 but not Strep/Flag-kinase domain alone, indicating that LRRK2 self-associates but the kinase domain is not sufficient for the interaction [22]. A subsequent study reported LRRK2 self-interaction in the cell by co-immunoprecipitation of differentially tagged LRRK2 fragments and yeast two-hybrid (Y2H) analysis [23]. Both Y2H and co-immunoprecipitation experiments suggested that the ROC domain might act as a hub of interaction with multiple LRRK2 domains, including ROC itself, supporting the existence of ROC-ROC dimers in the cells.

Multiple studies provided additional evidence supporting the existence of dimeric LRRK2 in the cell by means of gel filtration of cell lysates expressing endogenous or ectopically expressed full-length LRRK2 as well as blue native gels [20, 23–25]. These experiments indicate that the dimer appears as the prevalent conformer, with monomers and higher-order oligomers present as a minor proportion. Of interest, dimeric LRRK2 undergoes autophosphorylation *in cis* [23], a finding that was later confirmed using an antibody against phosphorylated S1292, a LRRK2 autophosphorylation site [26]. Interestingly, kinase-inactive LRRK2 mutants form high molecular weight (HMW) complexes when analyzed from total cell lysates [20, 23]. These HMW species are likely to be hetero-complexes rather than higher-order oligomers, since pharmacological inhibition of LRRK2 kinase activity phenocopies the formation of HMW species in gel filtration only when LRRK2 is present in the cell extract but not isolated in solution [27]. Furthermore, LRRK2 is likely to be monomeric within the HMW complex, and the WD40 C-terminal domain is required for dimer formation and kinase activity [24, 28]. Of interest, the N-terminal region of LRRK2 isolated in solution distributes across monomers, dimers, and oligomers [29], suggesting that dimer formation could also be mediated by the N-terminus, as also supported by Y2H experiments [23].

In support of these findings, Sen and collaborators observed that, among the different LRRK2 species separated by gel filtration and enriched by immunoprecipitation, only the dimer is catalytically active [20]. If the dimer is the *bona fide* active conformation, interfering with LRRK2 dimerization may represent a strategy to switch off kinase activity.

There is however some controversy as to whether LRRK2 is truly a homodimer or instead a monomer, being the majority of the analytical methods employed strictly dependent upon the hydrodynamic radius of the molecule analyzed. To this regard, Ito and Iwastubo performed a series of experiments suggesting that the ~600 kDa complex previously observed in gel filtration and compatible with the size of a dimer might be a pool of monomeric LRRK2. They discuss this finding in the contest of other high molecular weight (HMW) proteins, which were previously shown to exhibit anomalous MW in blue native gels and gel filtration experiments than expected [30].

One important drawback of all these studies is that the oligomerization status of LRRK2 was analyzed from cell extracts and not in living cells. To overcome this potential limitation, James and collaborators performed number and brightness (N&B) analysis of fluorescence fluctuation spectroscopy data to measure LRRK2 oligomerization state in the cytoplasm and plasma membrane of living cells. N&B analysis provides a map of number of diffusing particles (N) and the intrinsic brightness (B) of each particle for every pixel in the image [31]. Using N&B analysis and normalizing the brightness levels for monomeric GFP, they found that the majority of cytoplasmic LRRK2 was monomeric, whereas LRRK2 oligomers (dimers, tetramers, or higher-order oligomers) were present near the plasma membrane [31]. Although this analysis could not conclusively prove that the membrane-associated oligomers are exclusively dimers, it does suggest the existence of LRRK2 homointeraction in living cells and indicates that this homotypic interaction is compartmentalized. It will be of great interest to apply N&B analysis to investigate whether any extracellular stimulus capable of activating LRRK2 can dynamically modulate the equilibrium between monomers and oligomers within defined subcellular compartments. If dimerization is truly associated with kinase activity, N&B analysis would also represent a powerful tool to record in live mode the effect kinase of inhibitors on LRRK2 oligomerization/compartmentalization and also provide a valuable platform for high-throughput pharmacodynamics screenings. To rule out artifactual mislocalization due to protein overexpression, future N&B analysis should be performed on endogenous proteins tagged with GFP via CRISPR/Cas-based systems.

Functional Implication of Dimerization

The N&B analysis of LRRK2 oligomerization in living cells has provided key indications that the kinase may shuttle from a diffuse monomeric state to a compartmentalized oligomeric state [31].

In support of this model, there is evidence that links LRRK2 activation with relocalization at the cytoskeleton. For instance, treatment of cells with arsenite induces LRRK2 dimerization and relocalization to centrosomes [32]. Also, multiple pathogenic mutations that display increased kinase activity and/or decreased GTPase activity enhance LRRK2 oligomerization and colocalization with microtubules [33].

In addition, two studies from Lavoie's group showed that membrane-associated LRRK2 is dimeric and catalytically more active than monomeric cytosolic LRRK2 [34], and lipopolysaccharide (LPS) stimulation of microglial cells results in relocalization of dimeric LRRK2 to membranes [35]. Moreover, membrane-recruited LRRK2 is predominantly dimeric, suggestive of an active state of the protein associated with a specific cellular function. In agreement with the proposed role of LRRK2 in autophagic processes [36–38], they found that membrane-associated LRRK2 dimer is implicated in the autophagic flux and clearance of aggregated proteins [35].

One apparent paradox is that LRRK2 kinase inhibitors cause LRRK2 cellular compartmentalization in skein-like structures [39], an event associated with LRRK2 activation, according to this model. Although there is no experimental evidence, one possibility could be that LRRK2 inhibitors stabilize a dimeric conformation associated with a kinase-independent LRRK2 active state that is capable of subcellular relocalization and activation of cognate partners through its scaffolding activity. Alternatively, at subsaturating concentrations of inhibitor, drug-bound LRRK2 may adopt a conformation prone to dimerization, recruiting and trans-activating drug-free LRRK2 monomers, which are competent to phosphorylate specific substrates. The latter mechanism has been described for RAF proteins. A number of RAF inhibitors can paradoxically activate downstream MEK-ERK signaling when applied to cells, facilitating RAF dimerization and promoting autoactivation of the dimerized RAF molecule not bound by the inhibitor (reviewed in [40]). Whether the mechanism of LRRK2 dimerization is analogous to that described for RAF is not clear at this time, and future research should address these opened questions.

14-3-3 proteins are probably the best characterized LRRK2 interactors. Phosphorylation of a cluster of N-terminal serine residues (S910, S935, S955, and S973) preceding the LRR domain of LRRK2 is required for 14-3-3 s binding, and dephosphorylation of these residues results in LRRK2:14-3-3 complex dissociation [39, 41, 42]. 14-3-3s can control the localization, activity, or stability of their target proteins preventing protein interactions by steric hindrance [43]. As an example, auto-inhibited RAF is stabilized by 14-3-3 proteins bound to phosphorylated S259 at the RAF N terminus, which maintain RAF in an inactive state in the cytosol. Upon activation by extracellular mitogens, GTP-bound RAS interacts with RAF causing S259 dephosphorylation, 14-3-3 s dissociation, and recruitment of RAF to

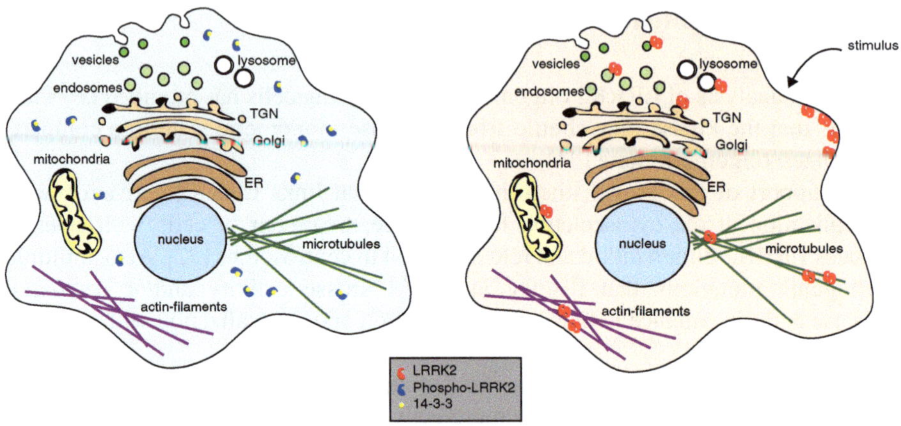

Fig. 6.2 Hypothetical representation of the subcellular distribution, quaternary structure, and phosphorylation state of LRRK2 in resting (*left*) and stimulated (*right*) cells

nanoclusters in the plasma membrane where RAF dimerizes (reviewed in [2]). Alternatively, 14-3-3s could direct their partners to specific cellular compartments where the kinases are competent to form macro-complexes able to transduce signals, as previously suggested for PAK4 [44].

In analogy to RAF, monomeric LRRK2 could be maintained in a latent state by inhibitory association with 14-3-3 in the cytosol, and there is evidence that 14-3-3-bound LRRK2 is cytosolic [39, 41]. Upon stimulation (e.g., arsenite), LRRK2 becomes dephosphorylated at S910/S935 possibly through activation of PP1 phosphatases and/or inhibition of kinases such as CK1α [45, 46]. This would cause dissociation of 14-3-3 and relocalization of LRRK2 to defined subcellular compartments where it dimerizes, a process required for kinase activation (Fig. 6.2). However, LRRK2 phosphorylation at the N-terminal serine cluster does not always parallel with protein monomerization/inactivation. Indeed, upon LPS stimulation, dimerization on membranes correlates with increased cellular phosphorylation at Ser935 in peripheral and brain macrophages [35]. Whether a specific 14-3-3 isoform, distinct from that responsible for maintaining LRRK2 in the cytosol, directs LRRK2 to the membranes in macrophages is not known at this time.

While it is possible that different activation and relocalization mechanisms occur in different cell types, additional studies are clearly necessary to better understand the complex interplay between phosphorylation and dimerization in LRRK2 activation and signaling, with a major focus on neuronal systems.

Pathological Implication of Dimerization

As discussed in the previous paragraphs, multiple lines of evidence suggest that the active conformation of LRRK2 is the dimer. Mutations in LRRK2 are linked to autosomal, dominant Parkinson's disease, with clinical and pathological presentation similar to the sporadic syndrome [47]. Of interest, mutations that segregate with disease are clustered within the enzymatic core of the protein [47]. The G2019S mutation in the kinase domain, which is the most prevalent among the pathogenic LRRK2 mutations, has a clear effect in augmenting kinase activity, whereas the effect of the neighbor I2020T mutation is less obvious [48, 49]. ROC-COR mutations (N1437H, R1441C/G/H, and Y1699C) have been suggested to decrease GTPase activity, which would result in a prolonged active state of the protein, and have variable effects on kinase activity [26, 48]. One striking feature of the majority of LRRK2 pathogenic mutations, with the notable exception of the kinase hyperactive G2019S, is that they all display moderate to severe loss of S910/S935 phosphorylation and 14-3-3 binding [39, 42, 50]. If dephosphorylated LRRK2 represents the active state of the protein, this should also correlate with more compartmentalization, according to the model proposed above. Indeed, mutant LRRK2, with the exception of the G2019S, is more clustered in the cell, and there seems to be a good correlation between the degree of dephosphorylation and cellular clustering [39]. LRRK2 associates with BAG5, GAK, and Rab7L1 to promote autophagic clearance of trans-Golgi network (TNG)-derived vesicles [51]. Strikingly, overexpression of disease-associated mutations, but not kinase-dead or GTP-deficient binding mutants, in combination with Rab7L1, causes a massive decrease of TGN-derived vesicles, indicating abnormally high compartmentalization to TGN-derived vesicles [51].

A handful of studies interrogated the quaternary state of mutant LRRK2. Jorgensen and co-workers observed that the electrophoretic pattern of LRRK2 wild type, R1441C, and G2019S in blue native gels is almost identical, with the presence of one band around 600 kDa and a higher MW band [24]. Using the same technique, Sen and collaborators were instead able to identify an increased propensity of mutant LRRK2 to form dimers [20], a finding supporting the notion that pathogenic mutations are locked in an active, dimeric state. One study reported that the pathogenic mechanism behind the I2010T mutation involves the recruitment of the wild-type proteins and the formation of WT:I2020T heterodimers. The authors suggest that I2020T through a dominant negative effect leads to accelerated degradation of LRRK2 wild type [52, 53]. While not addressed by the authors, it can be speculated that the I2020T mutation may act similar to oncogenic BRAF mutations, which are also clustered in the activation loop, by stabilizing wild-type LRRK2 active conformation via dimerization.

Another study addressed the effect of the Y1699C mutation within the COR domain on human ROC-COR dimerization [16]. Using a competition assay, the authors found that the intramolecular ROC-COR interaction is favored over ROC-ROC dimerization. The pathogenic Y1699C mutation, situated at the ROC-COR interface, strengthens the intramolecular ROC-COR interaction, weakening the dimerization of LRRK2 at the ROC-COR tandem domain and resulting in decreased GTPase activity [16]. In *C. tepidum* ROC-COR, the Y804C PD mutation in the COR domain analogous to the Y1699C human mutation and the two L487V and Y558A mutations in the ROC domain, analogous to the I1371V and R1441G/C PD mutations, have much lower GTPase activity but similar dimerization properties

[14]. Taken together, the ROC-COR PD mutations seem to severely affect GTPase activity, whereas the effect on dimerization is unclear. Outside ROC-COR, a decreased GTPase activity by 20 % has also been reported for the G2019S mutant [54]. Overall, LRRK2 PD mutations may promote protein dimerization and/or decreased GTPase activity, which results in a prolonged active state of the protein. Additional studies are required to clarify the impact of LRRK2 mutations in protein dimerization.

Targeting Dimerization: A Therapeutic Opportunity?

Since the most common G2019S mutation increases LRRK2 kinase activity, much research has been focused toward the identification of potent, selective, and brain-permeable LRRK2 inhibitors (reviewed in [55]). However, recent studies highlighted the potential safety liabilities of LRRK2 inhibition in rodent and nonhuman primates [56]. In particular, LRRK2 inhibition causes peripheral side effects including lung pathology similar to what was observed in LRRK2 knockout mice [57]. One possible explanation is that LRRK2 inhibition destabilizes LRRK2 protein levels by increasing its ubiquitination and proteasomal degradation [58]. Of note, LRRK2 ubiquitination is preceded by dephosphorylation, as expected when kinase inhibition is applied [58]. Therefore, it is reasonable to explore alternative therapeutic approaches other than kinase inhibition to reduce LRRK2 activation in the cell that do not disturb its phosphorylation levels. Compounds interfering with LRRK2 dimerization may result effective at keeping LRRK2 in the monomeric state, which may be associated with reduced activity and cytoplasmic localization, as discussed. One current limitation in designing such class of compounds resides in our limited knowledge of the structural requirements for protein dimerization. However, one study indicates that the ROCO domain exerts an inhibitory effect on LRRK2 kinase activity [25]. The authors show that ROCO is critical for LRRK2 dimerization, and ROCO fragments inhibit LRRK2 kinase activity, possibly disrupting dimerization [25]. These findings support the notion that dimerization occurs via ROCO and that peptides competing with ROCO binding may block LRRK2 self-association.

A number of cell penetrating peptides have been successfully developed to inhibit protein kinase activity or protein dimerization. For example, the kinase activity of PAK1 can be inhibited by the cell permeable TA-PAK18 peptide, which blocks the interaction between PAK1 and PIX, a GEF essential for PAK1 activation [59]. Also, the synthetic peptido-mimetic compound ST2825 is able to inhibit homodimerization of MyD88, an adaptor protein essential in the intracellular signaling elicited by IL-1R and TLRs [60].

High-throughput approaches could be useful to identify small peptides capable to disrupt or reduce LRRK2 dimer formation. Dimerization can be monitored by analytical approaches (gel filtration chromatography, blue native gels) or in live cells using N&B or FRET analysis. Alternatively, ad hoc peptides could be designed based on the crystallographic structures available (human ROC and bacterial ROC-COR), an approach that would also validate or reject these structures.

Although targeting LRRK2 dimerization is appealing, fundamental questions on the precise mechanism that correlates LRRK2 dimerization and activation in neuronal cells are still open, and therefore, more research is needed in this direction before envisaging a dimer-interfering approach as therapeutic strategy.

Conclusions

From the available literature, there is reasonable evidence that LRRK2 can form dimers both in isolation and in the cellular contest, although X-ray crystals, cryo-EM, or dynamic SAXS analysis of LRRK2 structure, if available in the future, will conclusively shed light into the nature of these dimers and their physiological and pathological implications.

Acknowledgments This research was supported by the Italian Telethon Foundation (grant n. GGP12237) and the Michael J Fox Foundation for Parkinson's disease research.

Conflict of Interest The author declares no conflicts of interest.

References

1. Lemmon MA, Schlessinger J (2010) Cell signaling by receptor tyrosine kinases. Cell 141(7):1117–1134. doi:10.1016/j.cell.2010.06.011
2. Lavoie H, Therrien M (2015) Regulation of RAF protein kinases in ERK signalling. Nat Rev Mol Cell Biol 16(5):281–298. doi:10.1038/nrm3979
3. Parrini MC, Lei M, Harrison SC, Mayer BJ (2002) Pak1 kinase homodimers are autoinhibited in trans and dissociated upon activation by Cdc42 and Rac1. Mol Cell 9(1):73–83
4. Civiero L, Dihanich S, Lewis PA, Greggio E (2014) Genetic, structural, and molecular insights into the function of ras of complex proteins domains. Chem Biol 21(7):809–818. doi:10.1016/j.chembiol.2014.05.010
5. Bosgraaf L, Van Haastert PJ (2003) Roc, a Ras/GTPase domain in complex proteins. Biochim Biophys Acta 1643(1–3):5–10
6. Gilsbach BK, Kortholt A (2014) Structural biology of the LRRK2 GTPase and kinase domains: implications for regulation. Front Mol Neurosci 7:32. doi:10.3389/fnmol.2014.00032
7. Gasper R, Meyer S, Gotthardt K, Sirajuddin M, Wittinghofer A (2009) It takes two to tango: regulation of G proteins by dimerization. Nat Rev Mol Cell Biol 10(6):423–429. doi:10.1038/nrm2689
8. Gilsbach BK, Ho FY, Vetter IR, van Haastert PJ, Wittinghofer A, Kortholt A (2012) Roco kinase structures give insights into the mechanism of Parkinson disease-related leucine-rich-repeat kinase 2 mutations. Proc Natl Acad Sci U S A 109(26):10322–10327. doi:10.1073/pnas.1203223109
9. Liu Z, Mobley JA, DeLucas LJ, Kahn RA, West AB (2015) LRRK2 autophosphorylation enhances its GTPase activity. FASEB J 30:336–347. doi:10.1096/fj.15-277095

10. Gloeckner CJ, Schumacher A, Boldt K, Ueffing M (2009) The Parkinson disease-associated protein kinase LRRK2 exhibits MAPKKK activity and phosphorylates MKK3/6 and MKK4/7, in vitro. J Neurochem 109(4):959–968. doi:10.1111/j.1471-4159.2009.06024.x
11. Civiero L, Vancraenenbroeck R, Belluzzi E, Beilina A, Lobbestael E, Reyniers L, Gao F, Micetic I, De Maeyer M, Bubacco L, Baekelandt V, Cookson MR, Greggio E, Taymans JM (2012) Biochemical characterization of highly purified leucine-rich repeat kinases 1 and 2 demonstrates formation of homodimers. PLoS One 7(8), e43472. doi:10.1371/journal.pone.0043472
12. Deng J, Lewis PA, Greggio E, Sluch E, Beilina A, Cookson MR (2008) Structure of the ROC domain from the Parkinson's disease-associated leucine-rich repeat kinase 2 reveals a dimeric GTPase. Proc Natl Acad Sci U S A 105(5):1499–1504. doi:10.1073/pnas.0709098105
13. Liao J, Wu CX, Burlak C, Zhang S, Sahm H, Wang M, Zhang ZY, Vogel KW, Federici M, Riddle SM, Nichols RJ, Liu D, Cookson MR, Stone TA, Hoang QQ (2014) Parkinson disease-associated mutation R1441H in LRRK2 prolongs the "active state" of its GTPase domain. Proc Natl Acad Sci U S A 111(11):4055–4060. doi:10.1073/pnas.1323285111
14. Gotthardt K, Weyand M, Kortholt A, Van Haastert PJ, Wittinghofer A (2008) Structure of the Roc-COR domain tandem of *C. tepidum*, a prokaryotic homologue of the human LRRK2 Parkinson kinase. EMBO J 27(16):2239–2249. doi:10.1038/emboj.2008.150
15. Rudi K, Ho FY, Gilsbach BK, Pots H, Wittinghofer A, Kortholt A, Klare JP (2015) Conformational heterogeneity of the Roc domains in *C. tepidum* Roc-COR and implications for human LRRK2 Parkinson mutations. Biosci Rep 35(5), e00254. doi:10.1042/BSR20150128
16. Daniels V, Vancraenenbroeck R, Law BM, Greggio E, Lobbestael E, Gao F, De Maeyer M, Cookson MR, Harvey K, Baekelandt V, Taymans JM (2011) Insight into the mode of action of the LRRK2 Y1699C pathogenic mutant. J Neurochem 116(2):304–315. doi:10.1111/j.1471-4159.2010.07105.x
17. Lewis PA, Greggio E, Beilina A, Jain S, Baker A, Cookson MR (2007) The R1441C mutation of LRRK2 disrupts GTP hydrolysis. Biochem Biophys Res Commun 357(3):668–671. doi:10.1016/j.bbrc.2007.04.006
18. Li Y, Dunn L, Greggio E, Krumm B, Jackson GS, Cookson MR, Lewis PA, Deng J (2009) The R1441C mutation alters the folding properties of the ROC domain of LRRK2. Biochim Biophys Acta 1792(12):1194–1197. doi:10.1016/j.bbadis.2009.09.010
19. Terheyden S, Ho FY, Gilsbach BK, Wittinghofer A, Kortholt A (2015) Revisiting the Roco G-protein cycle. Biochem J 465(1):139–147. doi:10.1042/BJ20141095
20. Sen S, Webber PJ, West AB (2009) Dependence of leucine-rich repeat kinase 2 (LRRK2) kinase activity on dimerization. J Biol Chem 284(52):36346–36356. doi:10.1074/jbc.M109.025437
21. Meyer S, Bohme S, Kruger A, Steinhoff HJ, Klare JP, Wittinghofer A (2009) Kissing G domains of MnmE monitored by X-ray crystallography and pulse electron paramagnetic resonance spectroscopy. PLoS Biol 7(10), e1000212. doi:10.1371/journal.pbio.1000212
22. Gloeckner CJ, Kinkl N, Schumacher A, Braun RJ, O'Neill E, Meitinger T, Kolch W, Prokisch H, Ueffing M (2006) The Parkinson disease causing LRRK2 mutation I2020T is associated with increased kinase activity. Hum Mol Genet 15(2):223–232. doi:10.1093/hmg/ddi439
23. Greggio E, Zambrano I, Kaganovich A, Beilina A, Taymans JM, Daniels V, Lewis P, Jain S, Ding J, Syed A, Thomas KJ, Baekelandt V, Cookson MR (2008) The Parkinson disease-associated leucine-rich repeat kinase 2 (LRRK2) is a dimer that undergoes intramolecular autophosphorylation. J Biol Chem 283(24):16906–16914. doi:10.1074/jbc.M708718200
24. Jorgensen ND, Peng Y, Ho CC, Rideout HJ, Petrey D, Liu P, Dauer WT (2009) The WD40 domain is required for LRRK2 neurotoxicity. PLoS One 4(12), e8463. doi:10.1371/journal.pone.0008463
25. Klein CL, Rovelli G, Springer W, Schall C, Gasser T, Kahle PJ (2009) Homo—and heterodimerization of ROCO kinases: LRRK2 kinase inhibition by the LRRK2 ROCO fragment. J Neurochem 111(3):703–715. doi:10.1111/j.1471-4159.2009.06358.x

26. Sheng Z, Zhang S, Bustos D, Kleinheinz T, Le Pichon CE, Dominguez SL, Solanoy HO, Drummond J, Zhang X, Ding X, Cai F, Song Q, Li X, Yue Z, van der Brug MP, Burdick DJ, Gunzner-Toste J, Chen H, Liu X, Estrada AA, Sweeney ZK, Scearce-Levie K, Moffat JG, Kirkpatrick DS, Zhu H (2012) Ser1292 autophosphorylation is an indicator of LRRK2 kinase activity and contributes to the cellular effects of PD mutations. Sci Transl Med 4(164), 164ra161. doi:10.1126/scitranslmed.3004485
27. Cirnaru MD, Marte A, Belluzzi E, Russo I, Gabrielli M, Longo F, Arcuri L, Murru L, Bubacco L, Matteoli M, Fedele E, Sala C, Passafaro M, Morari M, Greggio E, Onofri F, Piccoli G (2014) LRRK2 kinase activity regulates synaptic vesicle trafficking and neurotransmitter release through modulation of LRRK2 macro-molecular complex. Front Mol Neurosci 7:49. doi:10.3389/fnmol.2014.00049
28. Rudenko IN, Kaganovich A, Hauser DN, Beylina A, Chia R, Ding J, Maric D, Jaffe H, Cookson MR (2012) The G2385R variant of leucine-rich repeat kinase 2 associated with Parkinson's disease is a partial loss-of-function mutation. Biochem J 446(1):99–111. doi:10.1042/BJ20120637
29. Lu B, Zhai Y, Wu C, Pang X, Xu Z, Sun F (2010) Expression, purification and preliminary biochemical studies of the N-terminal domain of leucine-rich repeat kinase 2. Biochim Biophys Acta 1804(9):1780–1784. doi:10.1016/j.bbapap.2010.05.004
30. Ito G, Iwatsubo T (2012) Re-examination of the dimerization state of leucine-rich repeat kinase 2: predominance of the monomeric form. Biochem J 441(3):987–994. doi:10.1042/BJ20111215
31. James NG, Digman MA, Gratton E, Barylko B, Ding X, Albanesi JP, Goldberg MS, Jameson DM (2012) Number and brightness analysis of LRRK2 oligomerization in live cells. Biophys J 102(11):L41–L43. doi:10.1016/j.bpj.2012.04.046
32. Mamais A, Chia R, Beilina A, Hauser DN, Hall C, Lewis PA, Cookson MR, Bandopadhyay R (2014) Arsenite stress down-regulates phosphorylation and 14-3-3 binding of leucine-rich repeat kinase 2 (LRRK2), promoting self-association and cellular redistribution. J Biol Chem 289(31):21386–21400. doi:10.1074/jbc.M113.528463
33. Kett LR, Boassa D, Ho CC, Rideout HJ, Hu J, Terada M, Ellisman M, Dauer WT (2012) LRRK2 Parkinson disease mutations enhance its microtubule association. Hum Mol Genet 21(4):890–899. doi:10.1093/hmg/ddr526
34. Berger Z, Smith KA, Lavoie MJ (2010) Membrane localization of LRRK2 is associated with increased formation of the highly active LRRK2 dimer and changes in its phosphorylation. Biochemistry 49(26):5511–5523. doi:10.1021/bi100157u
35. Schapansky J, Nardozzi JD, Felizia F, LaVoie MJ (2014) Membrane recruitment of endogenous LRRK2 precedes its potent regulation of autophagy. Hum Mol Genet 23(16):4201–4214. doi:10.1093/hmg/ddu138
36. Tong Y, Yamaguchi H, Giaime E, Boyle S, Kopan R, Kelleher RJ 3rd, Shen J (2010) Loss of leucine-rich repeat kinase 2 causes impairment of protein degradation pathways, accumulation of alpha-synuclein, and apoptotic cell death in aged mice. Proc Natl Acad Sci U S A 107(21):9879–9884. doi:10.1073/pnas.1004676107
37. Tong Y, Giaime E, Yamaguchi H, Ichimura T, Liu Y, Si H, Cai H, Bonventre JV, Shen J (2012) Loss of leucine-rich repeat kinase 2 causes age-dependent bi-phasic alterations of the autophagy pathway. Mol Neurodegener 7:2. doi:10.1186/1750-1326-7-2
38. Manzoni C, Mamais A, Dihanich S, Abeti R, Soutar MP, Plun-Favreau H, Giunti P, Tooze SA, Bandopadhyay R, Lewis PA (2013) Inhibition of LRRK2 kinase activity stimulates macroautophagy. Biochim Biophys Acta 1833(12):2900–2910. doi:10.1016/j.bbamcr.2013.07.020
39. Nichols RJ, Dzamko N, Morrice NA, Campbell DG, Deak M, Ordureau A, Macartney T, Tong Y, Shen J, Prescott AR, Alessi DR (2010) 14-3-3 binding to LRRK2 is disrupted by multiple Parkinson's disease-associated mutations and regulates cytoplasmic localization. Biochem J 430(3):393–404. doi:10.1042/BJ20100483

40. Holderfield M, Nagel TE, Stuart DD (2014) Mechanism and consequences of RAF kinase activation by small-molecule inhibitors. Br J Cancer 111(4):640–645. doi:10.1038/bjc.2014.139
41. Dzamko N, Deak M, Hentati F, Reith AD, Prescott AR, Alessi DR, Nichols RJ (2010) Inhibition of LRRK2 kinase activity leads to dephosphorylation of Ser(910)/Ser(935), disruption of 14-3-3 binding and altered cytoplasmic localization. Biochem J 430(3):405–413. doi:10.1042/BJ20100784
42. Li X, Wang QJ, Pan N, Lee S, Zhao Y, Chait BT, Yue Z (2011) Phosphorylation-dependent 14-3-3 binding to LRRK2 is impaired by common mutations of familial Parkinson's disease. PLoS One 6(3), e17153. doi:10.1371/journal.pone.0017153
43. Fu H, Subramanian RR, Masters SC (2000) 14-3-3 proteins: structure, function, and regulation. Annu Rev Pharmacol Toxicol 40:617–647. doi:10.1146/annurev.pharmtox.40.1.617
44. Bastea LI, Doppler H, Pearce SE, Durand N, Spratley SJ, Storz P (2013) Protein kinase D-mediated phosphorylation at Ser99 regulates localization of p21-activated kinase 4. Biochem J 455(2):251–260. doi:10.1042/BJ20130281
45. Lobbestael E, Zhao J, Rudenko IN, Beylina A, Gao F, Wetter J, Beullens M, Bollen M, Cookson MR, Baekelandt V, Nichols RJ, Taymans JM (2013) Identification of protein phosphatase 1 as a regulator of the LRRK2 phosphorylation cycle. Biochem J 456(1):119–128. doi:10.1042/BJ20121772
46. Chia R, Haddock S, Beilina A, Rudenko IN, Mamais A, Kaganovich A, Li Y, Kumaran R, Nalls MA, Cookson MR (2014) Phosphorylation of LRRK2 by casein kinase 1alpha regulates trans-Golgi clustering via differential interaction with ARHGEF7. Nat Commun 5:5827. doi:10.1038/ncomms6827
47. Cookson MR (2010) The role of leucine-rich repeat kinase 2 (LRRK2) in Parkinson's disease. Nat Rev Neurosci 11(12):791–797. doi:10.1038/nrn2935
48. Greggio E, Cookson MR (2009) Leucine-rich repeat kinase 2 mutations and Parkinson's disease: three questions. ASN Neuro 1(1):13–24. doi:10.1042/AN20090007
49. Ray S, Bender S, Kang S, Lin R, Glicksman MA, Liu M (2014) The Parkinson disease-linked LRRK2 protein mutation I2020T stabilizes an active state conformation leading to increased kinase activity. J Biol Chem 289(19):13042–13053. doi:10.1074/jbc.M113.537811
50. Doggett EA, Zhao J, Mork CN, Hu D, Nichols RJ (2012) Phosphorylation of LRRK2 serines 955 and 973 is disrupted by Parkinson's disease mutations and LRRK2 pharmacological inhibition. J Neurochem 120(1):37–45. doi:10.1111/j.1471-4159.2011.07537.x
51. Beilina A, Rudenko IN, Kaganovich A, Civiero L, Chau H, Kalia SK, Kalia LV, Lobbestael E, Chia R, Ndukwe K, Ding J, Nalls MA, Olszewski M, Hauser DN, Kumaran R, Lozano AM, Baekelandt V, Greene LE, Taymans JM, Greggio E, Cookson MR (2014) Unbiased screen for interactors of leucine-rich repeat kinase 2 supports a common pathway for sporadic and familial Parkinson disease. Proc Natl Acad Sci U S A 111(7):2626–2631. doi:10.1073/pnas.1318306111
52. Ohta E, Kawakami F, Kubo M, Obata F (2013) Dominant-negative effects of LRRK2 heterodimers: a possible mechanism of neurodegeneration in Parkinson's disease caused by LRRK2 I2020T mutation. Biochem Biophys Res Commun 430(2):560–566. doi:10.1016/j.bbrc.2012.11.113
53. Miyajima T, Ohta E, Kawada H, Maekawa T, Obata F (2015) The mouse/human cross-species heterodimer of leucine-rich repeat kinase 2: possible significance in the transgenic model mouse of Parkinson's disease. Neurosci Lett 588:142–146. doi:10.1016/j.neulet.2015.01.003
54. Liu M, Kang S, Ray S, Jackson J, Zaitsev AD, Gerber SA, Cuny GD, Glicksman MA (2011) Kinetic, mechanistic, and structural modeling studies of truncated wild-type leucine-rich repeat kinase 2 and the G2019S mutant. Biochemistry 50(43):9399–9408. doi:10.1021/bi201173d
55. Gilligan PJ (2015) Inhibitors of leucine-rich repeat kinase 2 (LRRK2): progress and promise for the treatment of Parkinson's disease. Curr Top Med Chem 15(10):927–938
56. Fuji RN, Flagella M, Baca M, Baptista MA, Brodbeck J, Chan BK, Fiske BK, Honigberg L, Jubb AM, Katavolos P, Lee DW, Lewin-Koh SC, Lin T, Liu X, Liu S, Lyssikatos JP, O'Mahony

J, Reichelt M, Roose-Girma M, Sheng Z, Sherer T, Smith A, Solon M, Sweeney ZK, Tarrant J, Urkowitz A, Warming S, Yaylaoglu M, Zhang S, Zhu H, Estrada AA, Watts RJ (2015) Effect of selective LRRK2 kinase inhibition on nonhuman primate lung. Sci Transl Med 7(273):273ra215. doi:10.1126/scitranslmed.aaa3634

57. Herzig MC, Kolly C, Persohn E, Theil D, Schweizer T, Hafner T, Stemmelen C, Troxler TJ, Schmid P, Danner S, Schnell CR, Mueller M, Kinzel B, Grevot A, Bolognani F, Stirn M, Kuhn RR, Kaupmann K, van der Putten PH, Rovelli G, Shimshek DR (2011) LRRK2 protein levels are determined by kinase function and are crucial for kidney and lung homeostasis in mice. Hum Mol Genet 20(21):4209–4223. doi:10.1093/hmg/ddr348
58. Zhao J, Molitor TP, Langston JW, Nichols RJ (2015) LRRK2 dephosphorylation increases its ubiquitination. Biochem J 469(1):107–120. doi:10.1042/BJ20141305
59. Daniels RH, Zenke FT, Bokoch GM (1999) alphaPix stimulates p21-activated kinase activity through exchange factor-dependent and -independent mechanisms. J Biol Chem 274(10):6047–6050
60. Loiarro M, Capolunghi F, Fanto N, Gallo G, Campo S, Arseni B, Carsetti R, Carminati P, De Santis R, Ruggiero V, Sette C (2007) Pivotal advance: inhibition of MyD88 dimerization and recruitment of IRAK1 and IRAK4 by a novel peptidomimetic compound. J Leukoc Biol 82(4):801–810. doi:10.1189/jlb.1206746

Chapter 7
LRRK2 and the Immune System

Nicolas L. Dzamko

Abstract Polymorphisms in leucine-rich repeat kinase 2 (LRRK2) have been linked to familial Parkinson's disease, increased risk of sporadic Parkinson's disease, increased risk of Crohn's inflammatory bowel disease, and increased susceptibility to leprosy. As well as LRRK2 mutations, these diseases share in common immune dysfunction and inflammation. LRRK2 is highly expressed in particular immune cells and has been biochemically linked to the intertwined pathways regulating inflammation, mitochondrial function, and autophagy/lysosomal function. This review outlines what is currently understood about LRRK2 function in the immune system and the potential implications of LRRK2 dysfunction for diseases genetically linked to this enigmatic enzyme.

Keywords Inflammation • Parkinson's disease • Crohn's disease • Innate immunity • Cytokine • Kinase

Introduction

Leucine-rich repeat kinase 2 (LRRK2) is a 280 kDa, multi-domain protein that has dual catalytic GTPase and kinase activities, as well as number of protein-protein interaction domains. The GTPase domain of LRRK2 belongs to the Roco family of GTPases, characterized by a Ras of complex proteins (Roc) catalytic domain and a C-terminal of Ras (COR) regulatory domain [1]. The kinase domain of LRRK2 belongs to the serine/threonine family of protein kinases, with an apparent preference for threonine phosphorylation [2]. Intriguingly, both catalytic domains are seemingly linked, with functional GTPase activity required for kinase activity and a number of LRRK2 autophosphorylation sites located in the GTPase domain [3–6]. The linked catalytic activities of LRRK2 may be important, as a number of mutations in these domains are causative for Parkinson's disease (PD).

Missense mutations in LRRK2 were first identified as causal for autosomal dominantly inherited PD in 2004 [7, 8]. Six missense mutations across the COR/Roc/

N.L. Dzamko (✉)
School of Medical Sciences, University of NSW, Kensington, NSW 2052, Australia

Neuroscience Research Australia, Randwick, NSW 2031, Australia
e-mail: n.dzamko@neura.edu.au

H.J. Rideout (ed.), *Leucine-Rich Repeat Kinase 2 (LRRK2)*,
Advances in Neurobiology 14, DOI 10.1007/978-3-319-49969-7_7

kinase domains have subsequently been confirmed as pathogenic for PD, and collectively LRRK2 mutations are the most common cause of familial PD [9]. The most prevalent pathogenic LRRK2 mutation results in the substitution of the glycine at position 2019 to serine (G2019S mutation). This glycine is located in the "DFG" motif of the LRRK2 kinase domain, a highly conserved structural motif that marks the beginning of the activation loop that is important for regulating kinase activity. Indeed, the LRRK2 G2019S mutation increases the catalytic kinase activity of LRRK2 by two- to threefold [10, 11]. This close association between genetic risk of PD and increased LRRK2 activity has stimulated the development of potent and selective kinase inhibitors of LRRK2. Some initial preclinical studies are suggestive of a protective effect for LRRK2 kinase inhibitors [12–15]; however, distinct mechanisms of action are lacking, largely because knowledge regarding the biological functions of LRRK2 is still somewhat rudimentary. One emerging area of interest in LRRK2 biology, however, involves the potential role of the LRRK2 protein in inflammation and innate immunity.

This interest stems from a combination of findings, including a predominate expression of LRRK2 in immune cells, the close association between inflammation and PD pathogenesis, and that common genetic variation in the LRRK2 gene is also linked to increased risk of inflammatory bowel disease, as well as an increase in the susceptibility to infection by the leprosy-causing bacteria, *Mycobacterium leprae* [16, 17]. Animal modeling has further established a role for LRRK2 in inflammatory bowel disease and Crohn's disease [18], suggesting that LRRK2 function and/or dysfunction may underlie a number of inflammatory disorders. This book chapter will outline what is currently understood about LRRK2 and the immune system and how LRRK2 may contribute to disease pathogenesis.

LRRK2 Is a Member of the Receptor-Interacting Protein Kinase (RIPK) Family

LRRK2 is located in the tyrosine kinase-like (TKL) branch of the human kinome, where it shows greatest homology to the similarly named LRRK1 and also homology to the receptor-interacting protein kinases (RIPK1–RIPK5) [19]. The best-studied member of this family is RIPK2, which plays an important role in innate immunity. RIPK2 acts as the effector kinase for the intracellular pattern recognition receptor, NOD2 (nucleotide-binding oligomerization domain-containing protein 2) [20]. NOD2, like other pattern recognition receptors, recognizes distinct molecules from foreign pathogens, which bind to the receptors leucine-rich repeat domain. In particular, NOD2 binds the bacterial peptidoglycan, muramyl dipeptide, which upon binding to the leucine-rich repeat domain of NOD2 results in the recruitment of RIPK2 to the complex and the potent activation of NFκB-mediated pro-inflammatory cytokines [21]. Mutations in NOD2 are the greatest genetic risk factor for Crohn's inflammatory bowel disease [22, 23], and kinase inhibitors of RIPK2 may be protective against NOD2-driven inflammatory diseases [24, 25]. Common genetic variants in NOD2, RIPK2, and LRRK2 are also associated with increased susceptibility to leprosy [16, 17]. Moreover, LRRK2 and RIPK2 are both upregulated in skin biopsies from leprosy patients compared to controls [26]. Finally, LRRK2 has also been shown to co-immunoprecipitate with RIPK1, along with death domain-containing proteins FADD and TRADD, which can mediate cell death downstream of RIPK1 activation [27]. Moreover, PD-causing mutations increase the interaction of LRRK2 with FADD, but not RIPK1 or TRADD, resulting in a potentiated caspase-8-dependent cell death response in culture.

Collectively these studies implicate LRRK2 in known functions of the RIPK family. The interaction of LRRK2 with other RIPKs may be important, for example, the interaction of RIPK1 and RIPK3 is required for mediating the specific mode of programmed cell death called necroptosis [28–30]. Moreover, it is currently unknown whether LRRK2 further parallels NOD2/RIPK2 and acts as a pathogen-recognition receptor itself.

The Expression of LRRK2 in Immune Cells

Virtually concomitant with the advent of antibodies capable of detecting endogenous LRRK2 was the realization that the enzyme is particularly highly expressed in cells of the immune system (Fig. 7.1). Studies aimed at elucidating which immune

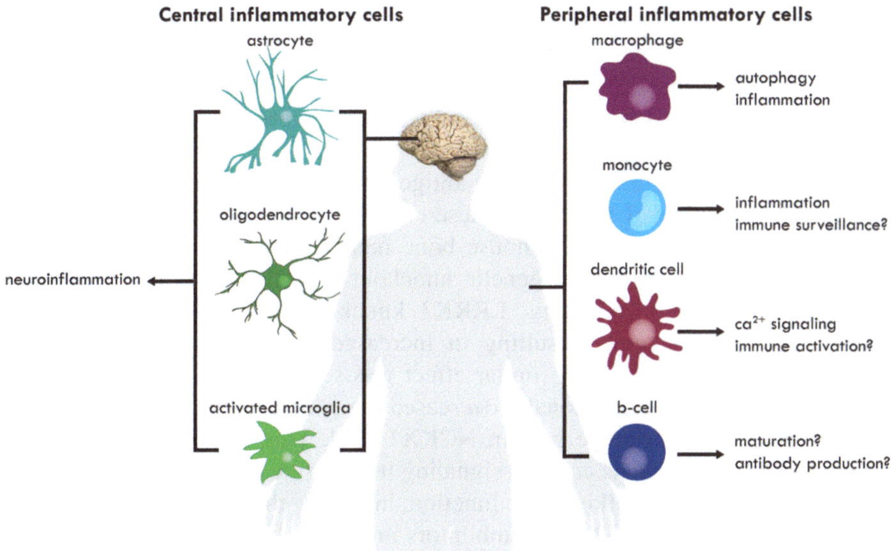

Fig. 7.1 The expression of LRRK2 in immune cells. LRRK2 is highly expressed in cells of the peripheral immune system including monocytes, B lymphocytes, and dendritic cells. Functional studies in these cell types indicate an important role for LRRK2 in the regulation of aspects of both innate and adaptive immunity following immune cell activation. LRRK2 is also expressed in glial cells in the brain where it may play an important role in regulating neuroinflammation

cell types express LRRK2, and if expression is altered in disease states, may be important for understanding the elusive biological functions of LRRK2.

Immune Cells from Myeloid Progenitors

The highest expression of LRRK2 in peripheral immune mononuclear cells is in cells of myeloid lineage, monocytes, and dendritic cells [31, 32]. Within the monocyte population, the pro-inflammatory $CD14^{+}CD16^{+}$ monocyte subgroup has the greatest expression of LRRK2 protein [33]. LRRK2 protein is also readily detectable in mouse bone marrow-derived macrophages and mouse RAW264.7 macrophage cells [32, 34], as well as human monocyte-derived macrophages and differentiated human THP-1 macrophage-like cells [31, 35]. Moreover, the expression of LRRK2 is further increased in monocyte/macrophage cells following stimulation with interferon gamma (IFNϒ) [31, 33, 35]. As type 2 IFNϒ is a classic activator of monocytes and macrophages, this result is consistent with the increased expression of LRRK2 observed in activated $CD14^{+}CD16^{+}$ monocytes and suggests that a high expression of LRRK2 will also be observed in pro-inflammatory M1 macrophages. Although complete mechanisms are unknown, a recent study demonstrated a requirement for extracellular signal-related kinase 5 (ERK5) activity in the induction of LRRK2 by IFNϒ [35]. This result followed diligent observations that the dual LRRK2/ERK5 kinase inhibitor LRRK2-IN1 [36] could suppress IFNϒ-mediated LRRK2 induction but other inhibitors of LRRK2 could not. The effect of LRRK2-IN1, however, was reproduced with other selective inhibitors of ERK5, as well as silencing of the ERK5 gene. In summary, the expression profile of LRRK2 in activated monocytes and macrophage subtypes is suggestive of a role in innate immunity and pro-inflammatory states.

LRRK2 is also expressed in human antigen-presenting dendritic cells [31, 32], but functional studies of LRRK2 in these cells are lacking. One study shows the presence of LRRK2 protein in mouse bone marrow-derived dendritic cells and demonstrates an effect of both genetic knockout of LRRK2 and LRRK2 kinase inhibition on calcium signaling. LRRK2 knockout decreased activity of the sodium-calcium exchanger resulting in increased cytosolic calcium following thapsigargin treatment [37]. A similar effect was seen with the LRRK2 inhibitor GSK2578215A, which additionally decreased the gene expression of the sodium/potassium/calcium exchange protein NCKX1 [37]. This study suggests a novel role for LRRK2 to regulate calcium signaling in dendritic cells, which is important for aspects of dendritic cell immune function, including the host response to pathogens [38]. Whether LRRK2 kinase inhibitors impact on human dendritic cell function, and to what extent, is currently unknown. It is also unknown to what extent LRRK2 is expressed in other myeloid immune cell types such as the eosinophils, neutrophils, or basophils.

Immune Cells from Lymphoid Progenitors

LRRK2 is also robustly expressed in $CD19^+$ B lymphocytes [31–33], and Epstein-Barr virus-immortalized B lymphoblasts derived from LRRK2 mutation carriers have been useful for screening LRRK2 kinase inhibitors in vitro [39]. In mice, the expression of LRRK2 is greatest in the IgG-secreting B-2 subset of B lymphocytes rather than the predominantly IgM-secreting B-1 cells [40]. Moreover, the expression of LRRK2 was greatest when B-2 lymphocytes were in the resting state, as activation with phorbol 12-myristate 13-acetate (PMA) or lipopolysaccharide (LPS) dramatically reduced LRRK2 expression [40]. Interestingly, LRRK2-deficient rats were shown to have reduced peripheral lymphocytes and reduced splenic B lymphocytes [41], suggesting a role for LRRK2 in the development and/or maintenance of B-lymphocyte populations, at least in rodents.

In contrast to B lymphocytes, the expression of LRRK2 in T lymphocytes appears low to negligible. In mouse splenic lymphocytes, LRRK2 protein could not be readily detected in $CD3^+$ T lymphocytes [42]. In human $CD3^+$ T lymphocytes isolated from human peripheral blood mononuclear cells, very low levels of LRRK2 mRNA and protein could be detected [32, 33]; however, like other immune cell types, LRRK2 expression in T lymphocytes could be increased with IFNϒ stimulation [31]. As the majority of studies have employed the pan T-lymphocyte marker CD3, it is unknown whether the low expression of LRRK2 occurs across all T-lymphocyte subpopulations or whether the expression of LRRK2 may be restricted to distinct T-lymphocyte types. For example, one study has noted a lower expression of IL-17 and IFNϒ following activation of splenic lymphocytes from LRRK2-deficient mice [43], suggesting a potential role in cytokine production by specialized Th17 cells. Whether LRRK2 expression is altered in activated T lymphocytes is unknown. Moreover, potential roles for LRRK2 in cell-mediated immunity are yet to be fully explored.

Glial Cells

Given its pathological links to neurodegenerative disease, there has been much interest in the expression of LRRK2 in brain cells. Of particular interest to the current review are microglia, the resident immune cells of the brain. Initial investigations into LRRK2 expression in wild-type rodent brain, however, suggested a predominantly neuronal localization, with little to no mention of LRRK2 immunoreactivity in glial cells [44–46]. A lack of LRRK2 expression in mouse microglia was also observed by Moehle and colleagues; however, when the resident microglia were activated by intracranial LPS injection, LRRK2 expression by microglia became evident [47]. Subsequently, a number of studies have demonstrated that LRRK2 knockdown and knockout or kinase inhibition can attenuate microglial-mediated inflammatory cytokine production induced by different stimuli [47–50].

These results suggest that LRRK2 expression is induced in rodent microglia as part of the inflammatory response, consistent with the previously mentioned increase in LRRK2 expression seen in activated monocytes and macrophages [31, 32]. Intriguingly, microglial activation and neuroinflammation are robustly associated with PD [51]; however, whether LRRK2 is expressed in human microglia under PD pathological conditions is less clear. The majority of studies to date demonstrate a predominant neuronal expression of LRRK2 in the human PD brain [52, 53], with only one early report suggesting a weak constitutive expression of LRRK2 in human microglia [54].

In addition to microglia, the macroglial cells, astrocytes, and oligodendrocytes can also contribute to brain inflammation. Two studies have used primary mouse astrocytes or astrocytoma-derived U87 cells to demonstrate an effect of LRRK2 inhibitors on lysosome function and inflammation, respectively [15, 55], suggesting a potential role of LRRK2 in this cell type. As is the case for microglia though, strong evidence that LRRK2 is expressed in human astrocytes is currently lacking. This is largely due to a number of the commercially available LRRK2 antibodies being considered unsuitable for immunohistochemistry on the human brain [56]. However, with new studies suggesting that neuroprotective effects of LRRK2 kinase inhibitors in mouse models may be mediated via astrocytes or microglia [12, 15, 47], it will be important to better elucidate the expression of LRRK2 in human glial cells. In particular, the expression of LRRK2 under pathological conditions may be important. Evidence for this comes from studies of multiple system atrophy (MSA), an alpha-synucleinopathy in which inclusions form in oligodendrocytes, instead of neurons as seen for PD. In postmortem brain samples from MSA patients, LRRK2 could be detected in inclusion-containing oligodendrocytes but not in the oligodendrocytes from control or PD brain tissue [57]. Such an observation may again be consistent with the concept that LRRK2 expression is very low in glial cells but increased under certain pathological conditions.

LRRK2 as a Modulator of Inflammatory Cytokine Production

LRRK2 is increasingly being implicated in immunity, in particular the innate immune inflammatory response (Fig. 7.2). Although mechanistic insight into how LRRK2 modulates the inflammatory response is in its infancy, two major inflammatory pathways have been biochemically linked to LRRK2 action.

The TLR Pathway

The toll-like family of pathogen-associated molecular pattern receptors are major mediators of the innate immune inflammatory response. They detect a diverse array of viral and bacterial pathogens that bind to the leucine-rich repeat domain of

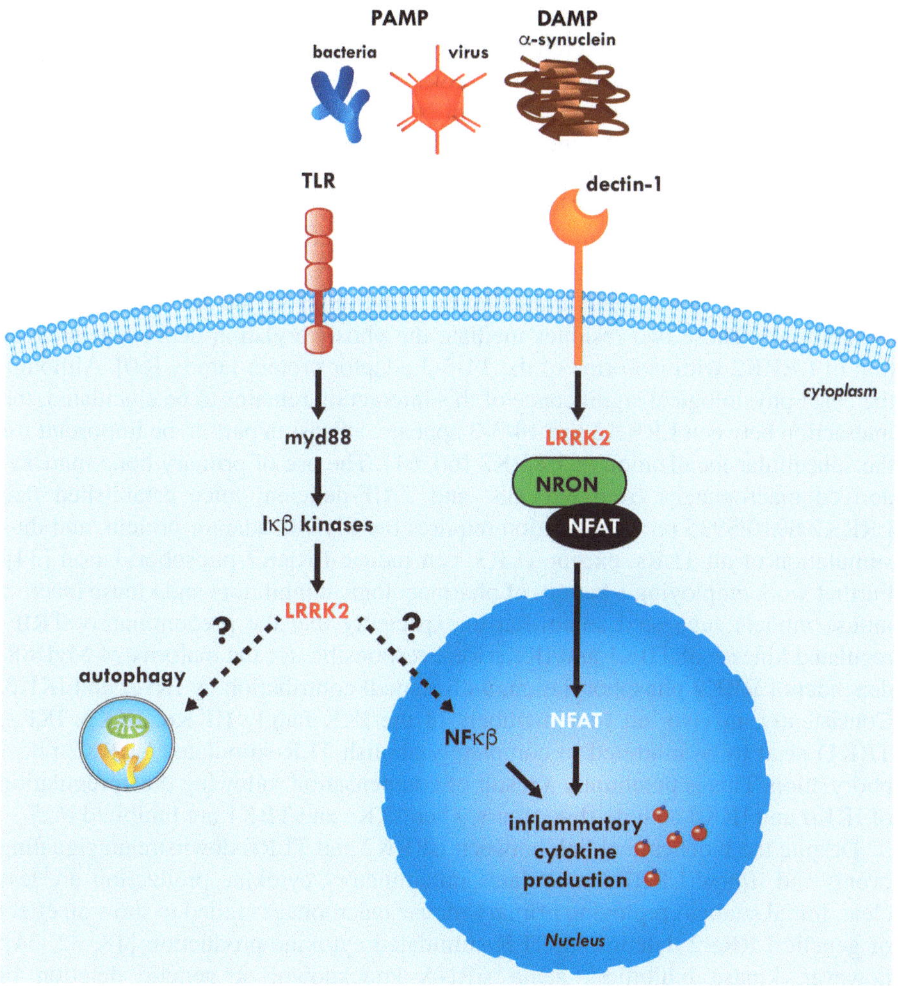

Fig. 7.2 LRRK2 contributes to the host immune response. LRRK2 has been biochemically linked to both the toll-like receptor (TLR) and dectin-1 signaling pathways that mediate the host immune response to pathogen-associated molecular patterns (PAMP), as well as self-originating danger-associated molecular patterns (DAMP). While exact details remain to be elucidated, LRRK2 appears to positively regulate inflammatory cytokine production and autophagy following TLR activation. In contrast, LRRK2 appears to negatively regulate inflammation following activation of dectin-1, in this case by acting as a scaffold helping to retain the pro-inflammatory cytokine producing nuclear factor of activated T-cell (NFAT) transcription factor in the cytoplasm

distinct TLRs to trigger downstream activation of signaling pathways and transcription factors that regulate inflammatory cytokine production (for review, see [58, 59]). The classical TLR signal transduction cascade has two different branches, known as the MyD88- and TRIF-dependent pathways. The majority of TLRs signal through the MyD88 adaptor protein, while TLR3 signals through the TRIF adaptor protein and TLR4, and the receptor for LPS signals through both MyD88 and TRIF pathways. Although complex cross talk and counter regulation exist, activation of the MyD88 pathway results predominantly in the increased kinase activity of p38 MAPK, IKKα, and IKKβ, which, respectively, activates the inflammatory cytokine-producing transcription factors AP-1 and NFκB. Activation of the TRIF pathway results predominantly in the increased kinase activity of IKKε and TBK1, which phosphorylate and activate the IRF3/7 transcription factors to initiate production of type 1 interferon effectors, IFNα and IFNβ.

Activation of the TLR pathway can also lead to increased phosphorylation of the LRRK2 protein in macrophages. At least two sites are phosphorylated, serines 910 and 935, which are located immediately N-terminal to the leucine-rich repeat domain [34]. These two residues mediate the phosphorylation-dependent interaction of LRRK2 with isoforms of the 14-3-3 adaptor protein family [60]. Although the exact physiological significance of this interaction remains to be elucidated, the interaction between LRRK2 and 14-3-3 appears, at least in part, to be important for the subcellular localization of LRRK2 [60, 61]. The use of primary bone marrow-derived macrophages from MyD88- and TRIF-deficient mice established that LRRK2 S910/S935 phosphorylation requires the MyD88 adaptor protein, and thus stimulation of all TLRs, except TLR3, can induce LRRK2 phosphorylation [34]. Further work employing a battery of pharmacological inhibitors and kinase inactive mouse models suggested somewhat unexpectedly that the predominantly TRIF-regulated kinases of TBK1 and IKKε were responsible for the majority of MyD88-dependent LRRK2 phosphorylation, with a small contribution by IKKα and IKKβ. Consistently, in vivo, all four members of the IKK family (IKKα, IKKβ, IKKε, TBK1) need to be inhibited to completely abolish TLR-stimulated LRRK2 phosphorylation. This is potentially a result of compensation following the upregulation of IKKα and IKKβ activity that occurs when IKKε and TBK1 are inhibited [62].

Despite the biochemical link between LRRK2 and TLRs, downstream signaling events and if/how LRRK2 modulates inflammatory cytokine production are less clear. Initial studies employing primary mouse macrophages failed to show an effect of genetic LRRK2 deletion on TLR-stimulated cytokine production [18, 32, 34]; however, kinase inhibition, acute siRNA knockdown, or genetic deletion of LRRK2 in mouse microglia impaired the inflammatory response to TLR4 activating LPS [47, 50] or HIV-1 Tat protein [48]. Such results could be explained by a fundamental difference in inflammatory signaling between mouse macrophages and microglia, or potentially compensatory effects due to LRRK2 deletion in macrophages, where unlike microglia, LRRK2 is robustly constitutively expressed. Further support of a role for LRRK2 in the regulation of TLR-stimulated inflammatory cytokine production comes from mouse microglia overexpressing LRRK2 with the PD-causing R1441G mutation, that have increased inflammatory cytokine production following TLR4 activation [63]. Mechanistically, LRRK2 has been linked to regulation of the NFκB transcription factor using reporter assays; however, results from these studies suggest that levels of LRRK2 protein are important for transcriptional regulation, rather than LRRK2 kinase activity or PD-causing LRRK2 mutations [31, 50].

Thus, evidence suggests a role for LRRK2 to modulate TLR-stimulated inflammatory cytokine production, at least in mouse microglial cells; however, further work is required to understand how. Moreover, studies translating these findings to human cells are currently lacking but may be important as differences in innate immune regulation between humans and mice are evident [64, 65].

The NFAT Pathway

The nuclear factor of activated T-cell (NFAT) family of transcription factors are also important mediators of the immune response (for review, see [66]). This five-member family classically regulates inflammatory cytokine production in a manner dependent on calcium signaling, in which events leading to an increase in intracellular calcium result in the nuclear translocation of NFAT and induction of NFAT-mediated gene expression. NFAT signaling is complex and, in addition to calcium signaling, involves a number of regulatory kinases and interacting proteins [67]. The nuclear translocation of NFAT can also be repressed by the noncoding RNA repressor of NFAT (NRON) that acts as a scaffold for the NRON complex [68, 69]. Co-immunoprecipitation experiments demonstrated that LRRK2 associated with five of the 11 proteins comprising the NRON complex [18] suggests a potential role in NFAT regulation.

To further explore this role, mouse macrophages deficient in LRRK2 were treated with the yeast cell wall component zymosan, which activates TLR2 but also NFAT via the dectin-1 pathway. A higher nuclear accumulation of NFAT was measured in knockout cells along with increased production of inflammatory cytokines [18]. This result was not observed with a TLR2-specific agonist and could be inhibited by blocking calcineurin, strongly suggesting a role for LRRK2 in the negative regulation of NFAT by helping to sequester the protein into the cytoplasm. Activation of the dectin-1 pathway with the specific agonist curdlan does not result in increased LRRK2 serine 910 and 935 phosphorylation [34], further demonstrating at least two distinct pathways by which LRRK2 can modulate inflammatory cytokine production. Cellular studies using overexpression systems and luciferase reporter assays show that LRRK2 helps sequester all members of the NFAT family, in a manner independent of kinase activity [18].

LRRK2 and Pathogen Clearance

Effective clearance of pathogens is essential to immune homeostasis and in avoiding chronic low-grade inflammation. Apart from inflammatory cytokine production, LRRK2 has been implicated in at least two other direct aspects of pathogen clearance, the production of reactive oxygen species and induction of autophagy.

Targeting Intracellular Pathogens with Reactive Oxygen Species

The targeting of reactive oxygen species (ROS) against intracellular pathogens is recognized as an important innate immune response to pathogen clearance [70]. In particular, mitochondria-produced ROS is important for suppressing infection by certain intracellular bacteria, such as the gastroenteritis causing *S. typhimurium* [71]. Intriguingly, both ROS production and the clearance of *S. typhimurium* were impaired in RAW 264.7 macrophages following siRNA knockdown of LRRK2 [31], suggesting a role for LRRK2 in pathogen clearance by mitochondrial ROS production. Evidence for a role of LRRK2 in mitochondrial ROS production is also provided by studies that have examined the effect of pathogenic LRRK2 mutations on mitochondrial function in neural cells in the context of PD. In particular, neural stem cells expressing the G2019S mutation have reduced oxygen consumption, altered mitochondrial morphology, and altered mitochondrial trafficking [13]. Further studies using rodents and human stem cell models suggest that neuronal mitochondrial dysfunction is due to either defects in mitochondrial fission [72, 73] or the clearance of defective mitochondria due to inefficient autophagy [74, 75]. Thus, LRRK2-mediated mitochondrial dysfunction may lead to impaired ROS production following pathogen detection, but exactly how and to what extent this is responsible for impaired pathogen clearance by immune cells remains to be elucidated.

Clearance of Pathogens by Autophagy

The clearance of pathogens by selective autophagy, termed xenophagy, is another important aspect of host defense (for review, see [76, 77]). Xenophagy, like most forms of selective autophagy, uses ubiquitination to recognize substrates for degradation. Intracellular pathogens, such as *S. typhimurium*, can become polyubiquitinated and bind autophagic receptors, such as p62 (SQSTM1), NDP52, and optineurin (OPTN), which facilitate autophagic clearance via lysosomes. Intriguingly, stimulation of RAW264.7 macrophage cells with LPS results in the translocation of LRRK2 to autophagosome membranes [78]. A relocalization of LRRK2 to bacterial membrane structures was also observed in macrophages infected with *S. typhimurium* [31]. Moreover, both siRNA silencing and pharmacological inhibition resulted in impaired rapamycin-induced autophagy in the RAW264.7 macrophage cell type [78], and LRRK2-deficient mice exhibit autophagic defects in peripheral tissues with the highest LRRK2 expression [79, 80]. While these studies are suggestive of an important role for LRRK2 in autophagy-mediated clearance of pathogens, this has not been widely studied. Additionally, studies in neuronal cell models have suggested that LRRK2 kinase inhibitors can promote rather than impair autophagy [74, 81], and further complicating this pathway is that protein levels of LRRK2 are also regulated by chaperone-mediated autophagy [82], as well as the ubiquitin-proteasome system [83]. Thus, LRRK2 is implicated in the highly intertwined mitochondrial function and autophagy pathways with potential implications for pathogen clearance, but the cell type and disease context under investigation will be important for interpretation of future results.

LRRK2 and Disease Pathogenesis

Protein kinase signaling pathways can often regulate a myriad of biological responses. This may also be true of LRRK2, with the enzyme linked to WNT signaling, neurogenesis, synaptic function, vesicle trafficking, and protein synthesis, among others, all in addition to the mentioned roles in mitochondrial function, autophagy and innate immunity. As LRRK2 is also genetically linked to human disease, in particular, Parkinson's and Crohn's diseases, it is important to understand how the role of LRRK2 in the immune system may contribute to pathogenicity.

Parkinson's Disease

PD is an increasingly common movement disorder defined pathologically by the progressive death of smooth motor-controlling dopaminergic neurons in the substantia nigra region of the brain. Inflammation has long been associated with the pathogenesis of PD (for reviews, see [84–87]). Increased numbers of activated microglial cells and reactive astrocytes are hallmark features surrounding the degenerating dopaminergic neurons in PD brain [51]. Increased microglial activation is also observed in other midbrain regions, particularly the striatum, and also in hippocampal and cortical regions, suggesting widespread microglial activation in PD [88, 89]. Levels of inflammatory cytokines are increased in brain homogenates and cerebrospinal fluid from PD patients [84–87], and anti-inflammatory therapies including nonsteroidal anti-inflammatory drugs, or the microglial inhibitor minocycline, have demonstrated some, albeit limited, efficacy in the treatment of PD [90, 91].

Mechanistically, it is unclear how inflammation manifests in PD brain; however, evidence suggests a role for "sterile inflammation" meditated via danger-associated molecular patterns (DAMPs), in particular alpha-synuclein. The 17 kDa presynaptic protein, alpha-synuclein, propagates and accumulates in PD brain to form insoluble inclusions called Lewy bodies [92]. The propagation of alpha-synuclein occurs in an organized or staged manner and associates with the progressive symptoms of PD [93, 94]. Like LRRK2, missense mutations in alpha-synuclein also cause autosomal-dominant inherited PD [95]. Intriguingly, recent data suggest that alpha-synuclein can either directly activate TLR2 [96, 97] or TLR4 [98] signaling in microglia and/or potentiate the microglial inflammatory response to TLR agonists [99, 100]. This is of interest as it places the two main genetic causes of autosomal-dominant PD in the same biological pathway. Indeed, LRRK2 deletion and pharmacological inhibition were both shown to prevent alpha-synuclein-mediated neurodegeneration in rodent models, via mechanisms that included reduced inflammation and recruitment/activation of microglia [12, 101].

In addition to the established role of neuroinflammation in the pathogenesis of PD, there is increasing evidence of an important role for peripheral inflammation. Indeed, the first clinical and pathological signs of PD occur peripherally, pre-dating the defining motor symptoms classically induced by loss of dopaminergic neurons [102, 103]. Particularly affected is the gastrointestinal system, with constipation as a common early PD symptom coupled with the presence of pathological alpha-synuclein inclusions throughout the gastrointestinal tract [104–106]. Inflammatory cytokines are also often increased in serum from PD patients [84–87], and peripheral administration of LPS results in a PD-like phenotype in rodents [107]. A lack of early diagnostic markers for PD makes it difficult to establish causes of disease from consequences; however, the study of genetic risk factors may give clues to what drives PD [85]. In this regard, the high expression of LRRK2 in immune cells and its established biochemical links to innate immune pathways that regulate inflammation could be important for understanding the initiating events of PD.

Inflammatory Bowel Disease

Inflammatory bowel diseases, including Crohn's disease and ulcerative colitis, affect the gastrointestinal tract with symptomology including diarrhea, bleeding, anemia, and subsequent weight loss. The exact etiology of inflammatory bowel diseases is unknown, but genetics seemingly underlies the host susceptibility to dysregulated immune homeostasis involving the commensal intestinal microbiota [108]. Both the adaptive and innate immune systems are implicated, and in particular mutations in the NOD2 pattern recognition receptor commonly underlie susceptibility [22, 23].

That genetic variability in the LRRK2-MUC19 locus could potentially increase the risk of Crohn's disease was first identified following a meta-analysis of three GWAS studies [109]. Although it is not decisive which of these two genes may be associated with the increased risk [110], follow-up studies using LRRK2 knockout mice have demonstrated an increased susceptibility to dextran sulfate sodium-induced colitis [18] and intestinal infection with *Listeria monocytogenes* [111]. Ulcerative colitis risk shares substantial genetic overlap with Crohn's disease [112], and LRRK2 risk variants are likely associated with both forms of inflammatory bowel diseases [113].

Mechanistically, LRRK2 appears to contribute to innate immune inflammation in inflammatory bowel disease via cytoplasmic sequestering of the NFAT transcription factor. LRRK2 knockout mice had increased levels of NFAT in the nucleus of macrophages leading to increased production of inflammatory cytokines and colitis symptomology [18]. In these studies, activation of toll-like receptors also reduced

LRRK2 protein expression, and LRRK2 was reduced in peripheral blood mononuclear cells from four subjects carrying a missense LRRK2 Crohn's disease risk allele (T2397M) [18]. This suggests that LRRK2 acts as a negative regulator of innate inflammatory cytokine production. This role for LRRK2 may depend on disease context; however, toll-like receptor activation does not always result in reduced LRRK2 protein [32, 34]. Moreover, intestinal biopsies from Crohn's disease patients show increased expression of LRRK2 in lamina propria macrophages, dendritic cells, and B lymphocytes [31].

Genetic studies have also indicated a key role for autophagy dysregulation in the pathogenesis of inflammatory bowel disease [112], suggesting that functional autophagic pathways are important for the efficient clearance of pathogenic bacteria, or microbiota homeostasis. As described above, LRRK2 knockout mice show impaired kidney autophagy, and a reduction in LRRK2 protein may contribute to inefficient clearance of bacteria [31, 80]. Indeed, LRRK2 knockout mice are also more susceptible to inflammatory bowel disease caused by intestinal *Listeria monocytogenes* infection [111]. Intriguingly, increased susceptibility in these studies was associated with increased lysosomal degradation of bactericidal lysozyme in LRRK2-deficient cells. Collectively, these studies suggest that LRRK2 may regulate and mediate cross talk between autophagy/lysosomal pathways and inflammatory pathways with a loss of LRRK2 function detrimental for intestinal health.

Summary

Although first identified through linkage studies as an inherited cause of PD, the potential role of LRRK2 mutations in the risk of human disease continues to rise. Genome-wide association studies have implicated LRRK2 polymorphisms in the increased risk of sporadic PD, as well as the somewhat related synucleinopathy and multiple system atrophy [114]. Given that protein aggregation, impaired autophagy/lysosomal function, and inflammation are prevalent in many neurological diseases, it will be of interest to determine if LRRK2 variation contributes to the risk of other neurological diseases. For example, variation in TBK1, the kinase that phosphorylates LRRK2 in response to TLR activation, was recently implicated in the risk of motor neuron disease and frontotemporal dementia [115, 116].

Genetic studies backed by animal modeling have also implicated LRRK2 in increased susceptibility to Crohn's inflammatory bowel disease. In contrast to PD, where missense mutations are thought to cause a toxic gain of function, increased risk of inflammatory bowel disease appears to result from a loss of LRRK2 function. This suggests an important role for LRRK2 in immune homeostasis, with potentially a loss of LRRK2 function leading to more severe inflammatory symptoms that can have a juvenile onset, and gain of LRRK2 function leading to potentially a chronic low-grade inflammation contributing to neurodegeneration with aging. Recent genetic studies have also suggested that as much as 70 % of the genes involved in inflammatory bowel disease overlap with other autoimmune diseases including rheumatoid arthritis, lupus, and multiple sclerosis [112]. Thus, determining the disease specificity of LRRK2 variation in autoimmune inflammatory diseases in general will be of future interest.

It will also be of importance to better understand the mechanism(s) by which LRRK2 polymorphisms contribute to disease. LRRK2 knockout mice or knock-in mice with PD-causing mutations do not display an overt phenotype under basal conditions, suggesting that LRRK2 function, at least in the immune system, may be responsive to environmental stimuli. Once involved, LRRK2 seems linked to the complex pathways involved in the cross talk and regulation of inflammatory cytokine production, mitochondrial function, and autophagy/lysosomal function. Further identification of interacting proteins and LRRK2 substrates in these pathways would be a substantial forward step in elucidating the pathogenic function(s) of LRRK2.

Acknowledgments ND is funded by the Michael J. Fox Foundation, the Shake It Up Australia Foundation, Parkinson's NSW, and the National Health and Medical Research Council. I thank Heidi Cartwright for the figure preparation and Glenda Halliday for the helpful comments on the manuscript.

Conflict of Interest The author declares no conflicts of interest.

References

1. Bosgraaf L, Van Haastert PJ (2003) Roc, a Ras/GTPase domain in complex proteins. Biochim Biophys Acta 1643(1–3):5–10
2. Nichols RJ, Dzamko N, Hutti JE, Cantley LC, Deak M, Moran J, Bamborough P, Reith AD, Alessi DR (2009) Substrate specificity and inhibitors of LRRK2, a protein kinase mutated in Parkinson's disease. Biochem J 424(1):47–60. doi:10.1042/BJ20091035
3. Liu Z, Mobley JA, DeLucas LJ, Kahn RA, West AB (2015) LRRK2 autophosphorylation enhances its GTPase activity. FASEB J 30:336–347. doi:10.1096/fj.15-277095
4. Li T, Yang D, Zhong S, Thomas JM, Xue F, Liu J, Kong L, Voulalas P, Hassan HE, Park JS, MacKerell AD Jr, Smith WW (2014) Novel LRRK2 GTP-binding inhibitors reduced degeneration in Parkinson's disease cell and mouse models. Hum Mol Genet 23(23):6212–6222. doi:10.1093/hmg/ddu341
5. Gilsbach BK, Kortholt A (2014) Structural biology of the LRRK2 GTPase and kinase domains: implications for regulation. Front Mol Neurosci 7:32. doi:10.3389/fnmol.2014.00032
6. Biosa A, Trancikova A, Civiero L, Glauser L, Bubacco L, Greggio E, Moore DJ (2013) GTPase activity regulates kinase activity and cellular phenotypes of Parkinson's disease-associated LRRK2. Hum Mol Genet 22(6):1140–1156. doi:10.1093/hmg/dds522
7. Zimprich A, Biskup S, Leitner P, Lichtner P, Farrer M, Lincoln S, Kachergus J, Hulihan M, Uitti RJ, Calne DB, Stoessl AJ, Pfeiffer RF, Patenge N, Carbajal IC, Vieregge P, Asmus F, Muller-Myhsok B, Dickson DW, Meitinger T, Strom TM, Wszolek ZK, Gasser T (2004) Mutations in LRRK2 cause autosomal-dominant parkinsonism with pleomorphic pathology. Neuron 44(4):601–607. doi:10.1016/j.neuron.2004.11.005
8. Paisan-Ruiz C, Jain S, Evans EW, Gilks WP, Simon J, van der Brug M, Lopez de Munain A, Aparicio S, Gil AM, Khan N, Johnson J, Martinez JR, Nicholl D, Carrera IM, Pena AS, de Silva R, Lees A, Marti-Masso JF, Perez-Tur J, Wood NW, Singleton AB (2004) Cloning of the gene containing mutations that cause PARK8-linked Parkinson's disease. Neuron 44(4):595–600. doi:10.1016/j.neuron.2004.10.023

9. Paisan-Ruiz C (2009) LRRK2 gene variation and its contribution to Parkinson disease. Hum Mutat 30(8):1153–1160. doi:10.1002/humu.21038
10. West AB, Moore DJ, Biskup S, Bugayenko A, Smith WW, Ross CA, Dawson VL, Dawson TM (2005) Parkinson's disease-associated mutations in leucine-rich repeat kinase 2 augment kinase activity. Proc Natl Acad Sci U S A 102(46):16842–16847. doi:10.1073/pnas.0507360102
11. Jaleel M, Nichols RJ, Deak M, Campbell DG, Gillardon F, Knebel A, Alessi DR (2007) LRRK2 phosphorylates moesin at threonine-558: characterization of how Parkinson's disease mutants affect kinase activity. Biochem J 405(2):307–317. doi:10.1042/BJ20070209
12. Daher JP, Abdelmotilib HA, Hu X, Volpicelli-Daley LA, Moehle MS, Fraser KB, Needle E, Chen Y, Steyn SJ, Galatsis P, Hirst WD, West AB (2015) Leucine-rich repeat kinase 2 (LRRK2) pharmacological inhibition abates alpha-synuclein gene-induced neurodegeneration. J Biol Chem 290(32):19433–19444. doi:10.1074/jbc.M115.660001
13. Cooper O, Seo H, Andrabi S, Guardia-Laguarta C, Graziotto J, Sundberg M, McLean JR, Carrillo-Reid L, Xie Z, Osborn T, Hargus G, Deleidi M, Lawson T, Bogetofte H, Perez-Torres E, Clark L, Moskowitz C, Mazzulli J, Chen L, Volpicelli-Daley L, Romero N, Jiang H, Uitti RJ, Huang Z, Opala G, Scarffe LA, Dawson VL, Klein C, Feng J, Ross OA, Trojanowski JQ, Lee VM, Marder K, Surmeier DJ, Wszolek ZK, Przedborski S, Krainc D, Dawson TM, Isacson O (2012) Pharmacological rescue of mitochondrial deficits in iPSC-derived neural cells from patients with familial Parkinson's disease. Sci Transl Med 4(141), 141ra190. doi:10.1126/scitranslmed.3003985
14. Lee BD, Shin JH, VanKampen J, Petrucelli L, West AB, Ko HS, Lee YI, Maguire-Zeiss KA, Bowers WJ, Federoff HJ, Dawson VL, Dawson TM (2010) Inhibitors of leucine-rich repeat kinase-2 protect against models of Parkinson's disease. Nat Med 16(9):998–1000. doi:10.1038/nm.2199
15. Henry AG, Aghamohammadzadeh S, Samaroo H, Chen Y, Mou K, Needle E, Hirst WD (2015) Pathogenic LRRK2 mutations, through increased kinase activity, produce enlarged lysosomes with reduced degradative capacity and increase ATP13A2 expression. Hum Mol Genet 24(21):6013–6028. doi:10.1093/hmg/ddv314
16. Marcinek P, Jha AN, Shinde V, Sundaramoorthy A, Rajkumar R, Suryadevara NC, Neela SK, van Tong H, Balachander V, Valluri VL, Thangaraj K, Velavan TP (2013) LRRK2 and RIPK2 variants in the NOD 2-mediated signaling pathway are associated with susceptibility to Mycobacterium leprae in Indian populations. PLoS One 8(8), e73103. doi:10.1371/journal.pone.0073103
17. Zhang FR, Huang W, Chen SM, Sun LD, Liu H, Li Y, Cui Y, Yan XX, Yang HT, Yang RD, Chu TS, Zhang C, Zhang L, Han JW, Yu GQ, Quan C, Yu YX, Zhang Z, Shi BQ, Zhang LH, Cheng H, Wang CY, Lin Y, Zheng HF, Fu XA, Zuo XB, Wang Q, Long H, Sun YP, Cheng YL, Tian HQ, Zhou FS, Liu HX, Lu WS, He SM, Du WL, Shen M, Jin QY, Wang Y, Low HQ, Erwin T, Yang NH, Li JY, Zhao X, Jiao YL, Mao LG, Yin G, Jiang ZX, Wang XD, Yu JP, Hu ZH, Gong CH, Liu YQ, Liu RY, Wang DM, Wei D, Liu JX, Cao WK, Cao HZ, Li YP, Yan WG, Wei SY, Wang KJ, Hibberd ML, Yang S, Zhang XJ, Liu JJ (2009) Genomewide association study of leprosy. N Engl J Med 361(27):2609–2618. doi:10.1056/NEJMoa0903753
18. Liu Z, Lee J, Krummey S, Lu W, Cai H, Lenardo MJ (2011) The kinase LRRK2 is a regulator of the transcription factor NFAT that modulates the severity of inflammatory bowel disease. Nat Immunol 12(11):1063–1070. doi:10.1038/ni.2113
19. Zhang D, Lin J, Han J (2010) Receptor-interacting protein (RIP) kinase family. Cell Mol Immunol 7(4):243–249. doi:10.1038/cmi.2010.10
20. Park JH, Kim YG, McDonald C, Kanneganti TD, Hasegawa M, Body-Malapel M, Inohara N, Nunez G (2007) RICK/RIP2 mediates innate immune responses induced through Nod1 and Nod2 but not TLRs. J Immunol 178(4):2380–2386
21. Inohara N, Ogura Y, Fontalba A, Gutierrez O, Pons F, Crespo J, Fukase K, Inamura S, Kusumoto S, Hashimoto M, Foster SJ, Moran AP, Fernandez-Luna JL, Nunez G (2003) Host recognition of bacterial muramyl dipeptide mediated through NOD2. Implications for Crohn's disease. J Biol Chem 278(8):5509–5512. doi:10.1074/jbc.C200673200

22. Ogura Y, Bonen DK, Inohara N, Nicolae DL, Chen FF, Ramos R, Britton H, Moran T, Karaliuskas R, Duerr RH, Achkar JP, Brant SR, Bayless TM, Kirschner BS, Hanauer SB, Nunez G, Cho JH (2001) A frameshift mutation in NOD2 associated with susceptibility to Crohn's disease. Nature 411(6837):603–606. doi:10.1038/35079114
23. Hugot JP, Chamaillard M, Zouali H, Lesage S, Cezard JP, Belaiche J, Almer S, Tysk C, O'Morain CA, Gassull M, Binder V, Finkel Y, Cortot A, Modigliani R, Laurent-Puig P, Gower-Rousseau C, Macry J, Colombel JF, Sahbatou M, Thomas G (2001) Association of NOD2 leucine-rich repeat variants with susceptibility to Crohn's disease. Nature 411(6837):599–603. doi:10.1038/35079107
24. Tigno-Aranjuez JT, Benderitter P, Rombouts F, Deroose F, Bai X, Mattioli B, Cominelli F, Pizarro TT, Hoflack J, Abbott DW (2014) In vivo inhibition of RIPK2 kinase alleviates inflammatory disease. J Biol Chem 289(43):29651–29664. doi:10.1074/jbc.M114.591388
25. Nachbur U, Stafford CA, Bankovacki A, Zhan Y, Lindqvist LM, Fiil BK, Khakham Y, Ko HJ, Sandow JJ, Falk H, Holien JK, Chau D, Hildebrand J, Vince JE, Sharp PP, Webb AI, Jackman KA, Muhlen S, Kennedy CL, Lowes KN, Murphy JM, Gyrd-Hansen M, Parker MW, Hartland EL, Lew AM, Huang DC, Lessene G, Silke J (2015) A RIPK2 inhibitor delays NOD signalling events yet prevents inflammatory cytokine production. Nat Commun 6:6442. doi:10.1038/ncomms7442
26. Sun Y, Liu H, Yu G, Chen X, Liu H, Tian H, Zhou G, Zhang F (2011) Gene expression analysis of leprosy by using a multiplex branched DNA assay. Exp Dermatol 20(6):520–522. doi:10.1111/j.1600-0625.2011.01270.x
27. Ho CC, Rideout HJ, Ribe E, Troy CM, Dauer WT (2009) The Parkinson disease protein leucine-rich repeat kinase 2 transduces death signals via Fas-associated protein with death domain and caspase-8 in a cellular model of neurodegeneration. J Neurosci 29(4):1011–1016. doi:10.1523/JNEUROSCI.5175-08.2009
28. Cho YS, Challa S, Moquin D, Genga R, Ray TD, Guildford M, Chan FK (2009) Phosphorylation-driven assembly of the RIP1-RIP3 complex regulates programmed necrosis and virus-induced inflammation. Cell 137(6):1112–1123. doi:10.1016/j.cell.2009.05.037
29. He S, Wang L, Miao L, Wang T, Du F, Zhao L, Wang X (2009) Receptor interacting protein kinase-3 determines cellular necrotic response to TNF-alpha. Cell 137(6):1100–1111. doi:10.1016/j.cell.2009.05.021
30. Zhang DW, Shao J, Lin J, Zhang N, Lu BJ, Lin SC, Dong MQ, Han J (2009) RIP3, an energy metabolism regulator that switches TNF-induced cell death from apoptosis to necrosis. Science 325(5938):332–336. doi:10.1126/science.1172308
31. Gardet A, Benita Y, Li C, Sands BE, Ballester I, Stevens C, Korzenik JR, Rioux JD, Daly MJ, Xavier RJ, Podolsky DK (2010) LRRK2 is involved in the IFN-gamma response and host response to pathogens. J Immunol 185(9):5577–5585. doi:10.4049/jimmunol.1000548
32. Hakimi M, Selvanantham T, Swinton E, Padmore RF, Tong Y, Kabbach G, Venderova K, Girardin SE, Bulman DE, Scherzer CR, LaVoie MJ, Gris D, Park DS, Angel JB, Shen J, Philpott DJ, Schlossmacher MG (2011) Parkinson's disease-linked LRRK2 is expressed in circulating and tissue immune cells and upregulated following recognition of microbial structures. J Neural Transm 118(5):795–808. doi:10.1007/s00702-011-0653-2
33. Thevenet J, Pescini Gobert R, Hooft van Huijsduijnen R, Wiessner C, Sagot YJ (2011) Regulation of LRRK2 expression points to a functional role in human monocyte maturation. PLoS One 6(6), e21519. doi:10.1371/journal.pone.0021519
34. Dzamko N, Inesta-Vaquera F, Zhang J, Xie C, Cai H, Arthur S, Tan L, Choi H, Gray N, Cohen P, Pedrioli P, Clark K, Alessi DR (2012) The IkappaB kinase family phosphorylates the Parkinson's disease kinase LRRK2 at Ser935 and Ser910 during Toll-like receptor signaling. PLoS One 7(6), e39132. doi:10.1371/journal.pone.0039132
35. Kuss M, Adamopoulou E, Kahle PJ (2014) Interferon-gamma induces leucine-rich repeat kinase LRRK2 via extracellular signal-regulated kinase ERK5 in macrophages. J Neurochem 129(6):980–987. doi:10.1111/jnc.12668

36. Deng X, Dzamko N, Prescott A, Davies P, Liu Q, Yang Q, Lee JD, Patricelli MP, Nomanbhoy TK, Alessi DR, Gray NS (2011) Characterization of a selective inhibitor of the Parkinson's disease kinase LRRK2. Nat Chem Biol 7(4):203–205. doi:10.1038/nchembio.538
37. Yan J, Almilaji A, Schmid E, Elvira B, Shimshek DR, van der Putten H, Wagner CA, Shumilina E, Lang F (2015) Leucine-rich repeat kinase 2-sensitive $Na^{+}/Ca2^{+}$ exchanger activity in dendritic cells. FASEB J 29(5):1701–1710. doi:10.1096/fj.14-264028
38. Connolly SF, Kusner DJ (2007) The regulation of dendritic cell function by calcium-signaling and its inhibition by microbial pathogens. Immunol Res 39(1–3):115–127
39. Deng X, Choi HG, Buhrlage SJ, Gray NS (2012) Leucine-rich repeat kinase 2 inhibitors: a patent review (2006–2011). Expert Opin Ther Pat 22(12):1415–1426. doi:10.1517/13543776.2012.729041
40. Kubo M, Kamiya Y, Nagashima R, Maekawa T, Eshima K, Azuma S, Ohta E, Obata F (2010) LRRK2 is expressed in B-2 but not in B-1 B cells, and downregulated by cellular activation. J Neuroimmunol 229(1–2):123–128. doi:10.1016/j.jneuroim.2010.07.021
41. Ness D, Ren Z, Gardai S, Sharpnack D, Johnson VJ, Brennan RJ, Brigham EF, Olaharski AJ (2013) Leucine-rich repeat kinase 2 (LRRK2)-deficient rats exhibit renal tubule injury and perturbations in metabolic and immunological homeostasis. PLoS One 8(6), e66164. doi:10.1371/journal.pone.0066164
42. Maekawa T, Kubo M, Yokoyama I, Ohta E, Obata F (2010) Age-dependent and cell-population-restricted LRRK2 expression in normal mouse spleen. Biochem Biophys Res Commun 392(3):431–435. doi:10.1016/j.bbrc.2010.01.041
43. Wandu WS, Tan C, Ogbeifun O, Vistica BP, Shi G, Hinshaw SJ, Xie C, Chen X, Klinman DM, Cai H, Gery I (2015) Leucine-rich repeat kinase 2 (Lrrk2) deficiency diminishes the development of experimental autoimmune uveitis (EAU) and the adaptive immune response. PLoS One 10(6), e0128906. doi:10.1371/journal.pone.0128906
44. Biskup S, Moore DJ, Celsi F, Higashi S, West AB, Andrabi SA, Kurkinen K, Yu SW, Savitt JM, Waldvogel HJ, Faull RL, Emson PC, Torp R, Ottersen OP, Dawson TM, Dawson VL (2006) Localization of LRRK2 to membranous and vesicular structures in mammalian brain. Ann Neurol 60(5):557–569. doi:10.1002/ana.21019
45. Melrose HL, Kent CB, Taylor JP, Dachsel JC, Hinkle KM, Lincoln SJ, Mok SS, Culvenor JG, Masters CL, Tyndall GM, Bass DI, Ahmed Z, Andorfer CA, Ross OA, Wszolek ZK, Delldonne A, Dickson DW, Farrer MJ (2007) A comparative analysis of leucine-rich repeat kinase 2 (Lrrk2) expression in mouse brain and Lewy body disease. Neuroscience 147(4):1047–1058. doi:10.1016/j.neuroscience.2007.05.027
46. Mandemakers W, Snellinx A, O'Neill MJ, de Strooper B (2012) LRRK2 expression is enriched in the striosomal compartment of mouse striatum. Neurobiol Dis 48(3):582–593. doi:10.1016/j.nbd.2012.07.017
47. Moehle MS, Webber PJ, Tse T, Sukar N, Standaert DG, DeSilva TM, Cowell RM, West AB (2012) LRRK2 inhibition attenuates microglial inflammatory responses. J Neurosci 32(5):1602–1611. doi:10.1523/JNEUROSCI.5601-11.2012
48. Puccini JM, Marker DF, Fitzgerald T, Barbieri J, Kim CS, Miller-Rhodes P, Lu SM, Dewhurst S, Gelbard HA (2015) Leucine-rich repeat kinase 2 modulates neuroinflammation and neurotoxicity in models of human immunodeficiency virus 1-associated neurocognitive disorders. J Neurosci 35(13):5271–5283. doi:10.1523/JNEUROSCI.0650-14.2015
49. Marker DF, Puccini JM, Mockus TE, Barbieri J, Lu SM, Gelbard HA (2012) LRRK2 kinase inhibition prevents pathological microglial phagocytosis in response to HIV-1 Tat protein. J Neuroinflammation 9:261. doi:10.1186/1742-2094-9-261
50. Kim B, Yang MS, Choi D, Kim JH, Kim HS, Seol W, Choi S, Jou I, Kim EY, Joe EH (2012) Impaired inflammatory responses in murine Lrrk2-knockdown brain microglia. PLoS One 7(4), e34693. doi:10.1371/journal.pone.0034693
51. McGeer PL, Itagaki S, Boyes BE, McGeer EG (1988) Reactive microglia are positive for HLA-DR in the substantia nigra of Parkinson's and Alzheimer's disease brains. Neurology 38(8):1285–1291

52. Guerreiro PS, Huang Y, Gysbers A, Cheng D, Gai WP, Outeiro TF, Halliday GM (2013) LRRK2 interactions with alpha-synuclein in Parkinson's disease brains and in cell models. J Mol Med 91(4):513–522. doi:10.1007/s00109-012-0984-y
53. Higashi S, Biskup S, West AB, Trinkaus D, Dawson VL, Faull RL, Waldvogel HJ, Arai H, Dawson TM, Moore DJ, Emson PC (2007) Localization of Parkinson's disease-associated LRRK2 in normal and pathological human brain. Brain Res 1155:208–219. doi:10.1016/j.brainres.2007.04.034
54. Miklossy J, Arai T, Guo JP, Klegeris A, Yu S, McGeer EG, McGeer PL (2006) LRRK2 expression in normal and pathologic human brain and in human cell lines. J Neuropathol Exp Neurol 65(10):953–963. doi:10.1097/01.jnen.0000235121.98052.54
55. Munoz L, Kavanagh ME, Phoa AF, Heng B, Dzamko N, Chen EJ, Doddareddy MR, Guillemin GJ, Kassiou M (2015) Optimisation of LRRK2 inhibitors and assessment of functional efficacy in cell-based models of neuroinflammation. Eur J Med Chem 95:29–34. doi:10.1016/j.ejmech.2015.03.003
56. Davies P, Hinkle KM, Sukar NN, Sepulveda B, Mesias R, Serrano G, Alessi DR, Beach TG, Benson DL, White CL, Cowell RM, Das SS, West AB, Melrose HL (2013) Comprehensive characterization and optimization of anti-LRRK2 (leucine-rich repeat kinase 2) monoclonal antibodies. Biochem J 453(1):101–113. doi:10.1042/BJ20121742
57. Huang Y, Song YJ, Murphy K, Holton JL, Lashley T, Revesz T, Gai WP, Halliday GM (2008) LRRK2 and parkin immunoreactivity in multiple system atrophy inclusions. Acta Neuropathol 116(6):639–646. doi:10.1007/s00401-008-0446-3
58. Kawai T, Akira S (2011) Toll-like receptors and their crosstalk with other innate receptors in infection and immunity. Immunity 34(5):637–650. doi:10.1016/j.immuni.2011.05.006
59. Kawasaki T, Kawai T (2014) Toll-like receptor signaling pathways. Front Immunol 5:461. doi:10.3389/fimmu.2014.00461
60. Nichols RJ, Dzamko N, Morrice NA, Campbell DG, Deak M, Ordureau A, Macartney T, Tong Y, Shen J, Prescott AR, Alessi DR (2010) 14-3-3 binding to LRRK2 is disrupted by multiple Parkinson's disease-associated mutations and regulates cytoplasmic localization. Biochem J 430(3):393–404. doi:10.1042/BJ20100483
61. Dzamko N, Deak M, Hentati F, Reith AD, Prescott AR, Alessi DR, Nichols RJ (2010) Inhibition of LRRK2 kinase activity leads to dephosphorylation of Ser (910)/Ser (935), disruption of 14-3-3 binding and altered cytoplasmic localization. Biochem J 430(3):405–413. doi:10.1042/BJ20100784
62. Clark K, Peggie M, Plater L, Sorcek RJ, Young ER, Madwed JB, Hough J, McIver EG, Cohen P (2011) Novel cross-talk within the IKK family controls innate immunity. Biochem J 434(1):93–104. doi:10.1042/BJ20101701
63. Gillardon F, Schmid R, Draheim H (2012) Parkinson's disease-linked leucine-rich repeat kinase 2(R1441G) mutation increases proinflammatory cytokine release from activated primary microglial cells and resultant neurotoxicity. Neuroscience 208:41–48. doi:10.1016/j.neuroscience.2012.02.001
64. Mestas J, Hughes CC (2004) Of mice and not men: differences between mouse and human immunology. J Immunol 172(5):2731–2738
65. Bryant CE, Monie TP (2012) Mice, men and the relatives: cross-species studies underpin innate immunity. Open Biol 2(4):120015. doi:10.1098/rsob.120015
66. Rao A, Luo C, Hogan PG (1997) Transcription factors of the NFAT family: regulation and function. Annu Rev Immunol 15:707–747. doi:10.1146/annurev.immunol.15.1.707
67. Macian F (2005) NFAT proteins: key regulators of T-cell development and function. Nat Rev Immunol 5(6):472–484. doi:10.1038/nri1632
68. Sharma S, Findlay GM, Bandukwala HS, Oberdoerffer S, Baust B, Li Z, Schmidt V, Hogan PG, Sacks DB, Rao A (2011) Dephosphorylation of the nuclear factor of activated T cells (NFAT) transcription factor is regulated by an RNA-protein scaffold complex. Proc Natl Acad Sci U S A 108(28):11381–11386. doi:10.1073/pnas.1019711108
69. Willingham AT, Orth AP, Batalov S, Peters EC, Wen BG, Aza-Blanc P, Hogenesch JB, Schultz PG (2005) A strategy for probing the function of noncoding RNAs finds a repressor of NFAT. Science 309(5740):1570–1573. doi:10.1126/science.1115901

70. Nathan C, Cunningham-Bussel A (2013) Beyond oxidative stress: an immunologist's guide to reactive oxygen species. Nat Rev Immunol 13(5):349–361. doi:10.1038/nri3423
71. West AP, Brodsky IE, Rahner C, Woo DK, Erdjument-Bromage H, Tempst P, Walsh MC, Choi Y, Shadel GS, Ghosh S (2011) TLR signalling augments macrophage bactericidal activity through mitochondrial ROS. Nature 472(7344):476–480. doi:10.1038/nature09973
72. Yue M, Hinkle KM, Davies P, Trushina E, Fiesel FC, Christenson TA, Schroeder AS, Zhang L, Bowles E, Behrouz B, Lincoln SJ, Beevers JE, Milnerwood AJ, Kurti A, McLean PJ, Fryer JD, Springer W, Dickson DW, Farrer MJ, Melrose HL (2015) Progressive dopaminergic alterations and mitochondrial abnormalities in LRRK2 G2019S knock-in mice. Neurobiol Dis 78:172–195. doi:10.1016/j.nbd.2015.02.031
73. Su YC, Qi X (2013) Inhibition of excessive mitochondrial fission reduced aberrant autophagy and neuronal damage caused by LRRK2 G2019S mutation. Hum Mol Genet 22(22):4545–4561. doi:10.1093/hmg/ddt301
74. Saez-Atienzar S, Bonet-Ponce L, Blesa JR, Romero FJ, Murphy MP, Jordan J, Galindo MF (2014) The LRRK2 inhibitor GSK2578215A induces protective autophagy in SH-SY5Y cells: involvement of Drp-1-mediated mitochondrial fission and mitochondrial-derived ROS signaling. Cell Death Dis 5, e1368. doi:10.1038/cddis.2014.320
75. Yang S, Xia C, Li S, Du L, Zhang L, Hu Y (2014) Mitochondrial dysfunction driven by the LRRK2-mediated pathway is associated with loss of Purkinje cells and motor coordination deficits in diabetic rat model. Cell Death Dis 5, e1217. doi:10.1038/cddis.2014.184
76. Castrejon-Jimenez NS, Leyva-Paredes K, Hernandez-Gonzalez JC, Luna-Herrera J, Garcia-Perez BE (2015) The role of autophagy in bacterial infections. Biosci Trends 9(3):149–159. doi:10.5582/bst.2015.01035
77. Knodler LA, Celli J (2011) Eating the strangers within: host control of intracellular bacteria via xenophagy. Cell Microbiol 13(9):1319–1327. doi:10.1111/j.1462-5822.2011.01632.x
78. Schapansky J, Nardozzi JD, Felizia F, Lavoie MJ (2014) Membrane recruitment of endogenous LRRK2 precedes its potent regulation of autophagy. Hum Mol Genet 23:4201–4214. doi:10.1093/hmg/ddu138
79. Tong Y, Yamaguchi H, Giaime E, Boyle S, Kopan R, Kelleher RJ 3rd, Shen J (2010) Loss of leucine-rich repeat kinase 2 causes impairment of protein degradation pathways, accumulation of alpha-synuclein, and apoptotic cell death in aged mice. Proc Natl Acad Sci U S A 107(21):9879–9884. doi:10.1073/pnas.1004676107
80. Tong Y, Giaime E, Yamaguchi H, Ichimura T, Liu Y, Si H, Cai H, Bonventre JV, Shen J (2012) Loss of leucine-rich repeat kinase 2 causes age-dependent bi-phasic alterations of the autophagy pathway. Mol Neurodegener 7:2. doi:10.1186/1750-1326-7-2
81. Manzoni C, Mamais A, Dihanich S, Abeti R, Soutar MP, Plun-Favreau H, Giunti P, Tooze SA, Bandopadhyay R, Lewis PA (2013) Inhibition of LRRK2 kinase activity stimulates macroautophagy. Biochim Biophys Acta 1833(12):2900–2910. doi:10.1016/j.bbamcr.2013.07.020
82. Orenstein SJ, Kuo SH, Tasset I, Arias E, Koga H, Fernandez-Carasa I, Cortes E, Honig LS, Dauer W, Consiglio A, Raya A, Sulzer D, Cuervo AM (2013) Interplay of LRRK2 with chaperone-mediated autophagy. Nat Neurosci 16(4):394–406. doi:10.1038/nn.3350
83. Wang L, Xie C, Greggio E, Parisiadou L, Shim H, Sun L, Chandran J, Lin X, Lai C, Yang WJ, Moore DJ, Dawson TM, Dawson VL, Chiosis G, Cookson MR, Cai H (2008) The chaperone activity of heat shock protein 90 is critical for maintaining the stability of leucine-rich repeat kinase 2. J Neurosci 28(13):3384–3391. doi:10.1523/JNEUROSCI.0185-08.2008
84. Orr CF, Rowe DB, Halliday GM (2002) An inflammatory review of Parkinson's disease. Prog Neurobiol 68(5):325–340
85. Dzamko N, Geczy CL, Halliday GM (2014) Inflammation is genetically implicated in Parkinson's disease. Neuroscience 302:89–102. doi:10.1016/j.neuroscience.2014.10.028
86. Kannarkat GT, Boss JM, Tansey MG (2013) The role of innate and adaptive immunity in Parkinson's disease. J Parkinson's Dis 3(4):493–514. doi:10.3233/JPD-130250
87. Deleidi M, Gasser T (2013) The role of inflammation in sporadic and familial Parkinson's disease. Cell Mol Life Sci 70(22):4259–4273. doi:10.1007/s00018-013-1352-y

88. Imamura K, Hishikawa N, Sawada M, Nagatsu T, Yoshida M, Hashizume Y (2003) Distribution of major histocompatibility complex class II-positive microglia and cytokine profile of Parkinson's disease brains. Acta Neuropathol 106(6):518–526. doi:10.1007/s00401-003-0766-2
89. Dubiel JS, Jaskiewicz J, Zmudka K, Brzostek T, Surdacki A (1992) Influence of verapamil on the coronary arteriovenous difference in long chain free fatty acids in hypertrophic cardiomyopathy. Eur J Clin Pharmacol 42(2):217–218
90. Kim HS, Suh YH (2009) Minocycline and neurodegenerative diseases. Behav Brain Res 196(2):168–179. doi:10.1016/j.bbr.2008.09.040
91. Gagne JJ, Power MC (2010) Anti-inflammatory drugs and risk of Parkinson disease: a meta-analysis. Neurology 74(12):995–1002. doi:10.1212/WNL.0b013e3181d5a4a3
92. Spillantini MG, Schmidt ML, Lee VM, Trojanowski JQ, Jakes R, Goedert M (1997) Alpha-synuclein in Lewy bodies. Nature 388(6645):839–840. doi:10.1038/42166
93. Halliday G, Hely M, Reid W, Morris J (2008) The progression of pathology in longitudinally followed patients with Parkinson's disease. Acta Neuropathol 115(4):409–415. doi:10.1007/s00401-008-0344-8
94. Braak H, Del Tredici K, Rub U, de Vos RA, Jansen Steur EN, Braak E (2003) Staging of brain pathology related to sporadic Parkinson's disease. Neurobiol Aging 24(2):197–211
95. Polymeropoulos MH, Lavedan C, Leroy E, Ide SE, Dehejia A, Dutra A, Pike B, Root H, Rubenstein J, Boyer R, Stenroos ES, Chandrasekharappa S, Athanassiadou A, Papapetropoulos T, Johnson WG, Lazzarini AM, Duvoisin RC, Di Iorio G, Golbe LI, Nussbaum RL (1997) Mutation in the alpha-synuclein gene identified in families with Parkinson's disease. Science 276(5321):2045–2047
96. Kim C, Ho DH, Suk JE, You S, Michael S, Kang J, Joong Lee S, Masliah E, Hwang D, Lee HJ, Lee SJ (2013) Neuron-released oligomeric alpha-synuclein is an endogenous agonist of TLR2 for paracrine activation of microglia. Nat Commun 4:1562. doi:10.1038/ncomms2534
97. Codolo G, Plotegher N, Pozzobon T, Brucale M, Tessari I, Bubacco L, de Bernard M (2013) Triggering of inflammasome by aggregated alpha-synuclein, an inflammatory response in synucleinopathies. PLoS One 8(1), e55375. doi:10.1371/journal.pone.0055375
98. Fellner L, Irschick R, Schanda K, Reindl M, Klimaschewski L, Poewe W, Wenning GK, Stefanova N (2013) Toll-like receptor 4 is required for alpha-synuclein dependent activation of microglia and astroglia. Glia 61(3):349–360. doi:10.1002/glia.22437
99. Couch Y, Alvarez-Erviti L, Sibson NR, Wood MJ, Anthony DC (2011) The acute inflammatory response to intranigral alpha-synuclein differs significantly from intranigral lipopolysaccharide and is exacerbated by peripheral inflammation. J Neuroinflammation 8:166. doi:10.1186/1742-2094-8-166
100. Roodveldt C, Labrador-Garrido A, Gonzalez-Rey E, Lachaud CC, Guilliams T, Fernandez-Montesinos R, Benitez-Rondan A, Robledo G, Hmadcha A, Delgado M, Dobson CM, Pozo D (2013) Preconditioning of microglia by alpha-synuclein strongly affects the response induced by toll-like receptor (TLR) stimulation. PLoS One 8(11), e79160. doi:10.1371/journal.pone.0079160
101. Daher JP, Volpicelli-Daley LA, Blackburn JP, Moehle MS, West AB (2014) Abrogation of alpha-synuclein-mediated dopaminergic neurodegeneration in LRRK2-deficient rats. Proc Natl Acad Sci U S A 111(25):9289–9294. doi:10.1073/pnas.1403215111
102. Hawkes CH, Del Tredici K, Braak H (2007) Parkinson's disease: a dual-hit hypothesis. Neuropathol Appl Neurobiol 33(6):599–614. doi:10.1111/j.1365-2990.2007.00874.x
103. Braak H, Del Tredici K (2009) Neuroanatomy and pathology of sporadic Parkinson's disease. Adv Anat Embryol Cell Biol 201:1–119
104. Sung HY, Park JW, Kim JS (2014) The frequency and severity of gastrointestinal symptoms in patients with early Parkinson's disease. J Mov Disord 7(1):7–12. doi:10.14802/jmd.14002
105. Hilton D, Stephens M, Kirk L, Edwards P, Potter R, Zajicek J, Broughton E, Hagan H, Carroll C (2014) Accumulation of alpha-synuclein in the bowel of patients in the pre-clinical phase of Parkinson's disease. Acta Neuropathol 127(2):235–241. doi:10.1007/s00401-013-1214-6

106. Cersosimo MG, Raina GB, Pecci C, Pellene A, Calandra CR, Gutierrez C, Micheli FE, Benarroch EE (2013) Gastrointestinal manifestations in Parkinson's disease: prevalence and occurrence before motor symptoms. J Neurol 260(5):1332–1338. doi:10.1007/s00415-012-6801-2
107. Qin L, Wu X, Block ML, Liu Y, Breese GR, Hong JS, Knapp DJ, Crews FT (2007) Systemic LPS causes chronic neuroinflammation and progressive neurodegeneration. Glia 55(5):453–462. doi:10.1002/glia.20467
108. Haag LM, Siegmund B (2015) Intestinal microbiota and the innate immune system—a cross-talk in Crohn's disease pathogenesis. Front Immunol 6:489. doi:10.3389/fimmu.2015.00489
109. Barrett JC, Hansoul S, Nicolae DL, Cho JH, Duerr RH, Rioux JD, Brant SR, Silverberg MS, Taylor KD, Barmada MM, Bitton A, Dassopoulos T, Datta LW, Green T, Griffiths AM, Kistner EO, Murtha MT, Regueiro MD, Rotter JI, Schumm LP, Steinhart AH, Targan SR, Xavier RJ, Consortium NIG, Libioulle C, Sandor C, Lathrop M, Belaiche J, Dewit O, Gut I, Heath S, Laukens D, Mni M, Rutgeerts P, Van Gossum A, Zelenika D, Franchimont D, Hugot JP, de Vos M, Vermeire S, Louis E, Belgian-French IBDC, Wellcome Trust Case Control C, Cardon LR, Anderson CA, Drummond H, Nimmo E, Ahmad T, Prescott NJ, Onnie CM, Fisher SA, Marchini J, Ghori J, Bumpstead S, Gwilliam R, Tremelling M, Deloukas P, Mansfield J, Jewell D, Satsangi J, Mathew CG, Parkes M, Georges M, Daly MJ (2008) Genome-wide association defines more than 30 distinct susceptibility loci for Crohn's disease. Nat Genet 40(8):955–962. doi:10.1038/ng.175
110. Kumar V, Mack DR, Marcil V, Israel D, Krupoves A, Costea I, Lambrette P, Grimard G, Dong J, Seidman EG, Amre DK, Levy E (2013) Genome-wide association study signal at the 12q12 locus for Crohn's disease may represent associations with the MUC19 gene. Inflamm Bowel Dis 19(6):1254–1259. doi:10.1097/MIB.0b013e318281f454
111. Zhang Q, Pan Y, Yan R, Zeng B, Wang H, Zhang X, Li W, Wei H, Liu Z (2015) Commensal bacteria direct selective cargo sorting to promote symbiosis. Nat Immunol 16(9):918–926. doi:10.1038/ni.3233
112. de Lange KM, Barrett JC (2015) Understanding inflammatory bowel disease via immunogenetics. J Autoimmun 64:91–100. doi:10.1016/j.jaut.2015.07.013
113. Umeno J, Asano K, Matsushita T, Matsumoto T, Kiyohara Y, Iida M, Nakamura Y, Kamatani N, Kubo M (2011) Meta-analysis of published studies identified eight additional common susceptibility loci for Crohn's disease and ulcerative colitis. Inflamm Bowel Dis 17(12):2407–2415. doi:10.1002/ibd.21651
114. Heckman MG, Schottlaender L, Soto-Ortolaza AI, Diehl NN, Rayaprolu S, Ogaki K, Fujioka S, Murray ME, Cheshire WP, Uitti RJ, Wszolek ZK, Farrer MJ, Sailer A, Singleton AB, Chinnery PF, Keogh MJ, Gentleman SM, Holton JL, Aoife K, Mann DM, Al-Sarraj S, Troakes C, Dickson DW, Houlden H, Ross OA (2014) LRRK2 exonic variants and risk of multiple system atrophy. Neurology 83(24):2256–2261. doi:10.1212/WNL.0000000000001078
115. Freischmidt A, Wieland T, Richter B, Ruf W, Schaeffer V, Muller K, Marroquin N, Nordin F, Hubers A, Weydt P, Pinto S, Press R, Millecamps S, Molko N, Bernard E, Desnuelle C, Soriani MH, Dorst J, Graf E, Nordstrom U, Feiler MS, Putz S, Boeckers TM, Meyer T, Winkler AS, Winkelman J, de Carvalho M, Thal DR, Otto M, Brannstrom T, Volk AE, Kursula P, Danzer KM, Lichtner P, Dikic I, Meitinger T, Ludolph AC, Strom TM, Andersen PM, Weishaupt JH (2015) Haploinsufficiency of TBK1 causes familial ALS and fronto-temporal dementia. Nat Neurosci 18(5):631–636. doi:10.1038/nn.4000
116. Pottier C, Bieniek KF, Finch N, van de Vorst M, Baker M, Perkersen R, Brown P, Ravenscroft T, van Blitterswijk M, Nicholson AM, DeTure M, Knopman DS, Josephs KA, Parisi JE, Petersen RC, Boylan KB, Boeve BF, Graff-Radford NR, Veltman JA, Gilissen C, Murray ME, Dickson DW, Rademakers R (2015) Whole-genome sequencing reveals important role for TBK1 and OPTN mutations in frontotemporal lobar degeneration without motor neuron disease. Acta Neuropathol 130(1):77–92. doi:10.1007/s00401-015-1436-x

Chapter 8
Regulation of LRRK2 by Phosphatases

Jean-Marc Taymans

Abstract LRRK2 is a highly phosphorylated protein, and evidence of a physiological role for LRRK2 phosphorylation has accumulated in recent years for cellular phosphosites, many of which are found in the ANK-LRR interdomain region, i.e., the S910/S935/S955/S973 sites as well as recently for autophosphorylation sites, at least one of which has been confirmed in cells, S1292. LRRK2 phosphorylation is modulated in several disease or potential therapy relevant conditions such as in disease mutant variants of LRRK2 or following LRRK2 kinase inhibitor treatment. This chapter will focus on the regulation of LRRK2 phosphorylation and more specifically the role of phosphatases in LRRK2 dephosphorylation. This will include reviewing the conditions in which LRRK2 is found to be dephosphorylated, the molecular partners and phosphatases involved in regulating LRRK2 phosphorylation, as well as discussing how LRRK2 phosphatases may be therapeutic targets or biomarkers in their own right.

Keywords Phosphatase • PP1 • PP2A • Phosphorylation • Biomarker

Introduction

Leucine-rich repeat kinase 2 (LRRK2) harbors several functions, such as GTPase and kinase functions, protein-scaffolding domains, or dimerization capacity. As is discussed elsewhere in this book, the study of LRRK2 biology points to Parkinson's disease mechanisms and therapeutic targeting opportunities. In particular, LRRK2's complexity offers several functions that may be monitored as biomarkers or targeted for therapy. For instance, LRRK2's kinase function has received much attention as a therapeutic targeting strategy with several compounds that have now been developed and that are valuable research tools as well as potential therapeutics [1–3]. Several other LRRK2 functions may also offer targeting or biomarker opportunities

J.-M. Taymans (✉)
Université de Lille, Inserm, CHU Lille, UMR-S 1172—JPARC Jean-Pierre Aubert Research Center, Neurosciences and Cancer, 1 Place de Verdun, Lille 59000, France
e-mail: jean-marc.taymans@inserm.fr

H.J. Rideout (ed.), *Leucine-Rich Repeat Kinase 2 (LRRK2)*,
Advances in Neurobiology 14, DOI 10.1007/978-3-319-49969-7_8

such as GTPase function [4–6], dimerization [7, 8], or phosphorylation regulation [9, 10].

LRRK2 is a highly phosphorylated protein, and evidence of a physiological role for LRRK2 phosphorylation has accumulated in recent years for cellular phosphosites, many of which are found in the ANK-LRR interdomain region, including the S910/S935/S955/S973 sites as well as recently for autophosphorylation sites, at least one of which has been confirmed in cells, S1292 [11]. Several LRRK2 pathogenic mutants display an overall reduced phosphorylation level at ANK-LRR interdomain sites or an increased phosphorylation level at the S1292 site [12–17], while both types of sites are dephosphorylated after cellular treatment with LRRK2 kinase inhibitors, although via different mechanisms. The shift of the phosphorylation equilibrium toward dephosphorylation of LRRK2 observed in disease indicates that phosphatases play an important role in LRRK2 cellular regulation.

This chapter will focus on the regulation of LRRK2 phosphorylation and more specifically the role of phosphatases in LRRK2 dephosphorylation. This will include reviewing the conditions in which LRRK2 is found to be dephosphorylated, the molecular partners and phosphatases involved in regulating LRRK2 phosphorylation, as well as discussing how LRRK2 phosphatases may be therapeutic targets or biomarkers in their own right.

LRRK2 Phosphorylation Sites: Many and Diverse

Phosphosite mapping studies have distinguished at least 20 phosphorylation sites in LRRK2, suggesting that close to 1 % of all amino acids are phosphorylated [13, 18–20]. Initially, the identified phosphorylation sites were subdivided into one of two notable classes: autophosphorylation sites and cellular phosphorylation sites (see Fig. 8.1). Further characterization of phosphosites has provided additional distinction as to the nature of LRRK2 phosphorylation sites, i.e., autophosphorylation sites and heterologous phosphorylation sites. Autophosphorylation sites were first identified after recombinant LRRK2 was submitted to in vitro autophosphorylation

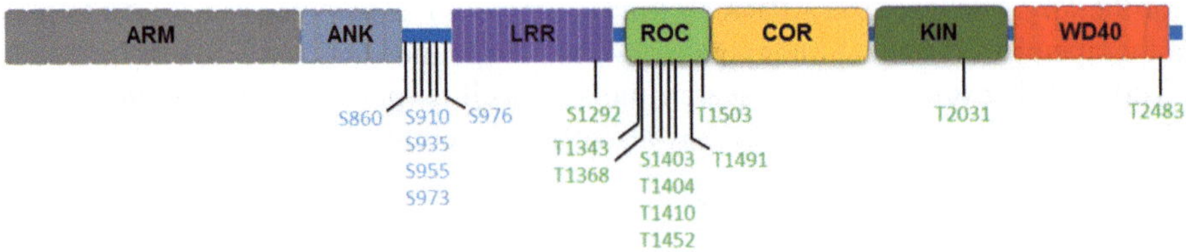

Fig. 8.1 Schematic of LRRK2 domain structure and phosphorylation sites. Phosphorylation sites are indicated below the LRRK2 schematic, with sites in orange indicating heterologous phosphorylation sites and in green in vitro autophosphorylation sites. Characteristics of these sites are summarized in Table 8.1

and have been found to reside predominantly in an autophosphorylation cluster in the ROC GTPase domain or as individual sites scattered outside of this cluster [20–23]. Thus far, only a handful of the autophosphorylation sites have been confirmed to exist under basal conditions in cells, including the S1292 or T1410 sites [17, 22]. The heterologous phosphorylation sites on the other hand were first identified in LRRK2 isolated from cells and therefore are classified as cellular phosphorylation sites. Again, these sites mostly exist in a cluster located in the ANK-LRR interdomain region, although additional heterologous phosphorylation sites exist scattered throughout the protein. For an extensive overview on LRRK2 phosphorylation sites, please refer to previously published reviews [11].

Of interest is the striking difference that exists between LRRK2's phosphorylation pattern and that of LRRK2's closest homolog LRRK1. LRRK1 resides in cells with levels of cellular phosphorylation that are similar to that of LRRK2 [19, 24]; however, the LRRK1 phosphorylation sites are not mapped to residues homologous to LRRK2 phosphorylation sites nor in regions of LRRK2 phosphorylation clusters [19]. This suggests that phosphorylation patterns of LRRK2 are LRRK2 specific and therefore that phosphorylation regulation and phosphorylation-dependent functions of LRRK2 are also LRRK2 specific, in line with the role of LRRK2 but not LRRK1 in Parkinson's disease. Given that this chapter focuses on phosphatase-mediated regulation of LRRK2 phosphorylation, the next sections will consider only phosphorylation sites which have been confirmed in cells.

Conditions in Which LRRK2 Phosphorylation Is Modified

LRRK2 Disease Mutants

Heterologous Phosphorylation Sites

Several LRRK2 mutants have been shown to display altered phosphorylation levels in cells under basal conditions. For the phosphorylation of sites in the ANK-LRR interdomain cluster, six of the seven mutants most consistently reported to segregate with disease have reduced basal phosphorylation levels. The outlier here is the G2019S variant of LRRK2 which displays phosphorylation levels comparable to wild type [12, 13, 15]. It should also be noted that of the six dephosphorylated variants of LRRK2, the extent of dephosphorylation varies, with five mutants showing extensive dephosphorylation (N1437H, R1441G, R1441H, Y1699C, I2020T), while the remaining mutant (R1441C) displays moderate dephosphorylation. Therefore, with the exception of G2019S, dephosphorylation of sites in the ANK-LRR interdomain region, and hence the intervention of phosphatases, is a characteristic of most disease variants of LRRK2. This is also in line with phosphorylation levels of LRRK2 G2385R, a common risk variant of LRRK2, which also displays a moderate reduction in phosphorylation relative to wild type.

Physiological Autophosphorylation Site (S1292)

Other data show a contrastingly different pattern of basal phosphorylation levels for autophosphorylation sites such as S1292 which is one of the few autophosphorylation sites that could be confirmed in vivo. On the contrary to the ANK-LRR interdomain cluster sites, S1292 phosphorylation is increased in most disease mutants, with the exception of the Y1699C and G2385R mutants [14, 17] that show levels comparable to WT. Interestingly, the increases observed in S1292 phosphorylation suggest that this event is not predominant in altering overall LRRK2 phosphorylation, as evidenced by overall phosphorylation of LRRK2 comparable to WT for G2019S or reduced for R1441C when measured by metabolic labeling with radioactive phosphate [12].

These data suggest that phosphorylation of LRRK2 is altered in the presence of LRRK2 disease mutants and that therefore LRRK2 phosphorylation may be involved in the disease process and may also be a disease biomarker. However, further work is required in this area, in the first instance, to understand why most but not all disease mutants display abovementioned phosphorylation changes.

LRRK2 Functional Mutants

Heterologous Phosphorylation Sites

Several studies have shown that LRRK2 phosphorylation is altered when using functional mutants affecting LRRK2 catalytic activities. For instance, at least some of the functional mutants that inhibit the GTP-binding capacity of LRRK2, such as K1347A or T1348N, display reduced phosphorylation levels [25, 26]. Interestingly, some authors found that kinase dead mutants of LRRK2 such as D1994A also led to decreases in LRRK2 phosphorylation levels; however, this was not uniformly the case for all kinase dead mutants, such as K1906A, K1906M, S2032A, and T2035A, which show phosphorylation levels comparable to WT or D2017A with intermediate phosphorylation levels [12, 13, 26, 27]. Inversely, other mutants that activate LRRK2 kinase activity, such as G2019S or T2031S, display phosphorylation levels comparable to WT [13, 27].

Physiological Autophosphorylation Site (S1292)

The phosphorylation level of the S1292 site is also modulated in the presence of several kinase-modifying mutants, such as reduced S1292 phosphorylation for the D1994A and D2017A kinase dead mutants [14, 17]. In addition, this site shows enhanced phosphorylation for at least one kinase-activating mutant, the G2019S.

Pharmacologically Induced Phosphorylation Changes

Heterologous Phosphorylation Sites

In a study by Dzamko, Nichols, Alessi and colleagues published in 2010, it was first found that compounds that inhibit LRRK2 kinase activity in vitro can induce dephosphorylation of LRRK2 at the S910-S935 sites in cells [28]. Subsequent studies showed that inhibitor-induced dephosphorylation of LRRK2 also affects the other phosphosites of the ANK-LRR interdomain cluster such as S955 and S973 [16]. Inhibitor treatment also results in a reduction of overall LRRK2 phosphorylation as measured by metabolic labeling with radioactive phosphates [12, 19], suggesting that the sites in the ANK-LRR cluster may be the dominant sites in overall inhibitor-induced LRRK2 dephosphorylation. The modulation of the cellular phosphorylation sites by LRRK2 inhibitors is also specific to LRRK2, and treatment with the same compounds does not lead to dephosphorylation in LRRK1 [19]. The potency of dephosphorylation is dependent on dose and potency of inhibitors applied [17, 29–33].

Physiological Autophosphorylation Site (S1292)

The S1292 autophosphorylation site is also dephosphorylated after cellular treatment with an LRRK2 kinase inhibitor [14, 17]. Similar to the inhibitor-induced dephosphorylation of the ANK-LRR interdomain sites, S1292 dephosphorylation is dependent on inhibitor potency and concentration.

LRRK2 Kinase Activity and LRRK2 Dephosphorylation: The Difference Between Kinase Activity Per Se and Pharmacological Inhibition

Heterologous Phosphorylation Sites

Although it appears at first glance to be logical that pharmacological inhibition of LRRK2 should induce dephosphorylation of LRRK2, it is actually counterintuitive for the heterologous phosphorylation sites given that S910-S935 are not autophosphorylation sites, nor are they sites that are regulated by the intensity of LRRK2 kinase activity. Indeed, studies testing several LRRK2 variants with a broad range of kinase activities have found that the phosphorylation state (at the S935 cluster) is not correlated with in vitro kinase activity of LRRK2 mutants [14, 27].

In contrast, reports thus far have shown a good correlation between in vitro activity of LRRK2 kinase inhibitors and their ability to dephosphorylate LRRK2 in cells. This apparent paradox was addressed by Vancraenenbroeck et al. who used a kinome-wide panel of kinase inhibitors with the initial intent to identify the classes of upstream kinases involved in phosphorylating LRRK2 in cells. In order to discern inhibitors that are acting on upstream kinases from those acting on LRRK2 itself, the in vitro potency of the kinase panel was tested in LRRK2 using the in vitro peptide (lrrktide)-based radiometric assay. Comparing the in vitro results to cellular dephosphorylation at S935 led to the conclusion that the compounds which most potently dephosphorylated LRRK2 are those which act on LRRK2 itself, a finding confirmed by in silico docking of the most potent compounds to the LRRK2 ATP-binding site [33]. Given that the S935 site is not an autophosphorylation site, this result was quite unexpected and suggests that binding of inhibitors to the LRRK2 kinase pocket induces dephosphorylation of LRRK2. As discussed below, this is likely due to the recruitment of phosphatases to dephosphorylate LRRK2 [12]. Although this suggests that the S935 cellular dephosphorylation is a good readout of cellular activity of LRRK2 compounds, it remains an indirect measure of LRRK2 cellular kinase activity, and future efforts should be directed toward strategies to complement the S935 dephosphorylation assay [34].

Physiological Autophosphorylation Site (S1292)

In contrast to the heterologous phosphorylation sites, the situation for the S1292 autophosphorylation site is more straightforward. Given that it is a site phosphorylated by LRRK2's own kinase activity, the dephosphorylation caused by inhibitor treatment is likely a consequence of the reduced phosphorylation of that site by LRRK2. For the disease mutants, the main caveat remains that the correlation between kinase activity and S1292 phosphorylation levels is not fully followed. For example, R1441C and G2019S show similar steady-state S1292 phosphorylation levels, while G2019S has a markedly higher kinase activity [14, 17]. Further research will be required to explain these discrepancies.

What We Know About the Phosphatases and Other Main Molecular Partners Regulating LRRK2 Phosphorylation

The most important partners of LRRK2 phosphoregulation are kinases and phosphatases. With regard to the regulation of the phosphorylation of LRRK2's heterologous sites, recent studies have begun to report kinases active at these sites. Studies in vitro and in COS-7 cells have suggested that protein kinase A (PKA) acts as a kinase for the S910-S935 sites (Muda et al. 2014); however, these findings are not confirmed in other cell types such as HEK293T cells [14]. This suggests cell-specific mechanisms of phosphorylation and is further supported by the work of Dzamko et al. (2012) who show that inhibitor of kappa B kinases (IKKs) phosphorylate LRRK2 in bone marrow-derived macrophages upon activation of toll-like receptor signaling which is specific to immune cells. Finally, casein kinase 1 alpha (CK1) was identified in a kinome-wide siRNA screen in HEK293T cells as a kinase for the ANK-LRR interdomain sites [35]. As tools such as site-specific phospho-antibodies become available, it may be expected that additional kinases be identified as LRRK2 phosphorylaters. As described above, autophosphorylation sites have been confirmed as such through studies with kinase inhibitors and functional mutants.

The search for LRRK2 phosphatases has also seen advances in recent years. First, for the heterologous phosphorylation sites, phosphoprotein phosphatase 1 catalytic subunit alpha (PP1alpha, HGNC symbol PPP1CA) has been shown as an important phosphatase at the LRRK2 ANK-LRR interdomain sites [12]. The study first shows that of a panel of recombinant serine/threonine phosphatases, only protein phosphatase 1 can efficiently dephosphorylate LRRK2 in vitro. In vitro dephosphorylation was demonstrated on purified LRRK2 protein carrying radioactive phosphates, showing that PP1 is responsible for dephosphorylation at the majority of LRRK2's phosphosites. This finding was confirmed for four sites with phospho-specific antibodies, i.e., S910, S935, S955, and S973. Of note, the recombinant LRRK2 protein can be dephosphorylated with excess amounts of recombinant alkaline phosphatase [12] or lambda phosphatase [13]; however, when tested in low concentrations, PP1 shows the highest dephosphorylation capacity [12].

The physiological relevance of PP1 as an LRRK2 phosphatase for the ANK-LRR interdomain sites could be demonstrated by pharmacological and molecular approaches. Upon pharmacological inhibition of cells with either PP1 or PP2A phosphatase inhibitors, it was observed that PP1 but not PP2A inhibition could reverse LRRK2 dephosphorylation. Interestingly, the effects of PP1 in LRRK2 phosphorylation appear to be broadly applicable across multiple cell types, including HEK293T, SH-SY5Y neuroblastoma cells, mouse primary cortical neurons, U2OS osteosarcoma cells, NIH3T3 mouse fibroblast cells, and A549 human lung cancer cells. This shows that PP1 is active as an LRRK2 phosphatase independent of the cell type tested, and it may be predicted that PP1 can dephosphorylate LRRK2 throughout multiple different tissues. In addition, conditions of LRRK2 dephosphorylation show an enhanced binding of PP1 to LRRK2. For example, treatment with LRRK2 kinase inhibitors that dephosphorylate LRRK2 or testing of LRRK2 mutants with low basal phosphorylation levels shows an enhanced association of PP1 to LRRK2 compared to basal levels [12]. Therefore, the mechanism of LRRK2 dephosphorylation involves the recruitment of PP1 to the LRRK2 complex.

Further evidence for the importance of the physiological role of PP1 in dephosphorylating has been reported in a study of the link between LRRK2 and arsenite stress [36]. Arsenite stress causes a reduction in S910/S935 phosphorylation of LRRK2 in cell culture, and this reduction is inhibited by the PP1/PP2A inhibitor calyculin A. Also, arsenite stress leads to the recruitment of PP1 to the LRRK2 complex, similar to what is observed with LRRK2 kinase inhibitor treatment [36].

PP1 class phosphatases are composed of a catalytic subunit, responsible for catalyzing dephosphorylation, and a regulatory subunit, responsible for substrate specificity. Together, the catalytic subunit and regulatory subunit are called the holoenzyme. There are more than 150 PP1 regulatory subunits reported, theoretically allowing more than 450 possible holoenzyme compositions (Bollen et al.

2010). This mode of functioning is necessary given that only three PP1 catalytic subunits are expressed in mammalian cells (PP1α, PP1β, and PP1γ; HGNC codes PPP1CA, PPP1CB, and PPP1CC) that on their own are insufficiently diverse to account for the specificity in the all phosphatase activity mediated by PP1. Indeed, PP1 together with PP2A (which is represented by only two catalytic subunits, PPP2CA and PPP2CB) accounts for more than 90 % of the protein phosphatase activity in eukaryotes [37, 38]. This contrasts with the large number of members in the kinase protein family, which includes ~400 serine/threonine kinases [39]. Therefore, a key issue is to identify the composition of the PP1 holoenzyme by identifying the LRRK2-specific PP1 regulatory subunit which associates with the PP1 catalytic subunit.

Concerning the regulation of LRRK2 phosphosites outside of the ANK-LRR interdomain region, little data is available; however, initial evidence suggests that other phosphatases are at play. Studies exploring dephosphorylation at the S1292 site show that inhibitor-induced dephosphorylation of LRRK2 at S1292 is insensitive to the phosphatase inhibitors calyculin A (mixed PP1 and PP2A inhibitor) and okadaic acid (selective PP2A inhibitor) [14], in contrast to what is observed at the S935 site where inhibitor-induced dephosphorylation is inhibited by calyculin A [12]. However, the low basal S1292 phosphorylation levels of the R1441G mutant are upregulated by both calyculin A and OA treatment [14], while the S935 phosphorylation levels of the same mutant are only upregulated by calyculin A [12]. These findings suggest the hypothesis that PP2A, rather than PP1, is the phosphatase system regulating R1441G LRRK2 at S1292. (See Fig. 8.2 for a schematic summary of phosphatases regulating LRRK2 phosphorylation sites).

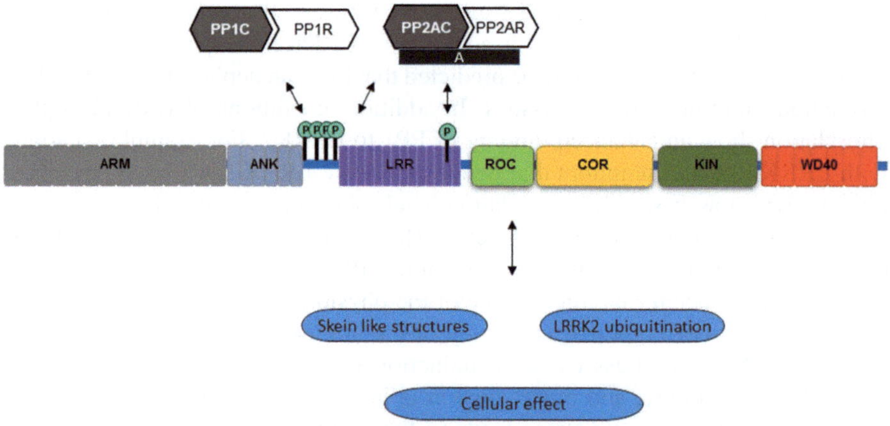

Fig. 8.2 Hypothetical schematic of LRRK2 phosphatases in relation to the ANK-LRR interdomain cluster and the S1292 autophosphorylation site. Shown are two phosphatase complexes that have been implicated in LRRK2 phosphoregulation, protein phosphatase 1 (catalytic subunit PP1C and regulatory subunit PP1R), and protein phosphatase 2A (catalytic subunit PP2AC and regulatory subunit PP2AR and scaffold subunit A). Of note, the precise subunit isoforms implicated remain to be confirmed. Downstream of LRRK2 phosphorylation, cellular effects may depend on the formation of LRRK2 positive skein-like structures or on effects of dephosphorylation on LRRK2 ubiquitination and stability

Functional Consequences of LRRK2 Dephosphorylation

Although still incompletely understood, it is now clear that changes in LRRK2 phosphorylation lead to several dynamic changes in the LRRK2 complex and in cellular phenotypes. On the molecular level, dephosphorylation at the ANK-LRR interdomain region (S910/S935) leads to the loss of 14-3-3 binding and recruitment of PP1 [12, 13, 15]. In addition, dephosphorylated LRRK2 displays enhanced self-association that is observed after LRRK2-IN1 treatment and to a greater extent after arsenite treatment [36]. Interestingly, phosphorylation may contribute to other biochemical changes in LRRK2, such as GTP binding. Indeed, both arsenite treatment and CK1 alpha inhibitor treatment led to a reduction in LRRK2 GTP binding, while LRRK2-IN1 treatment led to an increase in GTP binding [35, 36]. As all three of these treatments cause reductions in phosphorylation of the ANK-LRR interdomain sites, other factors are likely involved. For instance, these phenotypes are potentially additionally related to the interaction of LRRK2 with the guanine nucleotide exchange factor ARHGEF7, the degree of LRRK2 self-association, and/or with LRRK2 kinase activity. Indeed, dephosphorylation of LRRK2 with CK1 alpha inhibitor but not LRRK2-IN1 increases interaction with ARHGEF7, while the induced self-association of LRRK2 is significantly higher with arsenite treatment than with LRRK2-IN1 treatment, in line with the different effects on GTP binding. Thus far, the effect of arsenite on the LRRK2-ARHGEF7 binding has not been reported.

As far as cellular phenotypes go, a striking observation related to LRRK2 dephosphorylation is the relocalization of LRRK2 to discrete cytoplasmic pools upon dephosphorylation. Several different types of cytoplasmic accumulations of LRRK2 have been observed, including shuttling to filamentous skein-like structures [13, 19, 28] or the presence of amorphous and punctate accumulations [35]. With characterization still ongoing, at least some of the LRRK2 punctae are explained by the accumulation of LRRK2 in centrosomes [36], while skein-like filaments are at least partially located to microtubules [40].

Further phenotypic effects of dephosphorylation include the observation that LRRK2 dephosphorylation leads to ubiquitination, suggesting that dephosphorylation controls LRRK2 cellular expression levels [41]. Further research is required to fully define the phenotypic consequences of LRRK2 phosphorylation changes at the S935 cluster.

For the effects of dephosphorylation at the S1292 autophosphorylation site, fewer studies are available. In contrast to the ANK-LRR interdomain sites, there is no evidence suggesting that S1292 may govern LRRK2 subcellular organization. Experiments with the phospho-dead mutant S1292A show that negative effects of LRRK2 disease mutants on neurite outgrowth are dependent on the S1292 site, suggesting a role for this site in neuromorphology [17]. LRRK2 S1292A can also reverse the detrimental phenotype of enlarged lysosomes and reduce lysosomal capacity in cells [42]. However, it remains to be determined whether S1292 phosphorylation regulates neurotoxic effects.

Phosphatases of LRRK2, Potential Therapeutic Targets, or Disease Biomarkers in Their Own Right

As evidence accumulates that LRRK2 phosphorylation is important for LRRK2 biology and for LRRK2 pathogenic effects, the exploitation of LRRK2 phosphorylation as a therapeutic target or as disease or pharmacodynamic biomarker becomes warranted.

The identification of the LRRK2 phosphatase holoenzymes offers new targets to modulate LRRK2 function via the modulation of its phosphorylation. The targeting of phosphatases with modulators of phosphatase holoenzyme complexes is a novel theoretical approach (discussed in [10, 43]), the feasibility of which has begun to be illustrated, such as by the small molecule guanabenz or sephin1 that is a modulator of a PP1 holoenzyme through regulatory subunit 15A (PPP1R15A) [44, 45]. The feasibility of this approach for LRRK2 has been demonstrated using a short modulatory sequence directed at the PP1 phosphatase holoenzyme, i.e., the fragment of nuclear inhibitor of PP1 (NIPP1, fragment $NIPP1_{143-224}$) that potently inhibits PP1, to modulate LRRK2 phosphorylation [12]. Therefore, it can be hypothesized that the protein-protein interactions within PP1 phosphatase holoenzymes of LRRK2 and between LRRK2 and phosphatase regulatory subunits can be targeted to control LRRK2 phosphorylation and subsequent phosphorylation-dependent cellular phenotypes such as subcellular localization and ubiquitination. Based on the average differences in phosphorylation levels of LRRK2 disease mutants compared to wild type, targeting strategies may be directed at increasing phosphorylation at ANK-LRR interdomain sites and decreasing phosphorylation at the S1292 site. Further work is now warranted to better define the phosphatase holoenzymes active at these sites and to test LRRK2/holoenzyme complex targeting strategies.

There is also evidence pointing to the interest of measuring LRRK2 or LRRK2 phosphorylation as a disease marker or a marker of target engagement. LRRK2 is expressed in the central nervous system, both in the brain [46, 47] and in exosome isolates from cerebrospinal fluid (CSF) [48], indicating that detection of LRRK2 in CSF is feasible. Reports show increases in LRRK2 protein levels in the prefrontal cortex of PD patients relative to controls [49], suggesting that the increase in total LRRK2 protein expression is correlated with disease. LRRK2 phosphorylation is modulated in a majority of disease variants. For the S910-S935-S955-S973 phosphosites, levels are reduced for most mutants [13], while for phospho-S1292, levels are increased for most mutants (see Table 8.1). LRRK2 phosphorylatuon changes have begun to be studied in patients, with LRRK2-S935 phosphorylation reported as unaltered in patient peripheral blood mononuclear cells (PBMCs) in sporadic PD patients [34], and LRRK2-S1292 phosphorylation levels are increased in urinary exosomes of LRRK2 G2019S mutant carriers [50]. Also, all of these five sites are rapidly dephosphorylated and deubiquitinated upon LRRK2 inhibitor treatment [1, 41], considered potential therapeutics. Therefore, both LRRK2 levels and LRRK2 phosphorylation levels are promising markers for both disease and pharmacodynamic response. In line with this, proteins involved in LRRK2

Table 8.1 Overview of LRRK2 phosphorylation sites and phosphoregulation

	Effect on in vitro kinase activity relative to WT	Basal phosphorylation level of S910-S935 cluster in cells relative to WT	Basal phosphorylation level of autophosphorylation (S1292) in cells relative to WT	Effect of acute cellular treatment with LRRK2 kinase inhibitor on of S910-S935 cluster	Effect of acute cellular treatment with LRRK2 kinase inhibitor on autophosphorylation cluster (S1292)
LRRK2 WT	WT activity	WT level	WT level	Dephosphorylation	Dephosphorylation
Disease mutants					
N1437H	Variable: no to moderate enhancement	Low	Enhanced	Dephosphorylation	To be assessed
R1441C	Variable: no to moderate enhancement	Moderately lower	Enhanced	Moderate dephosphorylation	To be assessed
R1441G	Variable: no to moderate enhancement	Low	Enhanced	Dephosphorylation	Dephosphorylation
Y1699C	Variable: no to moderate enhancement	Low	WT level or slightly enhanced	Dephosphorylation	To be assessed
G2019S	Enhanced	WT level	Enhanced	Dephosphorylation	Dephosphorylation
I2020T	Variable: no to moderate enhancement	Low	Enhanced	Dephosphorylation	To be assessed
G2385R	Reduced	Moderately lower	WT level	Moderate dephosphorylation	To be assessed
Functional mutants					
GTP-binding deficient					
K1347A	Reduced	Low	Not done	Not applicable	To be assessed
T1348N	Reduced	Low	Not done	Not applicable	To be assessed

(continued)

Table 8.1 (continued)

	Effect on in vitro kinase activity relative to WT	Basal phosphorylation level of S910-S935 cluster in cells relative to WT	Basal phosphorylation level of autophosphorylation (S1292) in cells relative to WT	Effect of acute cellular treatment with LRRK2 kinase inhibitor on of S910-S935 cluster	Effect of acute cellular treatment with LRRK2 kinase inhibitor on autophosphorylation cluster (S1292)
Kinase dead					
K1906M	Reduced	WT levels	Not done	No change	To be assessed
D1994A	Reduced	Low	Reduced	No change	To be assessed
D2017A	Reduced	Moderately lower	To be assessed	No change	To be assessed
S2032A	Moderately reduced	WT levels	To be assessed	Dephosphorylated	To be assessed
T2035A	Reduced	WT levels	To be assessed	Dephosphorylated	To be assessed
Kinase activating					
T2031S	Enhanced	WT levels	To be assessed	Dephosphorylated	To be assessed
Kinase inhibitor insensitive					
A2016T	WT activity	Moderately reduced	To be assessed	No change	No change

Box 1: Phosphatases of LRRK2, Outstanding Questions
Phosphatases of LRRK2, outstanding issues

- Besides S1292, which autophosphorylation sites of LRRK2 are present under physiological conditions?
- What are the precise holoenzyme compositions of LRRK2 phosphatases?
- What are the structural determinants of LRRK2/phosphatase complexes?
- Are there other phosphatases besides PP1 that are involved in LRRK2 phosphoregulation?
- Is expression or activity of LRRK2 phosphatases altered in Parkinson's disease?

phosphoregulation are also potential biomarkers. To explore this, future research to analyze of the expression and activity levels of LRRK2 phosphatases would be justified.

Perspectives and Future Issues

Phosphoregulation of LRRK2 by phosphatases is emerging as an important phenomenon for LRRK2 biology offering new perspectives in developing LRRK2-targeting therapeutics or LRRK2-based biomarkers. While the state of advancement of the field has begun to show which phosphatases are involved and which phenotypes are associated with phosphorylation levels at specific phosphosites or clusters of phosphosites, some issues remain to be addressed (summarized in Box 1). For instance, few autophosphorylation sites have been observed to occur in cells, and specific phosphatases at these sites remain to be confirmed. Also, precise PP1 holoenzyme compositions active in dephosphorylating LRRK2 must be identified, in particular regulatory subunits. In order to develop small molecules capable of disrupting complexes of LRRK2 and LRRK2 phosphatase holoenzymes, structural details of these complexes should be examined. Finally, the role of additional phosphatases active on LRRK2 should be examined, and levels of expression or activation of LRRK2 phosphatases should be compared in healthy versus disease groups. Future studies addressing these issues will enable the development and assessment of LRRK2 phosphatase-based disease, pharmacodynamic biomarkers, and therapeutic strategies .

Conflict of Interest The author declares no conflicts of interest.

References

1. Taymans J (2014) Can the increasing number of newly developed leucine-rich repeat kinase 2 inhibitors validate or invalidate a potential disease-modifying therapeutic approach for Parkinson's disease? Expert Opin Ther Pat 24:727–730. doi:10.1517/13543776.2014.915945
2. Galatsis P, Henderson JL, Kormos BL, Hirst WD (2014) Leucine-rich repeat kinase 2 (LRRK2) inhibitors. Top Med Chem 18:111–148. doi:10.1007/7355_2014_69
3. Taymans JM, Greggio E (2016) LRRK2 kinase inhibition as a therapeutic strategy for Parkinson's disease, where do we stand? Curr Neuropharmacol 14(3):214–225
4. Tsika E, Moore DJ (2013) Contribution of GTPase activity to LRRK2-associated Parkinson disease. Small GTPases 4:164–170. doi:10.4161/sgtp.25130
5. Taymans J-M (2012) The GTPase function of LRRK2. Biochem Soc Trans 40:1063–1069. doi:10.1042/BST20120133
6. Hong L, Sklar LA (2014) Targeting GTPases in Parkinson's disease: comparison to the historic path of kinase drug discovery and perspectives. Front Mol Neurosci 7:1–10. doi:10.3389/fnmol.2014.00052
7. Greggio E, Zambrano I, Kaganovich A et al (2008) The Parkinson disease-associated leucine-rich repeat kinase 2 (LRRK2) is a dimer that undergoes intramolecular autophosphorylation. J Biol Chem 283:16906–16914. doi:10.1074/jbc.M708718200
8. Civiero L, Vancraenenbroeck R, Belluzzi E et al (2012) Biochemical characterization of highly purified leucine-rich repeat kinases 1 and 2 demonstrates formation of homodimers. PLoS One 7, e43472. doi:10.1371/journal.pone.0043472
9. Zhao J, Hermanson SB, Carlson CB et al (2012) Pharmacological inhibition of LRRK2 cellular phosphorylation sites provides insight into LRRK2 biology. Biochem Soc Trans 40:1158–1162. doi:10.1042/BST20120137
10. Taymans J-M, Baekelandt V (2014) Phosphatases of α-synuclein, LRRK2, and tau: important players in the phosphorylation-dependent pathology of Parkinsonism. Front Genet 5:382. doi:10.3389/fgene.2014.00382
11. Lobbestael E, Baekelandt V, Taymans J-M (2012) Phosphorylation of LRRK2: from kinase to substrate. Biochem Soc Trans 40:1102–1110. doi:10.1042/BST20120128
12. Lobbestael E, Zhao J, Rudenko IN et al (2013) Identification of protein phosphatase 1 as a regulator of the LRRK2 phosphorylation cycle. Biochem J 456:119–128. doi:10.1042/BJ20121772
13. Nichols RJ, Dzamko N, Morrice NA et al (2010) 14-3-3 binding to LRRK2 is disrupted by multiple Parkinson's disease-associated mutations and regulates cytoplasmic localization. Biochem J 430:393–404. doi:10.1042/BJ20100483
14. Reynolds A, Doggett EA, Riddle SM et al (2014) LRRK2 kinase activity and biology are not uniformly predicted by its autophosphorylation and cellular phosphorylation site status. Front Mol Neurosci 7:54. doi:10.3389/fnmol.2014.00054
15. Li X, Wang QJ, Pan N et al (2011) Phosphorylation-dependent 14-3-3 binding to LRRK2 is impaired by common mutations of familial Parkinson's disease. PLoS One 6, e17153. doi:10.1371/journal.pone.0017153
16. Doggett EA, Zhao J, Mork CN et al (2012) Phosphorylation of LRRK2 serines 955 and 973 is disrupted by Parkinson's disease mutations and LRRK2 pharmacological inhibition. J Neurochem 120:37–45. doi:10.1111/j.1471-4159.2011.07537.x
17. Sheng Z, Zhang S, Bustos D et al (2012) Ser1292 autophosphorylation is an indicator of LRRK2 kinase activity and contributes to the cellular effects of PD mutations. Sci Transl Med 4, 164ra161. doi:10.1126/scitranslmed.3004485
18. West AB, Moore DJ, Choi C et al (2007) Parkinson's disease-associated mutations in LRRK2 link enhanced GTP-binding and kinase activities to neuronal toxicity. Hum Mol Genet 16:223–232. doi:10.1093/hmg/ddl471

19. Reyniers L, Del Giudice MG, Civiero L et al (2014) Differential protein-protein interactions of LRRK1 and LRRK2 indicate roles in distinct cellular signaling pathways. J Neurochem 131:239–250. doi:10.1111/jnc.12798
20. Gloeckner CJ, Boldt K, von Zweydorf F et al (2010) Phosphopeptide analysis reveals two discrete clusters of phosphorylation in the N-terminus and the Roc domain of the Parkinson-disease associated protein kinase LRRK2. J Proteome Res 9:1738–1745. doi:10.1021/pr9008578
21. Greggio E, Taymans J, Zhen EY et al (2009) The Parkinson's disease kinase LRRK2 autophosphorylates its GTPase domain at multiple sites. Biochem Biophys Res Commun 389:449–454. doi:10.1016/j.bbrc.2009.08.163
22. Pungaliya PP, Bai Y, Lipinski K et al (2010) Identification and characterization of a leucine-rich repeat kinase 2 (LRRK2) consensus phosphorylation motif. PLoS One 5, e13672. doi:10.1371/journal.pone.0013672
23. Kamikawaji S, Ito G, Iwatsubo T, Accepted J (2009) Identification of the autophosphorylation sites of LRRK2. Biochemistry 48:10963–10975. doi:10.1021/bi9011379
24. Greggio E, Lewis PA, van der Brug MP et al (2007) Mutations in LRRK2/dardarin associated with Parkinson disease are more toxic than equivalent mutations in the homologous kinase LRRK1. J Neurochem 102:93–102. doi:10.1111/j.1471-4159.2007.04523.x
25. Taymans J-M, Vancraenenbroeck R, Ollikainen P et al (2011) LRRK2 kinase activity is dependent on LRRK2 GTP binding capacity but independent of LRRK2 GTP binding. PLoS One 6, e23207. doi:10.1371/journal.pone.0023207
26. Ito G, Okai T, Fujino G et al (2007) GTP binding is essential to the protein kinase activity of LRRK2, a causative gene product for familial Parkinson's disease. Biochemistry 46:1380–1388. doi:10.1021/bi061960m
27. Ito G, Fujimoto T, Kamikawaji S et al (2014) Lack of correlation between the kinase activity of LRRK2 harboring kinase-modifying mutations and its phosphorylation at Ser910, 935, and Ser955. PLoS One 9, e97988. doi:10.1371/journal.pone.0097988
28. Dzamko N, Deak M, Hentati F et al (2010) Inhibition of LRRK2 kinase activity leads to dephosphorylation of Ser(910)/Ser(935), disruption of 14-3-3 binding and altered cytoplasmic localization. Biochem J 430:405–413. doi:10.1042/BJ20100784
29. Choi HG, Zhang J, Deng X et al (2012) Brain penetrant LRRK2 inhibitor. ACS Med Chem Lett 3:658–662. doi:10.1021/ml300123a
30. Reith AD, Bamborough P, Jandu K et al (2012) GSK2578215A; a potent and highly selective 2-arylmethyloxy-5-substitutent-N-arylbenzamide LRRK2 kinase inhibitor. Bioorg Med Chem Lett 22:5625–5629. doi:10.1016/j.bmcl.2012.06.104
31. Deng X, Dzamko N, Prescott A et al (2011) Characterization of a selective inhibitor of the Parkinson's disease kinase LRRK2. Nat Chem Biol 7:203–205. doi:10.1038/nchembio.538
32. Zhang J, Deng X, Choi HG et al (2012) Characterization of TAE684 as a potent LRRK2 kinase inhibitor. Bioorg Med Chem Lett 22:1864–1869. doi:10.1016/j.bmcl.2012.01.084
33. Vancraenenbroeck R, De Raeymaecker J, Lobbestael E et al (2014) In silico, in vitro and cellular analysis with a kinome-wide inhibitor panel correlates cellular LRRK2 dephosphorylation to inhibitor activity on LRRK2. Front Mol Neurosci 7:51. doi:10.3389/fnmol.2014.00051
34. Dzamko N, Chua G, Ranola M et al (2013) Measurement of LRRK2 and Ser910/935 phosphorylated LRRK2 in peripheral blood mononuclear cells from idiopathic Parkinson's disease patients. J Parkinsons Dis 3:145–152. doi:10.3233/JPD-130174
35. Chia R, Haddock S, Beilina A et al (2014) Phosphorylation of LRRK2 by casein kinase 1α regulates trans-Golgi clustering via differential interaction with ARHGEF7. Nat Commun 5:5827. doi:10.1038/ncomms6827
36. Mamais A, Chia R, Beilina A et al (2014) Arsenite stress down-regulates phosphorylation and 14-3-3 binding of leucine-rich repeat kinase 2 (LRRK2), promoting self-association and cellular redistribution. J Biol Chem 289:21386–21400. doi:10.1074/jbc.M113.528463

37. Moorhead GBG, Trinkle-Mulcahy L, Ulke-Lemee A (2007) Emerging roles of nuclear protein phosphatases. Nat Rev Mol Cell Biol 8:234–244. doi:10.1038/nrm2126
38. Virshup DM, Shenolikar S (2009) From promiscuity to precision: protein phosphatases get a makeover. Mol Cell 33:537–545. doi:10.1016/j.molcel.2009.02.015
39. Manning G, Whyte DB, Martinez R et al (2002) The protein kinase complement of the human genome. Science 298:1912–1934. doi:10.1126/science.1075762
40. Kett LR, Boassa D, Ho CC-Y et al (2012) LRRK2 Parkinson disease mutations enhance its microtubule association. Hum Mol Genet 21:890–899. doi:10.1093/hmg/ddr526
41. Zhao J, Molitor TP, Langston JW, Nichols RJ (2015) LRRK2 dephosphorylation increases its ubiquitination. Biochem J 496:107–120. doi:10.1042/BJ20141305
42. Henry AG, Aghamohammadzadeh S, Samaroo H et al (2015) Pathogenic LRRK2 mutations, through increased kinase activity, produce enlarged lysosomes with reduced degradative capacity and increase ATP13A2 expression. Hum Mol Genet 24(21):6013–6028. doi:10.1093/hmg/ddv314
43. De Munter S, Köhn M, Bollen M (2013) Challenges and opportunities in the development of protein phosphatase-directed therapeutics. ACS Chem Biol 8:36–45. doi:10.1021/cb300597g
44. Tsaytler P, Harding HP, Ron D, Bertolotti A (2011) Selective inhibition of a regulatory subunit of protein phosphatase 1 restores proteostasis. Science 332:91–94. doi:10.1126/science.1201396
45. Das I, Krzyzosiak A, Schneider K et al (2015) Preventing proteostasis diseases by selective inhibition of a phosphatase regulatory subunit. Science 348:239–242. doi:10.1126/science.aaa4484
46. Galter D, Westerlund M, Carmine A et al (2006) LRRK2 expression linked to dopamine-innervated areas. Ann Neurol 59:714–719. doi:10.1002/ana.20808
47. Taymans J, Van den Haute C, Baekelandt V (2006) Distribution of PINK1 and LRRK2 in rat and mouse brain. J Neurochem 98:951–961. doi:10.1111/j.1471-4159.2006.03919.x
48. Fraser KB, Moehle MS, Daher JPL et al (2013) LRRK2 secretion in exosomes is regulated by 14-3-3. Hum Mol Genet 22(24):4988–5000. doi:10.1093/hmg/ddt346
49. Cho HJ, Liu G, Jin SM et al (2013) MicroRNA-205 regulates the expression of Parkinson's disease-related leucine-rich repeat kinase 2 protein. Hum Mol Genet 22:608–620. doi:10.1093/hmg/dds470
50. Fraser KB, Moehle MS, Alcalay RN et al (2016) Urinary LRRK2 phosphorylation predicts parkinsonian phenotypes in G2019S LRRK2 carriers. Neurology 86(11):994–999. doi:10.1212/WNL.0000000000002436

Part III
LRRK2 Neurodegeneration, Modeling, and Therapeutic Options

Chapter 9
Models of LRRK2-Associated Parkinson's Disease

Yulan Xiong, Ted M. Dawson, and Valina L. Dawson

Funding for a portion of Dr. Dawson's LRRK2 research in the past was provided by Merck KGAA. Under a licensing agreement between Merck KGAA and the Johns Hopkins University, Dr. Dawson and the University shared fees received by the University on licensing some of the reagents used in his research. Dr. Dawson also was a paid consultant to Merck KGAA.

Y. Xiong (✉)
Neuroregeneration and Stem Cell Programs, Institute for Cell Engineering, Johns Hopkins University School of Medicine, Baltimore, MD 21205, USA

Department of Neurology, Johns Hopkins University School of Medicine, Baltimore, MD 21205, USA

Department of Anatomy and Physiology, Kansas State University College of Veterinary Medicine, Manhattan, KS 66506, USA
e-mail: yulanxiong@ksu.edu

T.M. Dawson (✉)
Department of Pharmacology and Molecular Sciences, Johns Hopkins University School of Medicine, Baltimore, MD 21205, USA

Neuroregeneration and Stem Cell Programs, Institute for Cell Engineering, Johns Hopkins University School of Medicine, Baltimore, MD 21205, USA

Department of Neurology, Johns Hopkins University School of Medicine, Baltimore, MD 21205, USA

Solomon H. Snyder Department of Neuroscience, Johns Hopkins University School of Medicine, Baltimore, MD 21205, USA

Adrienne Helis Malvin Medical Research Foundation, New Orleans, LA 70130-2685, USA
e-mail: vdawson@jhmi.edu

V.L. Dawson (✉)
Solomon H. Snyder Department of Neuroscience, Johns Hopkins University School of Medicine, Baltimore, MD 21205, USA

Adrienne Helis Malvin Medical Research Foundation, New Orleans, LA 70130-2685, USA

Department of Physiology, Johns Hopkins University School of Medicine, Baltimore, MD 21205, USA

Neuroregeneration and Stem Cell Programs, Institute for Cell Engineering, Johns Hopkins University School of Medicine, Baltimore, MD 21205, USA

Department of Neurology, Johns Hopkins University School of Medicine, Baltimore, MD 21205, USA
e-mail: tdawson@jhmi.edu

H.J. Rideout (ed.), *Leucine-Rich Repeat Kinase 2 (LRRK2)*,
Advances in Neurobiology 14, DOI 10.1007/978-3-319-49969-7_9

Abstract Mutations in the leucine-rich repeat kinase 2 (LRRK2) gene are the most common genetic causes of Parkinson's disease (PD) and also one of the strongest genetic risk factors in sporadic PD. The LRRK2 protein contains a GTPase and a kinase domain and several protein-protein interaction domains. Both in vitro and in vivo assays in different model systems have provided tremendous insights into the molecular mechanisms underlying LRRK2-induced dopaminergic neurodegeneration. Among all the model systems, animal models are crucial tools to study the pathogenesis of human disease. How do the animal models recapitulate LRRK2-induced dopaminergic neuronal loss in human PD? To answer this question, this review focuses on the discussion of the animal models of LRRK2-associated PD including genetic- and viral-based models.

Keywords LRRK2 • Animal models • Parkinson's disease

Introduction

Parkinson's disease (PD) is recognized as the most common movement disorder, affecting up to 1% of the population above the age of 60 and 4–5% above the age of 85 [1]. Clinical symptoms in PD patients include akinesia, resting tremor, muscle rigidity, and postural imbalance [1]. The cardinal symptoms are caused by the progressive degeneration of dopaminergic (DA) neurons in the substantia nigra pars compacta (SNpc) [2]. Although the majority of PD cases appear to be sporadic, in the past couple of decades, several genes have been identified to be responsible for this progressive neurodegenerative disease [3]. To date, genes encoding LRRK2 (leucine-rich repeat kinase 2), α-synuclein, parkin, DJ-1, PINK1 (phosphatase and tensin homolog deleted on chromosome 10-induced putative kinase 1), VPS35 (vacuolar protein sorting 35), DNAJC13, GBA (glucocerebrosidase), and EIF4G1 (eukaryotic initiation factor 4G1), among others, are associated with genetic forms of PD [3–8]. Mutations in the *LRRK2* gene (PARK8, dardarin, OMIM 609007) cause late-onset, autosomal dominant PD and is the most frequent genetic cause of PD, accounting for 4% of familial PD and 1% of sporadic PD across all populations. Importantly, LRRK2-mediated PD is clinically and pathologically indistinguishable from sporadic PD [9, 10], suggesting that understanding LRRK2-associated PD may lead to an understanding of sporadic PD.

The LRRK2 protein contains two enzymatic domains, a GTPase and a kinase domain and multiple protein-protein interaction domains including a leucine-rich repeat (LRR), a WD40 repeat, and a LRRK2-specific repeat domain (Fig. 9.1) [11, 12]. LRRK2 interaction domains are thought to serve as protein-binding modules where LRRK2 acts as a signaling scaffold. LRRK2 GTPase and kinase enzyme activity are important in regulating LRRK2-dependent cellular signaling pathways and may reciprocally regulate each other to direct LRRK2's ultimate function [13]. Pathogenic mutations of LRRK2 are centered on LRRK2 enzymatic domains

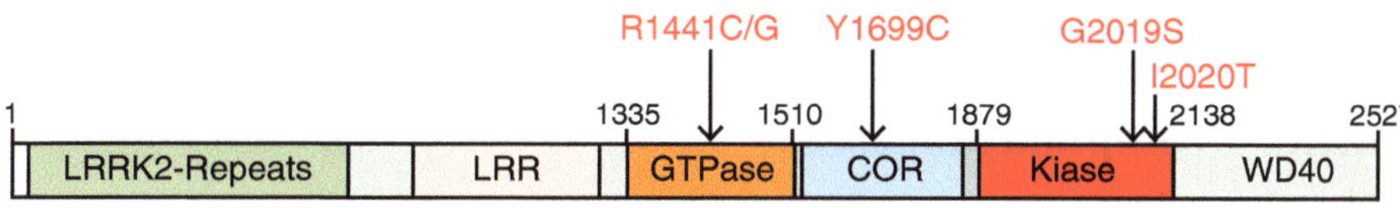

Fig. 9.1 Schematic showing the domain structure of LRRK2 protein and the position of pathogenic mutations. Residues 1–660 encode LRRK2-specific repeat sequences, 984–1278 encode the leucine-rich repeat (LRR), 1335–1510 encode the Roc GTPase domain, 1519–1795 encode the C-terminal of Ras (COR) domain, and 1879–2138 encode the kinase domain. Five confirmed LRRK2 pathogenic mutations: R1441C/G, Y1699C, G2019S, I2020T

(Fig. 9.1). Thus, LRRK2 enzymatic activity is important in PD. The most prevalent LRRK2 mutation, G2019S, is within the kinase domain. It accounts for 5–6% of autosomal dominant PD patients and ~1% of sporadic late-onset PD. Patients with the G2019S mutation exhibit Lewy bodies in most cases [1]. However, mutations in the GTPase domain and COR domain, such as R1441 C/R1441 G and Y1669C often vary on Lewy body pathology [10, 14]. This raises the possibility that these mutations cause disease via distinct pathogenic mechanisms.

Tremendous work in both in vitro and in vivo systems suggests that LRRK2 is involved in diverse pathways and cellular signaling including regulation of protein translation, vesicle trafficking, neurite outgrowth, autophagy, and cytoskeletal dynamics [15–17]. Several model systems have been developed to study LRRK2 function from yeast to invertebrates such as *Drosophila* and *C. elegans*, rodents, and patient-derived induced pluripotent stem cells (iPScs) [18]. Yeast, a eukaryotic single-cell organism, has been widely used to uncover the fundamental pathobiology of proteins associated with neurodegenerative diseases including PD. The first LRRK2 yeast model that revealed LRRK2 GTPase function plays a key role in LRRK2 pathobiology [19, 20]. The toxicity is closely associated with GTPase activity and defects in endocytic vesicular trafficking and autophagy [19]. More importantly, using this yeast model, the first GTPase-activating protein (GAP) for LRRK2, ArfGAP1, was identified and characterized [20]. Patient-derived iPScs provide highly relevant models for PD studies as the well-developed capacity to generate iPSc-derived DA neurons. Several LRRK2 iPSc models have been developed. DA neurons derived from LRRK2 iPScs display reduced neurite length, accumulation of α-synuclein and tau, increased vulnerability to cellular stress, and impaired autophagy and mitochondrial function [18]. IPSc models allow us to study LRRK2 pathobiology directly in human context. However, both yeast and iPSc models cannot recapitulate the physiological cell diversity in the intact mammalian brain and the complexity of brain circuits.

While the eukaryotic yeast and iPS cell models provide an important yet complementary insights to animal models on understanding disease mechanisms, this review focuses on the animal models of LRRK2-associated PD and discusses the advantages and disadvantages of each model and how each of these different models have contributed to understanding the role of LRRK2 in PD pathogenesis

Genetic Animal Models of LRRK2

LRRK2 Drosophila *Models*

Animal models are crucial tools for LRRK2 research. Invertebrate animals, especially *Drosophila*, have proven to play an important role in studying LRRK2 pathogenic mechanisms and developing therapeutics. *Drosophila* has well-defined nerve systems, which share similar neuronal transmitters with mammals. Importantly, *Drosophila* has DA neuronal clusters and can perform complicated behavioral tests, which mimic some DA-dependent behaviors in human. Several steps have been taken to establish and utilize LRRK2 *Drosophila* models:

1. Generation of *Drosophila* strains carrying mutations in genes linked to disease
2. Determination of the *Drosophila* models to see if they recapitulate the pathogenesis of the disease and in turn are good models of the disease
3. Investigation of the detailed molecular mechanisms underlying the phenotypes
4. Identification of genetic modifiers to dissect the signaling pathways involved in pathogenesis
5. Drug candidate screening

Several LRRK2 *Drosophila* models have been generated and are listed in Table 9.1.

LRRK2 Knockout *Drosophila* Models

Drosophila has one human LRRK2 homolog dLRRK, and residues affected by PD-causing mutations in LRRK2 are conserved in *Drosophila* LRRK2. To study the function of endogenous wild-type (WT) LRRK2, *Drosophila* LRRK2 knockouts have been generated [21–24]. Several groups reported that the homozygous mutant fly develops normally with a normal life span as well as unchanged number and pattern of DA neurons [21, 23, 24], although one group reported LRRK loss-of-function mutants exhibited a severely impaired locomotive activity and a severe reduction in tyrosine hydroxylase immunostaining and shrunken morphology of DA neurons in LRRK mutants [22]. In addition, Wang et al. showed that mutant flies containing C-terminal kinase domain truncated dLRRK are selectively sensitive to H_2O_2, but not to paraquat, rotenone, or β-mercaptoethanol [23]. By contrast, Imai et al. showed that *dLRRK* null flies are relatively resistant to general oxidative stress, such as paraquat and H_2O_2 treatment, compared to WT flies [21]. Thus, the exact role of dLRRK in oxidative stress remains unclear. The different phenotypes are possibly due to the different genomic loci of insertion for gene disruption and the different genetic backgrounds. As the majority of the reports support that *dLRRK* is dispensable for survival of DA neurons in flies and this is consistent with the phenotypes in LRRK2 knockout mice, the general consensus is that LRRK2 toxicity is from a gain-of-function and not a loss-of-function mechanism.

Table 9.1 *Drosophila* models for LRRK2-associated PD

LRRK2 variants	Genetic manipulations	Motor deficits/ life span	DA neuronal loss	Nigrostriatal dysfunction	Sensitivity to oxidative stress	Other phenotypes	References
dLRRK	Loss of function	Locomotor activity ↓	No changes	TH staining ↓ DA neurons shrunken	ND	ND	[22]
dLRRK	Loss of function	Life span ↓	No changes	No changes	Sensitive to hydrogen peroxide, not to paraquat, rotenone, and β-mercaptoethanol	ND	[23]
dLRRK	Loss of function	Life span ↓ Fertility ↓ Malformed abdomen	No changes	DA content ↑	Hydrogen peroxide ↓ Paraquat ↓	ND	[21]
dLRRK	Loss of function	Locomotor activity ↓	ND	ND	ND	ND	[24]
dLRRK	O/E	No changes	No changes	No changes	ND	ND	[22]
hLRRK2	O/E	Locomotor activity ↓ Life span ↓ Response to L-Dopa	TH neurons ↓ No response to L-Dopa	ND	ND	Retinal degeneration	[25]
hG2019S		Locomotor activity ↓ Life span ↓ Response to L-Dopa	TH neurons ↓↓ No response to L-Dopa	ND	ND	Retinal degeneration	

(continued)

Table 9.1 (continued)

LRRK2 variants	Genetic manipulations	Motor deficits/ life span	DA neuronal loss	Nigrostriatal dysfunction	Sensitivity to oxidative stress	Other phenotypes	References
dLRRK	O/E	ND	No changes	No changes	No changes	ND	[21]
dY1383C dI1915T		ND	No changes	TH staining ↓ DA content ↓	Hydrogen peroxide ↑ Paraquat ↑	ND	
hLRRK2	O/E at 29°C	Locomotor activity: 10 days ↓ 20 days ↑ Life span ↑ Fertility ↑	TH neurons ↓	ND	Rotenone ↑	Retinal degeneration	[27]
hI1122V hY1699C hI2020T		Locomotor activity: 10 days ↓ 20 days ↑ Life span ↑ in hY1699C, hI2020T Fertility ↑ in hI1122V, hI2020T	TH neurons ↓ the most with I 2020 T	ND	Rotenone ↑	Retinal degeneration	
hLRRK2	O/E	No changes	No changes	ND	No changes	No changes	[26]
hG2019S hY1699C hG2385R		Locomotor activity ↓ Life span ↓	TH neurons ↓	ND	hG2019S, ↑ hG2385R ↑ hY1699C no change	No changes	

LRRK2 variants	Genetic manipulations	Motor deficits/ life span	DA neuronal loss	Nigrostriatal dysfunction	Sensitivity to oxidative stress	Other phenotypes	References
hLRRK2	O/E	Locomotor activity: no changes Life span ↓	ND	Dendritic ends ↓	ND	ND	[28]
hG2019S		Locomotor activity ↓↓ Life span ↓↓	TH neurons ↓	Dendritic ends ↓↓ Axon degeneration ↑	ND	ND	
hR1441C hG2385R		Locomotor activity ↓ Life span ↓	ND	Dendritic ends ↓	ND	ND	
hLRRK2	O/E	ND	ND	ND	ND	Visual function: no changes	[29]
hG2019S		ND	ND	ND	ND	Visual function ↓	
hI1122V hR1441C hY1383C hI1915T hI2020T hG2385R hG2019S/K1906M		ND	ND	ND	ND	Visual function: no changes	

(continued)

Table 9.1 (continued)

LRRK2 variants	Genetic manipulations	Motor deficits/ life span	DA neuronal loss	Nigrostriatal dysfunction	Sensitivity to oxidative stress	Other phenotypes	References
hLRRK2 hG2019S	O/E	Locomotor activity: no changes	ND	Axon transport: no changes	ND	ND	[30]
hR1441C hY1699C		Locomotor activity ↓	ND	Axon transport ↓	ND	ND	
dR1069C dY1383C		Locomotor activity ↓	ND	Axon transport ↓	ND	ND	

ND not determined, *O/E* overexpression, ↑ increased, ↓ decreased

LRRK2 Transgenic *Drosophila* Models

In contrast to *dLRRK* knockout *Drosophila*, overexpression of both human LRRK2 and dLRRK pathogenic mutations in *Drosophila* leads to age-dependent DA-responsive reductions in locomotor activity and loss of DA neurons (Table 9.1) [21, 25–28]. Interestingly, in addition to the DA neurodegeneration, different LRRK2 mutations cause different phenotypes related to the degeneration. One recent study showed that LRRK2 G2019S induced extensive neurodegeneration throughout the visual system [29]. This degeneration is LRRK2 G2019S mutation specific and occurs in a kinase-dependent manner. Dopaminergic expression of LRRK2 G2019S led to nonautonomous cell death reminiscent of that seen in PD [29]. Another report showed that LRRK2 R1441C or Y1699C mutations in the GTPase-COR domain preferentially associates with deacetylated microtubules and inhibits axonal transport in *Drosophila*, causing locomotor deficits in vivo. These features are not seen with the LRRK2 G2019S mutation, suggesting that these defects are GTPase activity dependent [30]. A previous study suggested that reduced axonal transport rates caused by α-synuclein mutants might contribute to accumulation of α-synuclein and hence Lewy body formation and neuritic abnormalities in PD brain [31]. Taken together, reduced axonal transport rates may contribute to the formation of Lewy bodies or Lewy neurites in some PD cases carrying R1441C or Y1669C mutations. These studies suggest that different LRRK2 pathogenic mutations act at distinct pathways and cause varied neuropathology in that accompanies DA neurodegeneration.

Using LRRK2 *Drosophila* Models to Study Molecular Mechanisms Underlying LRRK2-Associated PD

Do LRRK2 *Drosophila* models reveal the pathogenic mechanisms underlying LRRK2-induced DA neurodegeneration? To address this, *Drosophila* offers a wide variety of genetic tools including genetic screens, which allow genome-wide analyses of genetic interactions based on the modification of a given phenotype, and candidate gene approaches, in which only those suspected genes are analyzed for modifications of the phenotype. Both strategies allow identification of components of signaling pathways involved in PD pathogenesis. Using LRRK2 *Drosophila* models, several in vivo LRRK2 interactors have been identified and characterized in different signaling pathways.

LRRK2 Function in Protein Synthesis/Translation

Drosophila dLRRK was shown to regulate protein translational pathways. Imai et al. first provided evidence that both dLRRK and human LRRK2 can phosphorylate eukaryotic initiation factor 4E-binding protein (4E-BP), a negative regulator of eukaryotic initiation factor 4E-mediated protein translation and a key mediator of various stress responses [21]. A link between dLRRK and protein synthesis was further strengthened by the observation from the same group that LRRK2 interacts with the microRNA pathway to regulate protein synthesis [32]. However, these *Drosophila* studies have yet to be extended to mammalian systems. A recent study, using a combination of an LRRK2 *Drosophila* model and human dopamine neurons, demonstrated that LRRK2 phosphorylates ribosome protein s15 to enhance protein translation and mediate LRRK2-induced neurodegeneration [33]. Taken together, there is strong convergent evidence that LRRK2 regulates protein translation machinery in diverse species and tissues.

LRRK2 Function in Vesicular Trafficking

Studies using LRRK2 *Drosophila* models have revealed potential roles for LRRK2 in multiple aspects of vesicle trafficking including endolysosomal pathways, synaptic vesicle (SV) endocytosis, ER-Golgi, and retromer trafficking. First, dLRRK was reported to localize to the membranes of late endosomes and lysosomes, physically interacts with the crucial mediator of late endosomal transport Rab7, and negatively regulates Rab7-dependent perinuclear localization of lysosomes [34]. LRRK2 has been further shown to localize at endosomes and interacts with clathrin light chains (CLCs) to limit Rac1 activation. These data identify a new pathway in which CLCs function with LRRK2 to control Rac1 activation on endosomes [35]. The function of LRRK2 in endolysosomal pathways is further strengthened by a study on novel ethyl methanesulfonate (EMS)-induced nonsense alleles in dLRRK, which cause striking defects in the endolysosomal and autophagy pathways [36]. Second, a study in *Drosophila* shows that LRRK2 functions on SV endocytosis at the neuromuscular junctions by phosphorylating endophilin A (EndoA) at S75 and mediating EndoA-dependent membrane tubulation and membrane association [37]. In addition, dLRRK has been demonstrated to regulate Golgi outpost (GOP) dynamics in dendrites through the golgin Lava lamp [38]. Moreover, genetic interactions between VPS35, Rab7L1, ArfGAP1, and LRRK2 in *Drosophila* highlight LRRK2's role in retromer and ER-Golgi trafficking [20, 39, 40]. All data taken together strongly support that LRRK2 plays a crucial role in vesicular trafficking pathway, which may provide potential mechanisms for accumulation of α-synuclein in LRRK2-associated PD.

LRRK2 Function in Dendritic Degeneration and Synaptic Morphology

Expression of LRRK2 G2019S in *Drosophila* dendritic arborization neurons induces mislocalization of the axonal protein tau in dendrites and causes dendrite degeneration. This may act through a mechanism in which LRRK2 G2019S promotes tau phosphorylation by the glycogen synthase kinase 3β (GSK3β) [28]. In addition, LRRK2 regulates synaptic morphology through interacting with 4E-BP at the postsynaptic site and phosphorylating Futsch at the presynaptic compartments of the *Drosophila* neuromuscular junctions [41]. These studies point out a possible role for LRRK2 in dendrite degeneration and synaptic dysfunction.

LRRK2 Genetic Interaction with Other PD Genes

As the number of genetic alterations linked to PD pathogenesis increases, establishing functional pathways and whether these genes or risk factors interact with each other will be crucial. *Drosophila* as a classical genetic model provides powerful tools to study genetic interactions between different genes. Using LRRK2 *Drosophila* models, genetic dissection revealed that LRRK2 interacts with other PD genes or risk factors such as VPS35, RAB7L, parkin, DJ-1, and PINK1 [27, 39, 40] and implicates several potential LRRK2 functions. Genetic interaction between LRRK2, VPS35, and Rab7L implicates LRRK2 function in retromer and lysosomal pathways that contribute to PD [39, 40]. Coexpression of human parkin in LRRK2 G2019S-expressing flies provides significant protection against DA neurodegeneration that occurs with age or in response to rotenone, suggesting a potential link between LRRK2, parkin, and mitochondria in the pathogenesis of LRRK2-related parkinsonism [27]. Genetic interaction between LRRK2 and parkin, DJ-1 or PINK1 also suggests that dominant PD genes may act via common pathways with the recessive PD genes.

Using LRRK2 *Drosophila* Models to Identify Potentially Therapeutic Compounds

The genetic LRRK2 *Drosophila* model represents a promising platform for inhibitor identification and validation. Studies have shown that GW5074, curcumin, or sorafenib significantly suppressed LRRK2 PD-like phenotypes in *Drosophila* [42, 43]. Although candidate compounds have been used in these studies, they open the possibility of performing compound screens, which may be useful for finding new drugs for treatment of LRRK2-associated PD.

LRRK2 C. elegans *Models*

The nematode *Caenorhabditis elegans* has a well-defined and genetically tractable nervous system that offers an effective model to explore basic mechanistic pathways that might underpin complex human neurological diseases. *C. elegans* contains only one *lrk-1* gene encoding a LRRK-like protein. Lrk-1 is localized in the Golgi apparatus and is required for polarized localization of SV proteins. The loss of *lrk-1* causes SV protein mislocalization to both presynaptic and dendritic endings in neurons, which are dependent on the AP-1 clathrin adaptor UNC-101 [44]. The results raise the possibility that the LRK-1 functions on the *trans*-Golgi network

(TGN) to exclude SV proteins from the dendrite-specific transport mechanisms mediated by the AP-1 clathrin adaptor complex. This study suggests that LRRK2 might function in the Golgi network. Recent identification of ArfGAP1, a Golgi protein that reciprocally regulates LRRK2-induced toxicity both in vitro and in vivo, might provide a new insight into LRRK2 function in ER to Golgi trafficking [20]. Other loss of function studies in *C. elegans* revealed that LRRK2 acts to protect *C. elegans* DA neurons from the toxicity of 6-hydroxydopamine and/or human α-synuclein, possibly through the p38 pathway, by supporting upregulation of GRP78 [45]. The loss of lrk-1 renders animals hypersensitive to the endoplasmic reticulum stressor tunicamycin, which is rescued by *PINK1* [46]. These studies suggest a functional link between LRRK2 and ER stress [45, 46].

While loss of the LRRK2 homolog in *C. elegans* provided information of the biological function of LRRK2, overexpression of human LRRK2 in *C. elegans* established a model that recapitulates key features of PD. Overexpression of human LRRK2 WT, R1441C, or G2019S in DA neurons in *C. elegans* causes age-dependent DA neurodegeneration, behavioral deficits, and locomotor dysfunction that is accompanied by a reduction of dopamine levels in vivo [47, 48]. Several studies suggested that these phenotypes could be caused by mitochondrial dysfunction, autophagy inhibition, and ER stress. Expressing human LRRK2 WT increased nematode survival by protecting against mitochondrial stress, but mutant forms of LRRK2 (G2019S or R1441C) enhanced vulnerability to mitochondrial dysfunction and inhibition of autophagy [47, 49]. Although LRRK2 G2019S consistently inhibits autophagy in multiple studies, the effects of LRRK2 WT appear to vary between studies even from the same group [47, 49]. The explanation for this variation appears to depend on whether or not α-synuclein is present [50]. Coexpressing LRRK2 WT with α-synuclein produces a modest age-dependent inhibition of autophagy [50]. Since *C. elegans*, like *Drosophila*, does not express endogenous α-synuclein, caution needs to be taken in interpreting studies using *C. elegans* models.

The observations from LRRK2 *C. elegans* models support a role for LRRK2 kinase and GTPase activity as critical mediators of neurotoxicity induced by mutant LRRK2. Overexpression of the GTP-binding defective mutant, K1347A, prevents the LRRK2-induced neurodegeneration and behavioral abnormalities [48]. LRRK2 kinase inhibitors TTT-3002 and LRRK2-IN1 protect against LRRK2 R1441C- or LRRK2 G2019S-induced neurodegeneration [42, 48]. These studies suggested that both LRRK2 GTPase and kinase activity play crucial roles in LRRK2-induced neurodegeneration in *C. elegans*.

LRRK2 Zebrafish Models

Although zebrafish has been established as an excellent vertebrate model for the study of human disease, zebrafish LRRK2 (zLRRK2) models are not well developed. There is one human LRRK2 homolog in zebrafish, zLRRK2, which has a high degree conservation of amino acid sequences with human LRRK2 (hLRRK2) proteins and the highest conservation within the kinase domain. Two groups reported the generation of loss-of-function zLRRK2 models, but with conflicting results. Sheng et al. first reported that the deletion of the WD40 domain of zLRRK2 by morpholino-targeted splicing caused parkinsonism-like phenotypes, including loss of DA neurons in the diencephalon and locomotion defects [51]. These neurodegenerative and locomotion defects could be rescued by overexpressing zLRRK2 or hLRRK2 mRNA. The administration of L-Dopa could also rescue the locomotion defects, but not the neurodegeneration [51]. However, a later study reported by Ren et al. demonstrated that the deletion of the WD40 domain of zLRRK2 using the same methods does not cause the loss of DA neurons [52]. Given the opposite results from two similar studies, the loss-of-function zLRRK2 models need further evaluation. Transient co-overexpression of human WT or GS LRRK2 with GFP-tagged ubiquitin in WT zebrafish embryos causes impaired clearance of transiently expressed ubiquitin, suggesting of ubiquitin proteasome system disruption [53]. The characterization on DA system was not performed [53]. Taken together, LRRK2 zebrafish models are underdeveloped and need more evaluation and characterization.

LRRK2 Mouse Models

Whereas all the models are important and can be used in a variety of research directions, generally more effort is placed on developing mouse models to study human genetic disorders because mice possess similar neuronal networks and basal ganglia circuitry with high conservation of homologs with the human disease-causing genes. Then, what are the criteria for the effective modeling of human diseases in mice? A good model should recapitulate the genetic and pathological features of the disease in human patients while avoiding spurious phenotypes that are not involved in human diseases [54–56]. For PD, mouse models that faithfully recapitulate the characteristic neurodegeneration and motor deficits as well as other hallmarks of PD such as α-synuclein aggregation are necessary. They would provide in vivo platforms to validate pathogenic molecular pathways and therapeutic strategies in more controlled physiological systems [55].

LRRK2 Knockout Mouse Models

A question frequently raised is whether LRRK2 pathology could be the result of a loss of function. To address this question, several groups generated and analyzed LRRK2 knockout mice. Consistent among the knockouts is that observation that there is no DA neurodegeneration although some abnormalities are observed outside the nervous system (Table 9.2) [57–63]. Andres-Mateo et al. reported the first LRRK2 knockout mouse model showing an intact nigrostriatal DA pathway up to 2 years of age and no altered sensitivity to MPTP-induced neurotoxicity [57]. Tong

Table 9.2 Mouse models for LRRK2-associated PD

LRRK2 variants	Genetic manipulations	Motor deficits	DA neuronal loss	Nigrostriatal dysfunction	Other phenotypes	References
LRRK2–/–	Knockout	ND	No changes	No changes	Lack of hypersensitivity to MPTP	[57]
LRRK2–/–	Knockout	No changes	No changes	ND	ND	[60]
LRRK2–/–	Knockout	ND	No changes	No changes	Accumulation of alpha-synuclein and ubiquitinated proteins; impaired autophagy-lysosomal pathway; increased apoptotic cell death, inflammatory responses, and oxidative damage	[63]
LRRK2 –/–	Knockout	No changes	No changes	No changes	An early-onset increase in number and size of secondary lysosomes in the kidney; hypertension blood pressure	[58]
LRRK2 –/–	Knockout	Abnormal exploratory behavior at 7 and 16 months	No changes	No changes	Degeneration in the kidney, increased autophagic activity	[59]

LRRK2 variants	Genetic manipulations	Motor deficits	DA neuronal loss	Nigrostriatal dysfunction	Other phenotypes	References
WT	BAC transgenic	No changes	No changes	No changes	ND	[67]
R1441G		Decline in rearing starting from 3 to 6 months; Response to L-Dopa	No changes	Axonal and tau pathology; Impaired dopamine release	ND	
WT	BAC transgenic	Hyperactive at 6 months	No changes	No changes	No changes	[66]
G2019S		No changes at 12 months	No changes	Decreased DA content and DA release and uptake	Enhanced kinase activity and phospho-tau	
WT	BAC transgenic	No changes	No changes	Reduction of extracellular dopamine levels	No changes	[68]
G2019S		No changes	No changes	Reduction of extracellular dopamine levels	Increased phospho-tau	
WT	Tet-off/CamKII-tTA inducible transgenic	No changes	ND	ND	ND	[60]
G2019S		Increased ambulatory activities at 12 months	ND	ND	ND	

(continued)

Table 9.2 (continued)

LRRK2 variants	Genetic manipulations	Motor deficits	DA neuronal loss	Nigrostriatal dysfunction	Other phenotypes	References
WT	Tet-off/Pitx3-tTA inducible transgenic	No changes	No changes	No changes	Decreased expression of DA genes	[71]
G2019S		No changes	No changes	Impaired dopamine homeostasis and release	Decreased expression of DA genes	
G2019S	CMVE-PDGFβ drived transgenic	No changes	TH neurodegeneration at 20 months	No changes	Reduced neurite complexity and autophagic abnormalities	[69]
R1441C		Decreased at 15 months	No changes	No changes	ND	
WT	CMVE-PDGFβ drived transgenic	No changes	No changes	No changes	No changes	[65]
G2019S		Impaired motor activity, response to L-Dopa	TH neurodegeneration starting from 12 months	Decreased dopamine transporters or TH staining	Activated MKK4-JNK pathway	
R1441C	ROSA26 drived transgenic	No changes	No changes	No changes	Nuclear abnormalities	[70]
WT	Thy1.2 drived transgenic	ND	No changes	No changes	No changes on neurite outgrowth	[73]
G2019S		ND	No changes	No changes	No changes on neurite outgrowth	
R1441C	Knock-in	Reduced response to AMPH in locomotor activity	No changes	Reduced catecholamine release in cultured mutant chromaffin cells	ND	[74]

LRRK2 variants	Genetic manipulations	Motor deficits	DA neuronal loss	Nigrostriatal dysfunction	Other phenotypes	References
G2019S	Knock-in	No changes	No changes	No changes	No changes in autophagy	[58]
G2019S	Knock-in	Homo-G2019Smice travel longer distances at 12 months	No changes	Impaired dopamine release, altered DA metabolism	Mitochondrial abnormalities; elevated glutamate release; increased phosphor-tau	[72, 75]

et al. demonstrated that LRRK2 knockouts develop striking kidney pathology and impaired autophagy-lysosomal function [62, 63]. The kidney phenotype was observed in two other LRRK2 knockouts, although the defects in autophagy changes were not observed [58, 59]. A recent study using LRRK2 knockouts suggests that LRRK2 influences neurogenesis and particularly neuronal morphogenesis [61].

Since the majority of LRRK2 PD patients exhibit α-synuclein deposition, the role of LRRK2 in α-synuclein pathology has been explored. Lin et al. showed that knockout of LRRK2-rescued A53T α-synuclein overexpression induced Golgi fragmentation, α-synuclein accumulation and aggregation, microglial activation, and forebrain neuronal degeneration [60]. On the other hand, Tong et al. demonstrated that LRRK2 knockout mice develop striking accumulation and aggregation of α-synuclein and Daher et al. showed that deletion of LRRK2 had no influence on the lethal neurodegenerative phenotype of the A53T α-synuclein transgenic mice [63, 64]. The different findings between these studies could be due to the different levels of α-synuclein expression and or technical concerns. Whether inhibition of LRRK2 could be employed as a therapeutic strategy to attenuate α-synuclein-mediated neuronal damage relevant to PD needs further investigation.

All the observations from the LRRK2 knockout mice suggest that LRRK2 plays little if any role in the development and survival of DA neurons is under physiologic conditions. Thus, PD caused by LRRK2 mutations are likely not due to a loss of LRRK2 function.

LRRK2 Transgenic Mouse Models

Many groups have generated LRRK2-related PD mouse models expressing LRRK2 WT or PD-associated mutant LRRK2 G2019S or R1441C/G (Table 9.2) [58, 60, 65–75]. Several transgenic techniques for LRRK2-related PD modeling in mice have been utilized, including conventional [65, 69, 70, 73], BAC transgenic [66–68], tet-inducible transgenic [60, 71], and mutant LRRK2 knock-in techniques [58, 72, 74, 75]. However, to date only two of the LRRK2 models exhibit age-dependent SNpc DA neurodegeneration [65, 69]. Most LRRK2 transgenic animals manifest deficits in DA transmission and DA-responsive behavior. Between the two studies with SNpc DA neurodegeneration, both used conventional transgenic techniques utilizing the PDGF-β promoter to generate LRRK2 mutant G2019S mouse lines. Ramonet et al. show that LRRK2 G2019S mice developed about 20% SNpc DA neurodegeneration at 20 months of age [69], while Chen et al. demonstrated more robust degeneration in the SNpc starting from 12 months of age with about 50% degeneration at 16 months of age without a phenotype in LRRK2 WT transgenic mice [65]. The different degrees of the degeneration may be due to the different overexpression levels of the transgenes.

Why don't most LRRK2 transgenic models exhibit SNpc DA degeneration? One potential explanation could be a lack of robust transgene overexpression in SNpc DA neurons. The BAC and knock-in models express mutant LRRK2 during development, and thus there may be compensatory mechanisms in the mouse that prevent loss of DA neurons. Thus, conditional and selective expression of LRRK2 in SNpc DA neurons may overcome this problem. A recent study reported an LRRK2 G2019S conditional transgenic mouse model using the tet-off system and a PitX3-tTA driver line to drive transgene expression in DA neurons. However, no SNpc DA degeneration was observed in this model [71]. The reason is unclear but may be related to not aging the mice to 24 months of age, or perhaps expression of LRRK2 only in DA neurons is not sufficient for DA degeneration to occur given that the endogenous LRRK2 expression levels are comparatively low in SNpc DA neurons and LRRK2 is also expressed in other neurons. Thus, overexpression of LRRK2 in other neurons at the same time as in DA neurons or other genetic and/or environmental factors may be required for degeneration of DA neurons.

LRRK2 Rat Models

For the last several decades, investigators have chosen to use mouse models because of the technologies that were available. Now the same technologies are available in the rat. As a model of human disease, the rat offers many advantages over the mouse and other organisms. Physiology is easier to monitor in the rat. Moreover, in many cases, the physiology is more like the corresponding human condition. The rat is more intelligent than the mouse and is capable of learning a wider variety of tasks that are important in mimicking human behavioral symptoms. Recently, both LRRK2 knockout and transgenic rat models have been generated and characterized.

LRRK2 Knockout Rat Models

Like other LRRK2 animal models, LRRK2 knockout rats have no significant loss of SNpc neurons. Similar to LRRK2 knockout mice, the loss of LRRK2 in rats leads to abnormal phenotypes in peripheral organs. Two studies have observed abnormal kidneys [76, 77]. Besides the kidney phenotype, Ness et al. observed significant weight gain in the LRRK2 knockout rats accompanied by significant increases in insulin and insulin-like growth factors [77]. They also found significant alterations in the cellular composition of the spleen in LRRK2 knockout animals, which Baptista et al. did not observe [76, 77]. Instead, they found LRRK2 knockout rats displayed an abnormal lung and liver phenotype. Using LRRK2 knockout rats, the West group demonstrated resistance to DA neurodegeneration elicited by intracranial administration of LPS and protection from α-synuclein-induced DA neurodegeneration and rhabdomyolysis-induced kidney injury [78, 79]. The abnormal peripheral phenotype of the LRRK2 knockout rat is suggestive of a complex LRRK2 biology influencing metabolism, immune function, and kidney homeostasis. The phenotype of LRRK2 knockout rat is consistent with LRRK2 knockout in other organisms such as *Drosophila*, *C. elegans*, and mouse, supporting the concept that

LRRK2 plays little role in the development and survival of DA neurons under physiologic conditions.

LRRK2 Transgenic Rat Models

The first LRRK2 transgenic rat model was developed by Zhou et al. using an inducible system [80]. Temporal expression of human LRRK2 G2019S in rats did not lead to DA neurodegeneration, but enhanced locomotor activity with age accompanied with impaired dopamine reuptake by the dopamine transporter (DAT) was observed. As a result of compromised DAT activity, amphetamine-evoked dopamine release and amphetamine-elicited locomotor activity were reduced in LRRK2 G2019S transgenic rats [80]. Since only two copies of LRRK2 transgene were expressed in this model, there may have been insufficient protein to produce DA neurodegeneration. Human BAC-LRRK2 G2019S or R1441G rats were developed, and mutant LRRK2 expression was approximately 5 ~ 8 times higher than endogenous rat LRRK2. However, both BAC-LRRK2 R1441G and G2019S transgenic rats do not show signs of neurodegeneration and do not develop significant motor or cognitive deficits with age [81–83]. Instead, LRRK2 G2019S induced oxidative stress in the striatum and substantia nigra and increased inducible nitric oxide synthase expression in SNpc DA neurons and abnormal morphology of SNpc DA neurons [81, 83]. Although this model does not reproduce the key features of end-stage PD, it may be useful in studying gene-environment interactions. However, a recent study indicates that BAC-LRRK2 R1441G transgenic rats did not show increased vulnerability to sub-toxic doses of paraquat [82]. Since these studies lacked a wild-type human LRRK2 transgenic rat as a control, it is not possible to conclude that the phenotype induced by mutant LRRK2 is due to the LRRK2 PD mutation or overexpression of the LRRK2 protein. All results from different LRRK2 transgenic rats suggest that rats compensate and accommodate LRRK2's toxic effects.

Viral-Mediated Animal Models of LRRK2

While the genetic LRRK2 models shed light on LRRK2 cellular functions and pathogenic pathways, development of recombinant viral vectors for in vivo delivery of transgenes has opened up a new possibility to model diseases in the CNS. The viral-mediated gene transfer approach in adulthood bypasses the development of compensatory effects. This approach also allows researchers to target specific neuronal populations, such as SNpc DA neurons. Another advantage of the viral-mediated gene transfer approach is that it allows researchers to control transgene dosage by modulating copy number of the transgene. While viral models allow us to recapitulate some of the neurodegeneration processes observed in PD patients that have so far been difficult to show in other models, there are caveats of non-physiological doses of transgenes and potential alterations in RNA translation.

Due to large size of LRRK2 gene and the limited packaging capacity of different viral vectors, so far only two LRRK2 viral models have been developed and characterized.

Herpes Simplex Virus (HSV)-LRRK2 Viral Model

The first LRRK2 viral model was developed by Lee et al. by carrying LRRK2 into HSV amplicons expressing a CMV-driven GFP reporter [84]. One advantage of HSV is that it is injected in the striatum and retrogradely transported into SNpc DA neurons, which avoids nonspecific inflammatory damage to the substantia nigra. In this model, after 3 weeks injection, the HSV-WT-LRRK2 induced modest SNpc DA neurodegeneration of about 10–20%, whereas the HSV-LRRK2 G2019S induced up to 50% neuronal loss in SNpc DA neurons. Interestingly, the kinase-dead LRRK2 does not induce neuronal loss, which strongly suggested that kinase activity of LRRK2 mediates LRRK2-induced DA neurodegeneration. This notion is further supported by the protective effects of pharmacological inhibition of LRRK2 kinase activity in this HSV model [84].

Adenoviruses (rAd)-LRRK2 Viral Model

Second-generation E1, E3, and E2a-deleted recombinant human serotype 5 adenoviruses (rAd) carrying LRRK2 WT and G2019S were generated by Dusonchet et al. [85]. Similar to HSV, adenoviral particles can be efficiently retrogradely transported to DA neurons within the SNpc following intrastriatal injections. Injection of rAd-LRRK2 G2019S into rat striatum causes a progressive loss of TH-positive DA neurons in the SNpc, reaching about 21% at 42 days postinjection, but no cell loss is detected in the rAd-GFP- or rAd-LRRK2 WT injected groups. Abnormal transient hyperphosphorylation of tau in dystrophic SNpc neuritic processes was observed upon LRRK2 overexpression at 10 days [85]. Tsika et al. further characterized the striatal pathology in this model [86]. Expression of LRRK2 G2019S selectively induces the accumulation of neuronal ubiquitin-positive inclusions accompanied by neurite degeneration and the altered distribution of axonal phosphorylated neurofilaments in the striatum. The pathological phenotype is dependent on LRRK2 kinase activity as a kinase-inactive mutation (LRRK2 G2019S/D1994N) completely ameliorates the pathological effects of LRRK2 G2019S [86].

Another LRRK2 viral model has been briefly mentioned in another study. The authors delivered lentiviral vectors carrying enhanced green fluorescent protein (eGFP)-tagged LRRK2 G2019S in adult mouse striatum and observed LRRK2 function in TGN turnover [87]. However, there was no characterization of this model in terms of pathology in the nigrostriatal pathways.

Concluding Remarks

Modeling of LRRK2-associated PD in various animal models has provided unprecedented insights into the potential mechanisms of LRRK2-mediated neurodegeneration such as regulation of protein translation, vesicle trafficking, neurite outgrowth, autophagy, and cytoskeletal dynamics. However, none of the current LRRK2 animal models fulfills all the key features of PD. Different LRRK2 animal models recapitulate different clinical and neuropathological features of LRRK2-associated PD, including the degeneration of nigrostriatal DA neurons, neuropathology, α-synuclein accumulation, abnormal striatal DA neurotransmission, and behavioral deficits.

Why are the animal models "imperfect" for modeling LRRK2-associated PD? First, for the simple animal models such as *Drosophila* and *C. elegans*, they do not have α-synuclein homolog and a true human LRRK2 homolog. PD patients harboring LRRK2 mutations frequently exhibit α-synuclein neuropathology in the form of Lewy bodies. A question about whether α-synuclein is required for LRRK2 pathology or vice visa has been raised. The challenge remains to validate the mechanisms identified in these model systems in human PD. Second, for LRRK2 rodent models, perhaps rodent DA neurons are particularly resistant to LRRK2 toxicity. In addition, there may be compensatory mechanisms in the rodents that prevent loss of DA neuron. Third, the fact is that LRRK2 mutations in humans are partially penetrant, implicating that there may be additional factors such as genetic and/or environmental stressors that are required for degeneration of DA neurons. Indeed, in both LRRK2 *Drosophila* and *C. elegans* models, treatments with mitochondrial function inhibitors exacerbate neurodegeneration. Fourth, the HSV- or adenovirus-mediated LRRK2 rodent models induce robust DA neurodegeneration, supporting the notion that both non-cell-autonomous and cell-autonomous processes contribute to the degeneration of DA neurons. The transgene can be virally expressed in both neurons and glia to activate the inflammatory pathway in glial cells and elicit neurodegeneration in DA neurons, which is largely absent in the genetic LRRK2 models. Therefore, non-cell-autonomous effects may provide a promising mechanism for LRRK2-induced PD in humans. All these possibilities need to be taken into consideration in developing future LRRK2 animal models.

Acknowledgments This work was supported by grants from the NIH/NINDS NS38377 (VLD and TMD), the JPB Foundation (TMD), NIH/NIA K01-AG046366 (YX), and the William N. & Bernice E. Bumpus Foundation Innovation Awards (YX). TMD is the Leonard and Madlyn Abramson Professor in Neurodegenerative Diseases. The authors acknowledge the joint participation by the Adrienne Helis Malvin Medical Research Foundation through its direct engagement in the continuous active conduct of medical research in conjunction with the Johns Hopkins Hospital and the Johns Hopkins University School of Medicine and the Foundation's Parkinson's Disease Programs, M-1, M-2.

Conflict of Interest The author declares no conflicts of interest.

References

1. Lees AJ, Hardy J, Revesz T (2009) Parkinson's disease. Lancet 373(9680):2055–2066. doi:10.1016/S0140-6736(09)60492-X
2. Savitt JM, Dawson VL, Dawson TM (2006) Diagnosis and treatment of Parkinson disease: molecules to medicine. J Clin Invest 116(7):1744–1754. doi:10.1172/JCI29178
3. Martin I, Dawson VL, Dawson TM (2011) Recent advances in the genetics of Parkinson's disease. Annu Rev Genomics Hum Genet 12:301–325. doi:10.1146/annurev-genom-082410-101440
4. Zimprich A, Benet-Pages A, Struhal W, Graf E, Eck SH, Offman MN, Haubenberger D, Spielberger S, Schulte EC, Lichtner P, Rossle SC, Klopp N, Wolf E, Seppi K, Pirker W, Presslauer S, Mollenhauer B, Katzenschlager R, Foki T, Hotzy C, Reinthaler E, Harutyunyan A, Kralovics R, Peters A, Zimprich F, Brucke T, Poewe W, Auff E, Trenkwalder C, Rost B, Ransmayr G, Winkelmann J, Meitinger T, Strom TM (2011) A mutation in VPS35, encoding a subunit of the retromer complex, causes late-onset Parkinson disease. Am J Hum Genet 89(1):168–175. doi:10.1016/j.ajhg.2011.06.008
5. Chartier-Harlin MC, Dachsel JC, Vilarino-Guell C, Lincoln SJ, Lepretre F, Hulihan MM, Kachergus J, Milnerwood AJ, Tapia L, Song MS, Le Rhun E, Mutez E, Larvor L, Duflot A, Vanbesien-Mailliot C, Kreisler A, Ross OA, Nishioka K, Soto-Ortolaza AI, Cobb SA, Melrose HL, Behrouz B, Keeling BH, Bacon JA, Hentati E, Williams L, Yanagiya A, Sonenberg N, Lockhart PJ, Zubair AC, Uitti RJ, Aasly JO, Krygowska-Wajs A, Opala G, Wszolek ZK, Frigerio R, Maraganore DM, Gosal D, Lynch T, Hutchinson M, Bentivoglio AR, Valente EM, Nichols WC, Pankratz N, Foroud T, Gibson RA, Hentati F, Dickson DW, Destee A, Farrer MJ (2011) Translation initiator EIF4G1 mutations in familial Parkinson disease. Am J Hum Genet 89(3):398–406. doi:10.1016/j.ajhg.2011.08.009
6. Vilarino-Guell C, Wider C, Ross OA, Dachsel JC, Kachergus JM, Lincoln SJ, Soto-Ortolaza AI, Cobb SA, Wilhoite GJ, Bacon JA, Behrouz B, Melrose HL, Hentati E, Puschmann A, Evans DM, Conibear E, Wasserman WW, Aasly JO, Burkhard PR, Djaldetti R, Ghika J, Hentati F, Krygowska-Wajs A, Lynch T, Melamed E, Rajput A, Rajput AH, Solida A, Wu RM, Uitti RJ, Wszolek ZK, Vingerhoets F, Farrer MJ (2011) VPS35 mutations in Parkinson disease. Am J Hum Genet 89(1):162–167. doi:10.1016/j.ajhg.2011.06.001
7. Vilarino-Guell C, Rajput A, Milnerwood AJ, Shah B, Szu-Tu C, Trinh J, Yu I, Encarnacion M, Munsie LN, Tapia L, Gustavsson EK, Chou P, Tatarnikov I, Evans DM, Pishotta FT, Volta M, Beccano-Kelly D, Thompson C, Lin MK, Sherman HE, Han HJ, Guenther BL, Wasserman WW, Bernard V, Ross CJ, Appel-Cresswell S, Stoessl AJ, Robinson CA, Dickson DW, Ross OA, Wszolek ZK, Aasly JO, Wu RM, Hentati F, Gibson RA, McPherson PS, Girard M, Rajput M, Rajput AH, Farrer MJ (2014) DNAJC13 mutations in Parkinson disease. Hum Mol Genet 23(7):1794–1801. doi:10.1093/hmg/ddt570
8. Scholz SW, Jeon BS (2015) GBA mutations and Parkinson disease: when genotype meets phenotype. Neurology 84(9):866–867. doi:10.1212/WNL.0000000000001321
9. Paisan-Ruiz C, Jain S, Evans EW, Gilks WP, Simon J, van der Brug M, Lopez de Munain A, Aparicio S, Gil AM, Khan N, Johnson J, Martinez JR, Nicholl D, Carrera IM, Pena AS, de Silva R, Lees A, Marti-Masso JF, Perez-Tur J, Wood NW, Singleton AB (2004) Cloning of the gene containing mutations that cause PARK8-linked Parkinson's disease. Neuron 44(4):595–600. doi:10.1016/j.neuron.2004.10.023
10. Zimprich A, Biskup S, Leitner P, Lichtner P, Farrer M, Lincoln S, Kachergus J, Hulihan M, Uitti RJ, Calne DB, Stoessl AJ, Pfeiffer RF, Patenge N, Carbajal IC, Vieregge P, Asmus F, Muller-Myhsok B, Dickson DW, Meitinger T, Strom TM, Wszolek ZK, Gasser T (2004) Mutations in LRRK2 cause autosomal-dominant parkinsonism with pleomorphic pathology. Neuron 44(4):601–607. doi:10.1016/j.neuron.2004.11.005
11. Mata IF, Wedemeyer WJ, Farrer MJ, Taylor JP, Gallo KA (2006) LRRK2 in Parkinson's disease: protein domains and functional insights. Trends Neurosci 29(5):286–293. doi:10.1016/j.tins.2006.03.006

12. Cookson MR (2010) The role of leucine-rich repeat kinase 2 (LRRK2) in Parkinson's disease. Nat Rev Neurosci 11(12):791–797. doi:10.1038/nrn2935
13. Berwick DC, Harvey K (2011) LRRK2 signaling pathways: the key to unlocking neurodegeneration? Trends Cell Biol 21(5):257–265. doi:10.1016/j.tcb.2011.01.001
14. Xiong Y, Dawson VL, Dawson TM (2012) LRRK2 GTPase dysfunction in the pathogenesis of Parkinson's disease. Biochem Soc Trans 40(5):1074–1079. doi:10.1042/BST20120093
15. Cookson MR (2015) LRRK2 pathways leading to neurodegeneration. Curr Neurol Neurosci Rep 15(7):42. doi:10.1007/s11910-015-0564-y
16. Martin I, Kim JW, Dawson VL, Dawson TM (2014) LRRK2 pathobiology in Parkinson's disease. J Neurochem 131(5):554–565. doi:10.1111/jnc.12949
17. Nikonova EV, Xiong Y, Tanis KQ, Dawson VL, Vogel RL, Finney EM, Stone DJ, Reynolds IJ, Kern JT, Dawson TM (2012) Transcriptional responses to loss or gain of function of the leucine-rich repeat kinase 2 (LRRK2) gene uncover biological processes modulated by LRRK2 activity. Hum Mol Genet 21(1):163–174. doi:10.1093/hmg/ddr451
18. Daniel G, Moore DJ (2015) Modeling LRRK2 pathobiology in Parkinson's disease: from yeast to rodents. Curr Top Behav Neurosci 22:331–368. doi:10.1007/7854_2014_311
19. Xiong Y, Coombes CE, Kilaru A, Li X, Gitler AD, Bowers WJ, Dawson VL, Dawson TM, Moore DJ (2010) GTPase activity plays a key role in the pathobiology of LRRK2. PLoS Genet 6(4), e1000902. doi:10.1371/journal.pgen.1000902
20. Xiong Y, Yuan C, Chen R, Dawson TM, Dawson VL (2012) ArfGAP1 is a GTPase activating protein for LRRK2: reciprocal regulation of ArfGAP1 by LRRK2. J Neurosci 32(11):3877–3886. doi:10.1523/JNEUROSCI.4566-11.2012
21. Imai Y, Gehrke S, Wang HQ, Takahashi R, Hasegawa K, Oota E, Lu B (2008) Phosphorylation of 4E-BP by LRRK2 affects the maintenance of dopaminergic neurons in *Drosophila*. EMBO J 27(18):2432–2443. doi:10.1038/emboj.2008.163
22. Lee SB, Kim W, Lee S, Chung J (2007) Loss of LRRK2/PARK8 induces degeneration of dopaminergic neurons in *Drosophila*. Biochem Biophys Res Commun 358(2):534–539. doi:10.1016/j.bbrc.2007.04.156
23. Wang D, Tang B, Zhao G, Pan Q, Xia K, Bodmer R, Zhang Z (2008) Dispensable role of *Drosophila* ortholog of LRRK2 kinase activity in survival of dopaminergic neurons. Mol Neurodegener 3:3. doi:10.1186/1750-1326-3-3
24. Tain LS, Mortiboys H, Tao RN, Ziviani E, Bandmann O, Whitworth AJ (2009) Rapamycin activation of 4E-BP prevents parkinsonian dopaminergic neuron loss. Nat Neurosci 12(9):1129–1135. doi:10.1038/nn.2372
25. Liu Z, Wang X, Yu Y, Li X, Wang T, Jiang H, Ren Q, Jiao Y, Sawa A, Moran T, Ross CA, Montell C, Smith WW (2008) A *Drosophila* model for LRRK2-linked parkinsonism. Proc Natl Acad Sci U S A 105(7):2693–2698. doi:10.1073/pnas.0708452105
26. Ng CH, Mok SZ, Koh C, Ouyang X, Fivaz ML, Tan EK, Dawson VL, Dawson TM, Yu F, Lim KL (2009) Parkin protects against LRRK2 G2019S mutant-induced dopaminergic neurodegeneration in *Drosophila*. J Neurosci 29(36):11257–11262. doi:10.1523/JNEUROSCI.2375-09.2009
27. Venderova K, Kabbach G, Abdel-Messih E, Zhang Y, Parks RJ, Imai Y, Gehrke S, Ngsee J, Lavoie MJ, Slack RS, Rao Y, Zhang Z, Lu B, Haque ME, Park DS (2009) Leucine-Rich Repeat Kinase 2 interacts with Parkin, DJ-1 and PINK-1 in a *Drosophila melanogaster* model of Parkinson's disease. Hum Mol Genet 18(22):4390–4404. doi:10.1093/hmg/ddp394
28. Lin CH, Tsai PI, Wu RM, Chien CT (2010) LRRK2 G2019S mutation induces dendrite degeneration through mislocalization and phosphorylation of tau by recruiting autoactivated GSK3ss. J Neurosci 30(39):13138–13149. doi:10.1523/JNEUROSCI.1737-10.2010
29. Hindle S, Afsari F, Stark M, Middleton CA, Evans GJ, Sweeney ST, Elliott CJ (2013) Dopaminergic expression of the Parkinsonian gene LRRK2-G2019S leads to non-autonomous visual neurodegeneration, accelerated by increased neural demands for energy. Hum Mol Genet 22(11):2129–2140. doi:10.1093/hmg/ddt061

30. Godena VK, Brookes-Hocking N, Moller A, Shaw G, Oswald M, Sancho RM, Miller CC, Whitworth AJ, De Vos KJ (2014) Increasing microtubule acetylation rescues axonal transport and locomotor deficits caused by LRRK2 Roc-COR domain mutations. Nat Commun 5:5245. doi:10.1038/ncomms6245
31. Saha AR, Hill J, Utton MA, Asuni AA, Ackerley S, Grierson AJ, Miller CC, Davies AM, Buchman VL, Anderton BH, Hanger DP (2004) Parkinson's disease alpha-synuclein mutations exhibit defective axonal transport in cultured neurons. J Cell Sci 117(Pt 7):1017–1024. doi:10.1242/jcs.00967
32. Gehrke S, Imai Y, Sokol N, Lu B (2010) Pathogenic LRRK2 negatively regulates microRNA-mediated translational repression. Nature 466(7306):637–641. doi:10.1038/nature09191
33. Martin I, Kim JW, Lee BD, Kang HC, Xu JC, Jia H, Stankowski J, Kim MS, Zhong J, Kumar M, Andrabi SA, Xiong Y, Dickson DW, Wszolek ZK, Pandey A, Dawson TM, Dawson VL (2014) Ribosomal protein s15 phosphorylation mediates LRRK2 neurodegeneration in Parkinson's disease. Cell 157(2):472–485. doi:10.1016/j.cell.2014.01.064
34. Dodson MW, Zhang T, Jiang C, Chen S, Guo M (2012) Roles of the *Drosophila* LRRK2 homolog in Rab7-dependent lysosomal positioning. Hum Mol Genet 21(6):1350–1363. doi:10.1093/hmg/ddr573
35. Schreij AM, Chaineau M, Ruan W, Lin S, Barker PA, Fon EA, Pherson PS (2015) LRRK2 localizes to endosomes and interacts with clathrin-light chains to limit Rac1 activation. EMBO Rep 16(1):79–86. doi:10.15252/embr.201438714
36. Dodson MW, Leung LK, Lone M, Lizzio MA, Guo M (2014) Novel ethyl methanesulfonate (EMS)-induced null alleles of the *Drosophila* homolog of LRRK2 reveal a crucial role in endolysosomal functions and autophagy in vivo. Dis Model Mech 7(12):1351–1363. doi:10.1242/dmm.017020
37. Matta S, Van Kolen K, da Cunha R, van den Bogaart G, Mandemakers W, Miskiewicz K, De Bock PJ, Morais VA, Vilain S, Haddad D, Delbroek L, Swerts J, Chavez-Gutierrez L, Esposito G, Daneels G, Karran E, Holt M, Gevaert K, Moechars DW, De Strooper B, Verstreken P (2012) LRRK2 controls an EndoA phosphorylation cycle in synaptic endocytosis. Neuron 75(6):1008–1021. doi:10.1016/j.neuron.2012.08.022
38. Lin CH, Li H, Lee YN, Cheng YJ, Wu RM, Chien CT (2015) Lrrk regulates the dynamic profile of dendritic Golgi outposts through the golgin Lava lamp. J Cell Biol 210(3):471. doi:10.1083/jcb.201411033
39. Linhart R, Wong SA, Cao J, Tran M, Huynh A, Ardrey C, Park JM, Hsu C, Taha S, Peterson R, Shea S, Kurian J, Venderova K (2014) Vacuolar protein sorting 35 (Vps35) rescues locomotor deficits and shortened lifespan in *Drosophila* expressing a Parkinson's disease mutant of Leucine-Rich Repeat Kinase 2 (LRRK2). Mol Neurodegener 9:23. doi:10.1186/1750-1326-9-23
40. MacLeod DA, Rhinn H, Kuwahara T, Zolin A, Di Paolo G, McCabe BD, Marder KS, Honig LS, Clark LN, Small SA, Abeliovich A (2013) RAB7L1 interacts with LRRK2 to modify intraneuronal protein sorting and Parkinson's disease risk. Neuron 77(3):425–439. doi:10.1016/j.neuron.2012.11.033
41. Lee S, Liu HP, Lin WY, Guo H, Lu B (2010) LRRK2 kinase regulates synaptic morphology through distinct substrates at the presynaptic and postsynaptic compartments of the *Drosophila* neuromuscular junction. J Neurosci 30(50):16959–16969. doi:10.1523/JNEUROSCI.1807-10.2010
42. Liu Z, Hamamichi S, Lee BD, Yang D, Ray A, Caldwell GA, Caldwell KA, Dawson TM, Smith WW, Dawson VL (2011) Inhibitors of LRRK2 kinase attenuate neurodegeneration and Parkinson-like phenotypes in *Caenorhabditis elegans* and *Drosophila* Parkinson's disease models. Hum Mol Genet 20(20):3933–3942. doi:10.1093/hmg/ddr312
43. Yang D, Li T, Liu Z, Arbez N, Yan J, Moran TH, Ross CA, Smith WW (2012) LRRK2 kinase activity mediates toxic interactions between genetic mutation and oxidative stress in a *Drosophila* model: suppression by curcumin. Neurobiol Dis 47(3):385–392. doi:10.1016/j.nbd.2012.05.020

44. Sakaguchi-Nakashima A, Meir JY, Jin Y, Matsumoto K, Hisamoto N (2007) LRK-1, a *C. elegans* PARK8-related kinase, regulates axonal-dendritic polarity of SV proteins. Curr Biol 17(7):592–598. doi:10.1016/j.cub.2007.01.074
45. Yuan Y, Cao P, Smith MA, Kramp K, Huang Y, Hisamoto N, Matsumoto K, Hatzoglou M, Jin H, Feng Z (2011) Dysregulated LRRK2 signaling in response to endoplasmic reticulum stress leads to dopaminergic neuron degeneration in *C. elegans*. PLoS One 6(8), e22354. doi:10.1371/journal.pone.0022354
46. Samann J, Hegermann J, von Gromoff E, Eimer S, Baumeister R, Schmidt E (2009) *Caenorhabditits elegans* LRK-1 and PINK-1 act antagonistically in stress response and neurite outgrowth. J Biol Chem 284(24):16482–16491. doi:10.1074/jbc.M808255200
47. Saha S, Guillily MD, Ferree A, Lanceta J, Chan D, Ghosh J, Hsu CH, Segal L, Raghavan K, Matsumoto K, Hisamoto N, Kuwahara T, Iwatsubo T, Moore L, Goldstein L, Cookson M, Wolozin B (2009) LRRK2 modulates vulnerability to mitochondrial dysfunction in *Caenorhabditis elegans*. J Neurosci 29(29):9210–9218. doi:10.1523/JNEUROSCI.2281-09.2009
48. Yao C, El Khoury R, Wang W, Byrd TA, Pehek EA, Thacker C, Zhu X, Smith MA, Wilson-Delfosse AL, Chen SG (2010) LRRK2-mediated neurodegeneration and dysfunction of dopaminergic neurons in a *Caenorhabditis elegans* model of Parkinson's disease. Neurobiol Dis 40(1):73–81. doi:10.1016/j.nbd.2010.04.002
49. Saha S, Liu-Yesucevitz L, Wolozin B (2014) Regulation of autophagy by LRRK2 in *Caenorhabditis elegans*. Neurodegener Dis 13(2–3):110–113. doi:10.1159/000355654
50. Saha S, Ash PE, Gowda V, Liu L, Shirihai O, Wolozin B (2015) Mutations in LRRK2 potentiate age-related impairment of autophagic flux. Mol Neurodegener 10(1):26. doi:10.1186/s13024-015-0022-y
51. Sheng D, Qu D, Kwok KH, Ng SS, Lim AY, Aw SS, Lee CW, Sung WK, Tan EK, Lufkin T, Jesuthasan S, Sinnakaruppan M, Liu J (2010) Deletion of the WD40 domain of LRRK2 in Zebrafish causes Parkinsonism-like loss of neurons and locomotive defect. PLoS Genet 6(4), e1000914. doi:10.1371/journal.pgen.1000914
52. Ren G, Xin S, Li S, Zhong H, Lin S (2011) Disruption of LRRK2 does not cause specific loss of dopaminergic neurons in zebrafish. PLoS One 6(6), e20630. doi:10.1371/journal.pone.0020630
53. Lichtenberg M, Mansilla A, Zecchini VR, Fleming A, Rubinsztein DC (2011) The Parkinson's disease protein LRRK2 impairs proteasome substrate clearance without affecting proteasome catalytic activity. Cell Death Dis 2, e196. doi:10.1038/cddis.2011.81
54. Xu Q, Shenoy S, Li C (2012) Mouse models for LRRK2 Parkinson's disease. Parkinsonism Relat Disord 18(Suppl 1):S186–S189. doi:10.1016/S1353-8020(11)70058-X
55. Dawson TM, Ko HS, Dawson VL (2010) Genetic animal models of Parkinson's disease. Neuron 66(5):646–661. doi:10.1016/j.neuron.2010.04.034
56. Lee Y, Dawson VL, Dawson TM (2012) Animal models of Parkinson's disease: vertebrate genetics. Cold Spring Harb Perspect Med 2(10):a009324. doi:10.1101/cshperspect.a009324
57. Andres-Mateos E, Mejias R, Sasaki M, Li X, Lin BM, Biskup S, Zhang L, Banerjee R, Thomas B, Yang L, Liu G, Beal MF, Huso DL, Dawson TM, Dawson VL (2009) Unexpected lack of hypersensitivity in LRRK2 knock-out mice to MPTP (1-methyl-4-phenyl-1,2,3,6-tetrahydropyridine). J Neurosci 29(50):15846–15850. doi:10.1523/JNEUROSCI.4357-09.2009
58. Herzig MC, Kolly C, Persohn E, Theil D, Schweizer T, Hafner T, Stemmelen C, Troxler TJ, Schmid P, Danner S, Schnell CR, Mueller M, Kinzel B, Grevot A, Bolognani F, Stirn M, Kuhn RR, Kaupmann K, van der Putten PH, Rovelli G, Shimshek DR (2011) LRRK2 protein levels are determined by kinase function and are crucial for kidney and lung homeostasis in mice. Hum Mol Genet 20(21):4209–4223. doi:10.1093/hmg/ddr348
59. Hinkle KM, Yue M, Behrouz B, Dachsel JC, Lincoln SJ, Bowles EE, Beevers JE, Dugger B, Winner B, Prots I, Kent CB, Nishioka K, Lin WL, Dickson DW, Janus CJ, Farrer MJ, Melrose HL (2012) LRRK2 knockout mice have an intact dopaminergic system but display alterations

in exploratory and motor co-ordination behaviors. Mol Neurodegener 7:25. doi:10.1186/1750-1326-7-25
60. Lin X, Parisiadou L, Gu XL, Wang L, Shim H, Sun L, Xie C, Long CX, Yang WJ, Ding J, Chen ZZ, Gallant PE, Tao-Cheng JH, Rudow G, Troncoso JC, Liu Z, Li Z, Cai H (2009) Leucine-rich repeat kinase 2 regulates the progression of neuropathology induced by Parkinson's-disease-related mutant alpha-synuclein. Neuron 64(6):807–827. doi:10.1016/j.neuron.2009.11.006
61. Paus M, Kohl Z, Ben Abdallah NM, Galter D, Gillardon F, Winkler J (2013) Enhanced dendritogenesis and axogenesis in hippocampal neuroblasts of LRRK2 knockout mice. Brain Res 1497:85–100. doi:10.1016/j.brainres.2012.12.024
62. Tong Y, Giaime E, Yamaguchi H, Ichimura T, Liu Y, Si H, Cai H, Bonventre JV, Shen J (2012) Loss of leucine-rich repeat kinase 2 causes age-dependent bi-phasic alterations of the autophagy pathway. Mol Neurodegener 7:2. doi:10.1186/1750-1326-7-2
63. Tong Y, Yamaguchi H, Giaime E, Boyle S, Kopan R, Kelleher RJ 3rd, Shen J (2010) Loss of leucine-rich repeat kinase 2 causes impairment of protein degradation pathways, accumulation of alpha-synuclein, and apoptotic cell death in aged mice. Proc Natl Acad Sci U S A 107(21):9879–9884. doi:10.1073/pnas.1004676107
64. Daher JP, Pletnikova O, Biskup S, Musso A, Gellhaar S, Galter D, Troncoso JC, Lee MK, Dawson TM, Dawson VL, Moore DJ (2012) Neurodegenerative phenotypes in an A53T alpha-synuclein transgenic mouse model are independent of LRRK2. Hum Mol Genet 21(11):2420–2431. doi:10.1093/hmg/dds057
65. Chen CY, Weng YH, Chien KY, Lin KJ, Yeh TH, Cheng YP, Lu CS, Wang HL (2012) (G2019S) LRRK2 activates MKK4-JNK pathway and causes degeneration of SN dopaminergic neurons in a transgenic mouse model of PD. Cell Death Differ 19(10):1623–1633. doi:10.1038/cdd.2012.42
66. Li X, Patel JC, Wang J, Avshalumov MV, Nicholson C, Buxbaum JD, Elder GA, Rice ME, Yue Z (2010) Enhanced striatal dopamine transmission and motor performance with LRRK2 overexpression in mice is eliminated by familial Parkinson's disease mutation G2019S. J Neurosci 30(5):1788–1797. doi:10.1523/JNEUROSCI.5604-09.2010
67. Li Y, Liu W, Oo TF, Wang L, Tang Y, Jackson-Lewis V, Zhou C, Geghman K, Bogdanov M, Przedborski S, Beal MF, Burke RE, Li C (2009) Mutant LRRK2(R1441G) BAC transgenic mice recapitulate cardinal features of Parkinson's disease. Nat Neurosci 12(7):826–828. doi:10.1038/nn.2349
68. Melrose HL, Dachsel JC, Behrouz B, Lincoln SJ, Yue M, Hinkle KM, Kent CB, Korvatska E, Taylor JP, Witten L, Liang YQ, Beevers JE, Boules M, Dugger BN, Serna VA, Gaukhman A, Yu X, Castanedes-Casey M, Braithwaite AT, Ogholikhan S, Yu N, Bass D, Tyndall G, Schellenberg GD, Dickson DW, Janus C, Farrer MJ (2010) Impaired dopaminergic neurotransmission and microtubule-associated protein tau alterations in human LRRK2 transgenic mice. Neurobiol Dis 40(3):503–517. doi:10.1016/j.nbd.2010.07.010
69. Ramonet D, Daher JP, Lin BM, Stafa K, Kim J, Banerjee R, Westerlund M, Pletnikova O, Glauser L, Yang L, Liu Y, Swing DA, Beal MF, Troncoso JC, McCaffery JM, Jenkins NA, Copeland NG, Galter D, Thomas B, Lee MK, Dawson TM, Dawson VL, Moore DJ (2011) Dopaminergic neuronal loss, reduced neurite complexity and autophagic abnormalities in transgenic mice expressing G2019S mutant LRRK2. PLoS One 6(4), e18568. doi:10.1371/journal.pone.0018568
70. Tsika E, Kannan M, Foo CS, Dikeman D, Glauser L, Gellhaar S, Galter D, Knott GW, Dawson TM, Dawson VL, Moore DJ (2014) Conditional expression of Parkinson's disease-related R1441C LRRK2 in midbrain dopaminergic neurons of mice causes nuclear abnormalities without neurodegeneration. Neurobiol Dis 71:345–358. doi:10.1016/j.nbd.2014.08.027
71. Liu G, Sgobio C, Gu X, Sun L, Lin X, Yu J, Parisiadou L, Xie C, Sastry N, Ding J, Lohr KM, Miller GW, Mateo Y, Lovinger DM, Cai H (2015) Selective expression of Parkinson's disease-related Leucine-rich repeat kinase 2 G2019S missense mutation in midbrain dopaminergic

neurons impairs dopamine release and dopaminergic gene expression. Hum Mol Genet 24(18):5299–5312. doi:10.1093/hmg/ddv249
72. Beccano-Kelly DA, Volta M, Munsie LN, Paschall SA, Tatarnikov I, Co K, Chou P, Cao LP, Bergeron S, Mitchell E, Han H, Melrose HL, Tapia L, Raymond LA, Farrer MJ, Milnerwood AJ (2015) LRRK2 overexpression alters glutamatergic presynaptic plasticity, striatal dopamine tone, postsynaptic signal transduction, motor activity and memory. Hum Mol Genet 24(5):1336–1349. doi:10.1093/hmg/ddu543
73. Garcia-Miralles M, Coomaraswamy J, Habig K, Herzig MC, Funk N, Gillardon F, Maisel M, Jucker M, Gasser T, Galter D, Biskup S (2015) No dopamine cell loss or changes in cytoskeleton function in transgenic mice expressing physiological levels of wild type or G2019S mutant LRRK2 and in human fibroblasts. PLoS One 10(4), e0118947. doi:10.1371/journal.pone.0118947
74. Tong Y, Pisani A, Martella G, Karouani M, Yamaguchi H, Pothos EN, Shen J (2009) R1441C mutation in LRRK2 impairs dopaminergic neurotransmission in mice. Proc Natl Acad Sci U S A 106(34):14622–14627. doi:10.1073/pnas.0906334106
75. Yue M, Hinkle KM, Davies P, Trushina E, Fiesel FC, Christenson TA, Schroeder AS, Zhang L, Bowles E, Behrouz B, Lincoln SJ, Beevers JE, Milnerwood AJ, Kurti A, McLean PJ, Fryer JD, Springer W, Dickson DW, Farrer MJ, Melrose HL (2015) Progressive dopaminergic alterations and mitochondrial abnormalities in LRRK2 G2019S knock-in mice. Neurobiol Dis 78:172–195. doi:10.1016/j.nbd.2015.02.031
76. Baptista MA, Dave KD, Frasier MA, Sherer TB, Greeley M, Beck MJ, Varsho JS, Parker GA, Moore C, Churchill MJ, Meshul CK, Fiske BK (2013) Loss of leucine-rich repeat kinase 2 (LRRK2) in rats leads to progressive abnormal phenotypes in peripheral organs. PLoS One 8(11), e80705. doi:10.1371/journal.pone.0080705
77. Ness D, Ren Z, Gardai S, Sharpnack D, Johnson VJ, Brennan RJ, Brigham EF, Olaharski AJ (2013) Leucine-rich repeat kinase 2 (LRRK2)-deficient rats exhibit renal tubule injury and perturbations in metabolic and immunological homeostasis. PLoS One 8(6), e66164. doi:10.1371/journal.pone.0066164
78. Boddu R, Hull TD, Bolisetty S, Hu X, Moehle MS, Daher JP, Kamal AI, Joseph R, George JF, Agarwal A, Curtis LM, West AB (2015) Leucine-rich repeat kinase 2 deficiency is protective in rhabdomyolysis-induced kidney injury. Hum Mol Genet 24(14):4078–4093. doi:10.1093/hmg/ddv147
79. Daher JP, Volpicelli-Daley LA, Blackburn JP, Moehle MS, West AB (2014) Abrogation of alpha-synuclein-mediated dopaminergic neurodegeneration in LRRK2-deficient rats. Proc Natl Acad Sci U S A 111(25):9289–9294. doi:10.1073/pnas.1403215111
80. Zhou H, Huang C, Tong J, Hong WC, Liu YJ, Xia XG (2011) Temporal expression of mutant LRRK2 in adult rats impairs dopamine reuptake. Int J Biol Sci 7(6):753–761
81. Lee JW, Tapias V, Di Maio R, Greenamyre JT, Cannon JR (2015) Behavioral, neurochemical, and pathologic alterations in bacterial artificial chromosome transgenic G2019S leucine-rich repeated kinase 2 rats. Neurobiol Aging 36(1):505–518. doi:10.1016/j.neurobiolaging.2014.07.011
82. Shaikh KT, Yang A, Youshin E, Schmid S (2015) Transgenic LRRK2 (R1441G) rats-a model for Parkinson disease? Peer J 3, e945. doi:10.7717/peerj.945
83. Walker MD, Volta M, Cataldi S, Dinelle K, Beccano-Kelly D, Munsie L, Kornelsen R, Mah C, Chou P, Co K, Khinda J, Mroczek M, Bergeron S, Yu K, Cao LP, Funk N, Ott T, Galter D, Riess O, Biskup S, Milnerwood AJ, Stoessl AJ, Farrer MJ, Sossi V (2014) Behavioral deficits and striatal DA signaling in LRRK2 p.G2019S transgenic rats: a multimodal investigation including PET neuroimaging. J Parkinsons Dis 4(3):483–498. doi:10.3233/JPD-140344
84. Lee BD, Shin JH, VanKampen J, Petrucelli L, West AB, Ko HS, Lee YI, Maguire-Zeiss KA, Bowers WJ, Federoff HJ, Dawson VL, Dawson TM (2010) Inhibitors of leucine-rich repeat kinase-2 protect against models of Parkinson's disease. Nat Med 16(9):998–1000. doi:10.1038/nm.2199

85. Dusonchet J, Kochubey O, Stafa K, Young SM Jr, Zufferey R, Moore DJ, Schneider BL, Aebischer P (2011) A rat model of progressive nigral neurodegeneration induced by the Parkinson's disease-associated G2019S mutation in LRRK2. J Neurosci 31(3):907–912. doi:10.1523/JNEUROSCI.5092-10.2011
86. Tsika E, Nguyen AP, Dusonchet J, Colin P, Schneider BL, Moore DJ (2015) Adenoviral-mediated expression of G2019S LRRK2 induces striatal pathology in a kinase-dependent manner in a rat model of Parkinson's disease. Neurobiol Dis 77:49–61. doi:10.1016/j.nbd.2015.02.019
87. Beilina A, Rudenko IN, Kaganovich A, Civiero L, Chau H, Kalia SK, Kalia LV, Lobbestael E, Chia R, Ndukwe K, Ding J, Nalls MA, International Parkinson's Disease Genomics Consortium, North American Brain Expression Consortium, Olszewski M, Hauser DN, Kumaran R, Lozano AM, Baekelandt V, Greene LE, Taymans JM, Greggio E, Cookson MR (2014) Unbiased screen for interactors of leucine-rich repeat kinase 2 supports a common pathway for sporadic and familial Parkinson disease. Proc Natl Acad Sci U S A 111(7):2626–2631. doi:10.1073/pnas.1318306111

Chapter 10
LRRK2 and the "LRRKtosome" at the Crossroads of Programmed Cell Death: Clues from RIP Kinase Relatives

Hardy J. Rideout and Diane B. Re

Abstract Since its cloning and identification in 2004, considerable gains have been made in the understanding of the basic functionality of leucine-rich repeat kinase 2 (LRRK2), including its kinase and GTPase activities, its protein interactors and subcellular localization, and its expression in the CNS and peripheral tissues. However, the mechanism(s) by which expression of mutant forms of LRRK2 lead to the death of dopaminergic neurons of the ventral midbrain remains largely uncharacterized. Because of its complex domain structure, LRRK2 exhibits similarities with multiple protein families including ROCO proteins, as well as the RIP kinases. Cellular models in which mutant LRRK2 is overexpressed in neuronal-like cell lines or in primary neurons have found evidence of apoptotic cell death involving components of the extrinsic as well as intrinsic death pathways. However, since the expression of LRRK2 is comparatively quite low in ventral midbrain dopaminergic neurons, the possibility exists that non-cell autonomous signaling also contributes to the loss of these neurons. In this chapter, we will discuss the different neuronal death pathways that may be activated by mutant forms of LRRK2, guided in part by the behavior of other members of the RIP kinase protein family.

Keywords LRRK2 • Apoptosis • Necroptosis • RIP kinase • Neurodegeneration • Oligomerization • FADD

H.J. Rideout (✉)
Division of Basic Neurosciences, Biomedical Research Foundation of the Academy of Athens, Soranou Efessiou 4, Athens 115 27, Greece
e-mail: hrideout@bioacademy.gr

D.B. Re
EHS Department and Motor Neuron Center, Columbia University, 722 W 168th Street Suite 1107-b, New York, NY 10032, USA
e-mail: dr2240@cumc.columbia.edu

H.J. Rideout (ed.), *Leucine-Rich Repeat Kinase 2 (LRRK2)*, Advances in Neurobiology 14, DOI 10.1007/978-3-319-49969-7_10

Introduction

The symptomology of Parkinson's disease (PD) that ultimately brings people to the Neurologist's office is typically the motor dysfunctions that are the consequence of death of dopamine-producing neurons in the substantia nigra pars compacta (SNpc). Of course the pathogenesis of the disease is considerably more complex and, as it's generally accepted, has begun years prior to the emergence of the overt motor symptoms and in regions distant from the ventral midbrain.

In the case of PD caused by expression of mutant forms of a protein called leucine-rich repeat kinase 2 (LRRK2), the same question remains: what is the mechanism by which widespread expression of altered forms of this protein preferentially leads to the death of SNpc dopaminergic neurons? LRRK2 is a large protein with a complex domain architecture and has a broad range of cellular functions in a wide variety of cell types and tissues (some of which are discussed in detail in other chapters of this volume). While the mutations most strongly associated with developing PD span multiple domains, particularly the enzymatic core of the protein, they possess the single uniform phenotype of causing the death of this specific neuronal population. It is widely expressed, yet at very low levels in ventral midbrain dopaminergic neurons, raising at least the possibility of a non-cell autonomous contribution to the neurodegenerative phenotype.

This chapter will focus on the induction of neuronal death signaling pathways by mutant LRRK2, keeping in mind the broader context of a considerably more complex disease pathogenesis. Potential clues to the mechanism of neuronal death induced by mutant forms of LRRK2 can be seen in other members of one protein family to which it belongs, the receptor-interacting protein (RIP) kinases.

Evidence of Cell/Neuronal Death Induced by Mutant LRRK2

Very soon after the first mutations in LRRK2 were identified in families with genetic forms of PD, it was reported that the same disease-linked mutations in LRRK2 were neurotoxic in different cellular models [1, 2]. The degree of cell death in both neuroblastoma cell lines and primary neurons was robust, reaching close to 50 % after just 48 h of expression. The initial method of assessment of neuronal "death" was based on the apparent health of neurites, rather than a classical cell viability measure. This was complemented however by the demonstration of TUNEL-positive nuclei, indicating that the mode of neuronal death induced by mutant LRRK2 was a form of programmed cell death, likely an apoptotic one. The neuronal death induced in these models required intact kinase activity of LRRK2, as multiple kinase-inactivating mutations reversed the phenotype [1, 3]. The apoptotic mode of cell death was confirmed soon after, in a report by Iaccarino and colleagues in 2007. Here, also using the SH-SY5Y neuroblastoma cell line, a robust induction of apoptotic death (defined solely by nuclear morphology) was observed following expression of three different mutant forms of LRRK2 [4]. The involvement of caspase signaling in the death induced by mutant LRRK2 was confirmed in this study by the appearance of cleaved/activated caspase-3 in cells with abnormal condensed nuclei and the reduction of such staining upon treatment with broad-spectrum caspase inhibitors. Subsequent studies continued to report apoptotic neuronal death in primary cultured neurons transiently overexpressing mutant forms of LRRK2, albeit at lower levels than those initially reported [5, 6]. Differences in expression levels, which are difficult to assess and compare in such transient overexpression studies, likely contribute to the variance in the induction of neuronal death. In another study, here employing automated image analysis, the longitudinal survival (plotted as the cumulative risk of death) in primary cultured neurons expressing mutant LRRK2 was assessed [7]. The authors elegantly demonstrated that LRRK2 kinase activity, as well as α-synuclein, modulated the death of neurons via changes in expression of LRRK2 itself.

Evidence of the mechanism of neuronal death in in vivo models of LRRK2 neurodegeneration is lacking, largely owing to the relative paucity of such models that exhibit neuronal loss, as well as the more protracted time course involved in an in vivo setting (weeks or months compared to 2–3 days in most cellular models). In fact, where a neurodegenerative phenotype is observed, only the *loss* of SNpc dopaminergic neurons is reported [8–11]. Confirmation of the mechanism(s) of neuronal loss that are observed in an isolated cell culture system (i.e., devoid of other potentially contributing cell types), in an in vivo setting is urgently needed. Combining conditional transgenic mouse models with viral vectors employing cell-specific promoters could potentially be a powerful tool to address this question. This is especially true in light of evidence that stimulated microglia from transgenic mice expressing mutant (R1441C) LRRK2 release greater levels of pro-inflammatory cytokines and induce neuronal death, in comparison to their wild-type counterparts [12]. It's likely that in the greater complexity of an organism, as opposed to purified embryonic neuronal cultures, the dynamics of dopamine neuron survival are significantly modulated by multiple intracellular pathways as well as communication with other cell types.

RIP Kinase Signaling in Neurodegeneration and Parallels to LRRK2 in Parkinson's Disease

Based on a number of criteria, including domain architecture, common protein interactors, and signaling pathways, many authors have included LRRK2 (as well as the related LRRK1) in the receptor-interacting protein (RIP) kinase family ([13]; Chap. 7 in this volume by N. Dzamko; and see Fig. 10.1). We have previously shown that LRRK2 can interact with RIP1 [5], although whether this interaction lead to potential reciprocal changes in phosphorylation status was not determined at the time.

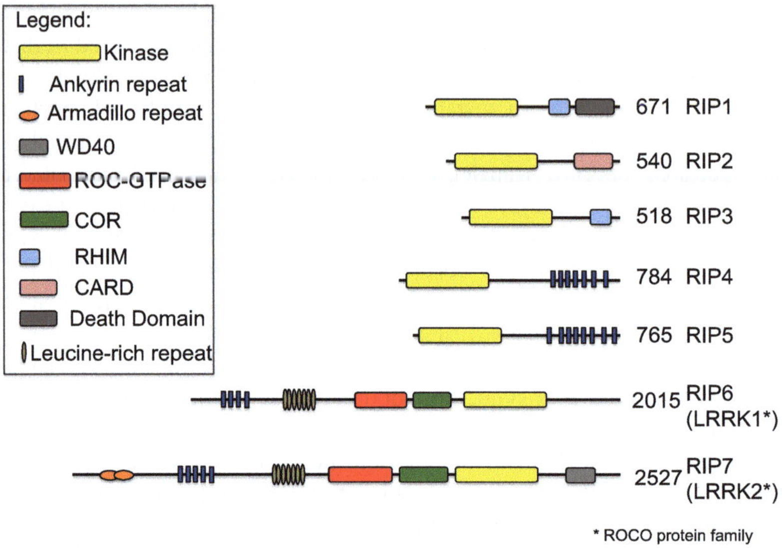

Fig. 10.1 Domain architecture of the RIP kinase family

Until 2005, when the term "necroptosis" was first applied [14], the modes of cell death were more broadly classified as either apoptotic, a regulated form of cellular demise, or necrotic, a passive uncontrolled type of cell death. With necroptosis, a type of cell death lacking the classical features of apoptosis, such as maintenance of plasma membrane integrity, but appearing to require the coordinated action of multiple signaling cascades, was beginning to be recognized. Among the effectors regulating the induction of necroptosis are RIP1, RIP3, and MLKL [15]; poly-ADP-ribose polymerase (PARP; [16]); the ubiquitination system [17]; as well as mitochondria and lysosomes [18, 19]. In addition to the TNF-α superfamily of plasma membrane death receptors typically associated with extrinsic apoptotic signaling [20, 21], necroptosis can also be induced by pathogen recognition receptors (e.g., Toll-like receptors (TLR)) expressed by cells of the innate immune system. For example, in bone marrow-derived dendritic cells, RIP1 and RIP3 are activated in response to treatment with the bacterial endotoxin LPS [22]. An important regulator of RIP- and RIP3-dependent necroptosis is caspase activity. Both of these kinases are cleavage substrates of caspase-8 [23, 24], leading to the inhibition of necroptosis and a shift to apoptotic cell death. It is worth noting here that selective activation of caspase-8 in brains of PD patients positive for mutations in LRRK2 has been demonstrated [5]. Additionally, RIP3 dimerization, which seeds further oligomer formation, is a requisite for necroptotic cell death and was dependent upon RIP1 kinase activity [25]. Multiple groups have shown that many of the pathogenic mutations in LRRK2 can re-localize within the cell into filamentous structures (e.g., [6, 26, 27]) resembling oligomeric death-effector filaments described by Siegel and colleagues [28]. At this point, however, it is unclear if such oligomerization of LRRK2 is necessary for its induction of neuronal death. Moreover, whether necroptotic death signaling, or a combined activation of apoptosis and necroptosis, plays any role in the loss of dopaminergic neurons in PD remains to be elucidated; in fact, only a single study thus far has implicated RIP1-dependent necroptotic death in a cell model of PD [29]. One potential scenario could be that an aborted apoptotic signaling cascade within a population of neurons reverts to a secondary necroptotic pathway in order to facilitate the death of the cell.

Stronger evidence exists, however, pointing to the involvement of RIP kinase signaling in other neurodegenerative diseases. The lysosomal storage disease, Gaucher's disease (GD), is a genetic metabolic disorder caused by mutations in the *GBA* gene, encoding lysosomal glucocerebrosidase. Among the many manifestations of this disease are often severe neurological symptoms, including PD. In fact, mutations in *GBA* are the most common genetic risk factor for PD [30]. A transgenic mouse model of GD, in which the endogenous mouse *Gba* is selectively deleted in neuronal cells (floxed *Gba* crossed with nestin-Cre mice), displays progressive cortical neuronal loss, without evidence of apoptotic cell death [31]. In the brains of these mice, the mRNA levels of both RIP1 and RIP3 were markedly elevated in comparison to control animals, suggesting that the neuronal loss was mediated by necroptosis. Remarkably, the life-span of mice lacking RIP3 (chosen because RIP1-null mice show embryonic lethality) treated with conduritol B epoxide (CBE; an inhibitor of GCase activity) was extended by almost threefold [31] and was accompanied by an abrogation of neuronal loss. This is salient for LRRK2-PD for a number of reasons. First, the link between lysosomal function, via GCase activity, and necroptotic signaling suggests another potential mechanism for neuronal loss in nigral dopamine neurons in PD, since lysosomal *dysfunction* is suspected as contributing at least in part to the pathology. Secondly, LRRK2 is closely related to RIP1 and RIP3 and is linked at multiple points to the regulation of autophagic/lysosomal degradation (the reader is referred to the chapter by Manzoni and Lewis in this volume for a more detailed discussion of LRRK2 and autophagy). Mice expressing a kinase dead form of LRRK2 from the endogenous locus have reduced levels of mTor in the kidney, coupled with increased levels of Akt [32]. Akt appears to be a phospho-substrate of RIP1 in neurons undergoing necroptosis, as is mTor [33]; and both proteins are required for the necroptotic death of neurons, downstream of formation of the RIP1/RIP3 "necrosome" complex. Moreover, caspase-8 mediates the cleavage of p62 in an RIP1-/RIP3-dependent manner [34] and is itself a substrate of lysosomal protein degradation [35], signaling the complex cross talk between different modes of cell death, controlled in part by the autophagic protein homeostasis machinery.

The motor neuron disease amyotrophic lateral sclerosis (ALS) is a devastating and fatal neurodegenerative disorder manifested by the loss of upper and lower motor neurons. Like PD, it is predominantly a sporadic disease; however about 10 % of the cases are familial (fALS), caused by mutations in several genes including superoxide dismutase 1 (SOD1), transactive response DNA-binding protein 43 (TDP-43), or chromosome 9 open reading frame 72 (C9ORF72). We and others have shown that in multiple cellular models of fALS, motor neuron death can be triggered by factors released by astrocytes [36–39]. Recently, however, in a "humanized" cellular model of ALS, we have demonstrated that primary astrocytes directly produced from patients with sporadic ALS (sALS) induced the death of human embryonic stem (ES) cell-derived motor neurons [40]. In this cell model of human ALS, and in the SOD1-linked murine cell model of fALS, the mechanism of motor neuron death was determined to be necroptotic, as the RIP1 inhibitor necrostatin-1 completely blocked the loss of motor neurons, as did downregulation of RIP1 by shRNA [40]. In these cells, caspase activity was detected early, yet inhibition with broad-spectrum inhibitors failed to protect motor neurons from astrocyte-dependent cell death, while Bax signaling, presumably at the level of the mitochondria, was a prerequisite. The fact that caspase inhibition is neuroprotective in cellular models of mutant LRRK2 neuronal death [4] does not entirely eliminate the possibility that necroptotic signaling also contributes to the loss of neuronal survival since caspase activity, particularly that of caspase-8, can shift the balance between apoptotic and necroptotic cell death depending on the setting and potential non-cell autonomous components. The majority of studies examining cell death induced by mutant LRRK2 have been in isolated cellular models (i.e., neuronal-like cell lines or highly enriched primary neuronal cultures) where the contributions of other cell types are removed. For example, in mixed primary cerebellar cultures, treatment with LPS induces neuronal death that is prevented by caspase inhibition-induced necroptosis of activated microglia [41]. When necrostatin-1 is added, microglia are protected, and neuronal death is then restored. While not a perfect correlate of models of mutant LRRK2-induced neurotoxicity, it is a good example of the complex signaling underlying necroptotic death and raises the possibility that the "picture" of neuronal death triggered by mutant LRRK2 may look completely different in a mixed culture setting. Moreover, in light of the reported interaction between RIP1 and LRRK2, two outstanding questions are immediately raised: (1) do RIP1 and LRRK2 phosphorylate each other, at baseline or in the context of a dying neuron? and (2) does necrostatin-1 protect against neuronal death induced by expression of mutant forms of LRRK2? .

In addition to GD and ALS, necroptosis linked to activation of a RIP kinase has also been recently described in a cellular model of Huntington's disease (HD). Zhu et al., using the ST14A striatal neuronal cell line expressing a fragment of Htt containing an expanded polyglutamine repeat, observed a marked increase in cell death upon treatment with pan-caspase- or caspase-8-specific inhibitors and that this death was blocked by the RIP1 kinase inhibitor necrostatin-1 [42]. Importantly, ICV delivered necrostatin-1 via an osmotic pump delayed the onset of motor dysfunction in the R6/2 mouse model of HD, improved performance on the rotarod test, delayed the decline in body weight, and provided a modest extension in survival. The authors concluded that the partial benefit provided by necrostatin-1 was attributable to the multiple modes of neuronal cell death, including apoptosis, observed in models of HD as well as in postmortem tissue of patients with late-stage HD [42]. This is a noteworthy example of a pathological condition with multiple forms of neuronal death coexisting and underlying the progression of the disease.

Perhaps the most intriguing link between RIP kinases and necroptotic signaling and LRRK2-mediated neuronal death comes from the phosphorylation of these

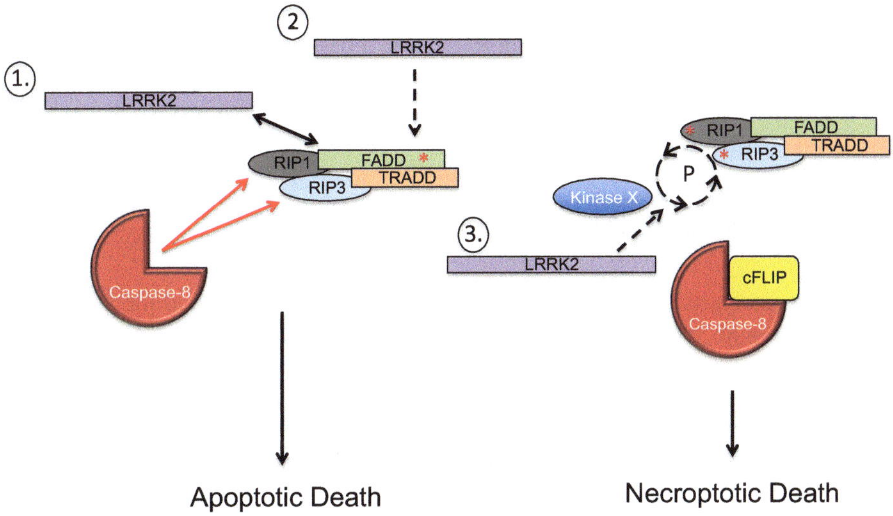

Fig. 10.2 LRRK2 in apoptotic and necroptotic signaling complexes. In response to a multitude of plasma membrane death receptor activation signals, the protein complex on the cytoplasmic face of the death receptor becomes internalized, recruiting RIP1 and RIP3 as well as caspase-8. The activation of an apoptotic cascade usually involves the caspase-8-dependent cleavage of RIP1 and RIP3. LRRK2 can potentially participate in this complex via its association with RIP1, TRADD, and FADD (Fig. 10.2, point *1*). Alternatively, mutations in LRRK2 may lead to cell death-associated phosphorylation of FADD (Fig. 10.2, point *2*), triggering an apoptotic cascade. In necroptotic death signaling, caspase-8 activity is inhibited either by its inactive homologue, cFLIP, or via caspase inhibitors and is unable to cleave and inactivate RIP1 and RIP3 kinases, which in turn are activated in a phosphorylation cycle with an upstream kinase. Whether this kinase is LRRK2 (Fig. 10.2, point *3*), thereby directly implicating its activity in necroptotic cell death, remains to be determined

kinases themselves. In cells treated with TNF-α, IKK-α/β can phosphorylate RIP1 and, in doing so, prevent it from being incorporated into the cytosolic necroptotic-inducing complex IIb [43]. Although the consequences are unclear at this point, LRRK2 is also phosphorylated by IKK-α/β, at the N-terminal cluster of Ser residues, particularly Ser910 and Ser935, but also potentially Ser955 [44]. It should be pointed out that multiple kinases, such as CK1α, have also been reported to phosphorylate LRRK2 at these sites [45], so the cell type, cellular environment, and genotype of LRRK2 may all potentially contribute to this regulation. Importantly, multiple pathogenic mutations in LRRK2 show a reduction in phosphorylation at these sites, at least in cellular overexpression studies [46–48]. However, whether this is linked to the capacity of mutant LRRK2 to induce neuronal death remains to be determined, especially since these studies were conducted in cells not susceptible to the death-inducing properties of LRRK2. In the case of RIP1, its phosphorylation by IKKα/IKKβ prevents its association with the death adaptor protein FADD [43]. It will thus be important to determine if the interaction of mutant LRRK2 with FADD [5] is modulated by its phosphorylation status. As WT LRRK2 also binds to FADD at baseline, perhaps an alteration in phosphorylation at one of the N-terminal

Ser residues of LRRK2 triggers the activation of FADD and the downstream death cascade. One possibility, shared with RIP1, is that baseline phosphorylation restricts LRRK2 in a pro-survival "mode," in which case the mutations associated with PD pathogenesis are, from this perspective, loss-of-function mutations.

Extrinsic Death Pathway Signaling and FADD

In part based on its similarity to other RIP kinases, we investigated the association of LRRK2 with members of the broader extrinsic death pathway. We found, in overexpression as well as at endogenous levels in mouse brain, that the death adaptor protein FADD interacted with WT LRRK2 [5]. This was also the case for another death domain containing protein TRADD, but only the interaction with FADD was altered significantly by pathogenic mutations in LRRK2. This led us to speculate whether FADD-dependent extrinsic death signaling played a role in the death of primary neurons expressing mutant forms of LRRK2. Indeed, overexpression of a dominant negative form of FADD, comprised of an enforced dimer of the FADD death domain, was protective in this cellular system. We found that this dominant negative, which lacks the death-effector domain required to recruit caspase-8, bound very strongly to LRRK2, whereas the isolated death-effector domain did not bind at all [5]. Downstream activation of caspase-8 was suspected in this model based on its participation in canonical FADD-dependent cell death; and in fact, downregulation of caspase-8 in primary neurons blocked the induction of neuronal death by mutant LRRK2.

Since LRRK2 itself is unable to bind a number of different plasma membrane death receptors (at least when co-expressed with the cytoplasmic domain of these receptors), the mechanism of activation of these components of the extrinsic pathway (i.e., FADD and caspase-8) diverges somewhat from the classical model involving an extracellular ligand. Here, occupation of the plasma membrane death receptor by ligand induces clustering of the receptor and recruitment of adaptor proteins such as FADD (see [49] for recent review). In many cellular models of overexpressed LRRK2, a redistribution of some pathogenic mutants into (presumably) oligomeric filamentous structures has been reported (e.g., [6, 26, 27]). A simplified model emerges then that mutant LRRK2 acts as a *soluble death receptor*, with the clustering at the plasma membrane mimicked by enhanced cytoplasmic oligomerization. Interestingly, RIP1 and RIP3 interaction and oligomerization in similar filamentous structures resembling β-amyloids have been described and demonstrated to be dependent on the RHIM domains of the two proteins [50]. Furthermore, mutations in the core RHIM residues of RIP1 and RIP3 have been shown to not only impair the formation of these β-amyloid-like filaments but also to prevent RIP kinase activation and necroptotic cell death to occur. LRRK2 does not possess similar RHIM domains but can also oligomerize (the reader is directed to Chapter X by E. Greggio and colleagues for a discussion on LRRK2 oligomerization), and with greater frequency with most pathogenic PD mutations.

It remains to be determined whether such structures are formed at endogenous levels in human brain. However, their presence in cellular models, particularly when also considering the formation of similar filamentous pools of LRRK2 dephosphorylated by pharmacological kinase inhibitors, and their shared dependence on binding of 14-3-3 to Ser910/935 can nevertheless still be informative for certain aspects of LRRK2 function.

The phosphorylation of FADD has multiple consequences, depending on the cell type, the kinase involved, and the residue modified. For example, phosphorylation at S194 by CK1a modulates cell cycle transition [51, 52], whereas phosphorylation at this residue by Plk1 in combination with phosphorylation at S203 by Aur-A leads to recruitment of caspase-8 and cell death signaling [53]. In our previous work [5], we did not determine whether the association between mutant LRRK2 and FADD led to alterations in the phosphorylation of FADD or whether this was required to induce neuronal death; however, an alteration in the phosphorylation levels of FADD, or specific residue, may be another explanation for the activation of FADD-dependent neuronal death despite a strong interaction with wild-type LRRK2. The two phosphorylation sites in FADD, S194 and S203, are located within the death domain of the protein, which we have previously shown to be the region of FADD that binds LRRK2 [5], suggesting that phosphorylation of FADD by LRRK2 is possible.

Intrinsic Death Pathway and Mitochondrial Signaling

The evidence that mutant LRRK2 activates components of the extrinsic death pathway such as FADD and caspase-8 does not necessarily imply that neuronal death induced by mutant LRRK2 *exclusively* relies on this pathway. In fact, there is much evidence to support that intrinsic, mitochondrial-dependent, cell death components also play a role. It is possible that the well-described cross talk between the extrinsic and intrinsic death pathways, involving the caspase-8-mediated cleavage and activation of the BH3-only domain protein Bid, is involved. RIP1-dependent necroptotic signaling can also recruit intrinsic pathway components, such as Bax, as in the astrocyte-mediated death of motor neurons described above [37, 40]. Moreover, there could also be regional or neuronal subtype-specific variations in the ways in which neurons respond to expression of mutant forms of LRRK2. Certainly, dopaminergic neurons of the SNpc are considerably more vulnerable to myriad insults in comparison to other neuronal populations and may also respond in different ways to changes in the local environment (e.g., altered signaling from infiltrating myeloid cells).

In terms of direct links between LRRK2 and classical intrinsic pathway proteins, two recent reports describe potential phospho-substrates of LRRK2. First, in 2015, Su et al. [54] reported that LRRK2 phosphorylates Bcl-2 at Thr56 and that this was required for the depolarization of mitochondria induced by G2019S-LRRK2, leading to the mitophagic removal of depolarized mitochondria. Previous work has shown that phosphorylation of Bcl-2 at Thr56, as well as Ser87, by p38 MAP kinase suppresses its antiapoptotic properties [55], raising the possibility that loss of neuronal survival could also be mediated at least in part by LRRK2-dependent phosphorylation of Bcl-2 at this residue. That Bcl-2 is indeed a true phospho-substrate of LRRK2 that awaits further validation; however, the fact that this residue seems to be important for the pro-survival, as well as the autophagic-suppressive, roles of Bcl-2 is in line with the known cellular functions of LRRK2.

The second proposed direct link between LRRK2 and a protein predominantly associated with cell death is the reported phosphorylation of p53 by LRRK2. In 2015, the group led by Seol and colleagues [56] identified Thr304/377 as candidate p53 phosphorylation sites modified by LRRK2. There has long been evidence that p53 may participate in neuronal death in PD, both from models as well as studies on postmortem PD brain tissue [57–59]; and other work has indeed shown that phosphorylation of p53 enhances its localization to the nucleus and activation of downstream death pathways [60]. Phosphorylation of recombinant p53 by purified LRRK2 (ΔN-LRRK2) was increased in the presence of the G2019S mutation and absent when D1994A kinase dead LRRK2 was present [56]. Additionally, while not indicative of a *direct* phosphorylation of p53 by LRRK2, elevated levels of phosphorylated p53 were observed in human neurons differentiated from iPS cells derived from G2019S-LRRK2-positive fibroblasts and in ventral midbrain tissue from transgenic mice overexpressing G2019S-LRRK2. In the differentiated human neurons, treatment with the LRRK2-IN1 inhibitor only partially reduced levels of phosphorylated p53, raising the possibility, as the authors rightly pointed out, that other kinases could also phosphorylate p53 at these sites. In cells and primary neurons overexpressing mutant G2019S-LRRK2, phosphorylated p53 was associated with an increase in p21, but not MDM2, expression, a finding recapitulated by overexpression of p53 with phosphomimetic substitutions at these sites. Taken together, these results are suggestive of phosphorylation of p53 by LRRK2. Whether this is a direct phosphorylation by LRRK2 in vivo awaits determination and, more specifically, whether p53-null mice are resistant to mutant LRRK2-induced neurodegeneration would be a strong confirmation of the importance of this posttranslational modification. Intriguingly, one transcriptional target of p53 is the recently identified p53-induced protein with a death domain (PIDD; [61]), which when in a complex with RIP1 activates NF-kB, but under conditions of cellular stress triggers cell death via a complex formed with caspase-2.

Other recent work has postulated a direct modulation of a critical mitochondrial membrane complex that controls the release of proapoptotic proteins such as AIF, cytochrome c, and SMAC/Diablo. Cui et al. [62] showed that LRRK2 interacts with the adenine nucleotide transporter (ANT), the voltage-dependent anion channel (VDAC), and the mitochondrial creatine kinase (mtCK), which together form the mitochondrial permeability transition pore complex. The localization of LRRK2 in this complex prevented the insertion of mtCK and increased the likelihood of opening of this pore by facilitating the direct interaction between ANT and VDAC, providing a vital clue to potential late-stage apoptotic signaling in mutant LRRK2-expressing neurons.

Is it Strictly a Cell Autonomous Mechanism?

Despite the fact that, according to multiple reports, expression of mutant LRRK2 in cultured primary neurons and neuroblastoma cells leads to cell death, this does not necessarily indicate that the same mechanism underlies the neurodegeneration in PD. The reader is directed to Chapter X for a more in-depth discussion of LRRK2 and the immune system. For example, LRRK2 is highly expressed in peripheral blood monocytes as well as cells of the myeloid cell lineage (macrophages and microglia; e.g., [63]), and there is compelling evidence to suggest a role for these cell types in the pathogenesis of PD. Primary microglia cells derived from transgenic mice expressing R1441C-LRRK2 release more IL-1β, IL-6, and TNF-α upon stimulation with LPS compared to non-transgenic control microglia and less IL-10 [12]. Importantly, conditioned medium from R1441C-LRRK2 microglia induced significantly greater toxicity when applied to primary cortical neurons [12], demonstrating in a cellular model at least, that non-cell autonomous mechanisms can also influence neuronal survival in the context of mutant LRRK2. In 2012, the West group nicely showed that LRRK2 levels and function increased in TLR4-stimulated microglia and, importantly, that cytokine release from stimulated microglia was attenuated in response to LRRK2 kinase inhibition [64].

TLR4 receptor stimulation can also recruit and activate RIP1 (see Ofengeim and Yuan for a recent review; [65]), leading under certain circumstances to the activation of NF-κB signaling. And inflammatory cytokine signaling by microglia in general is regulated by necroptotic processes within microglia themselves, under the control of RIP1 activity [66]. It is not known whether under baseline or activated conditions, there is an interaction between RIP1 and LRRK2 within microglia. Recall that these two kinases are known to interact [5], at least when co-expressed; therefore, it is tempting to speculate that their activity may jointly play a role in inflammatory signaling by microglia under pathological conditions.

Conclusions and Future Directions

While there still remains a considerable knowledge gap concerning the mechanism(s) by which expression of certain mutant forms of LRRK2 induces death of dopaminergic neurons of the SNpc in PD patients, there are several important findings in hand that can inform future studies. First, while the issues of expression level of mutant LRRK2 need to be addressed, from the cellular studies reported thus far, overexpression of pathogenic mutant forms of LRRK2 are sufficient to induce an apoptotic-like cell death in primary neurons. This death is caspase dependent and involves mitochondrial dysfunction, as well as upstream activation of extrinsic death pathway components. If these pathways are confirmed in animal models of LRRK2 neurodegeneration, several potential therapeutic targets then become available independently of approaches directed against LRRK2 kinase activity. Secondly, some of the known protein interactors of LRRK2, as well as several closely related protein kinases (of the RIP kinase family), participate in multiple well-characterized cell death pathways, including apoptosis, but also necroptosis. Studies moving forward can capitalize on the well-described components of necroptotic signaling complexes to determine (a) if LRRK2 is a constituent member of such complexes and (b) if manipulating these complexes, or the activity of other proteins contained therein, such as RIP1 or RIP3, can alter the course of mutant LRRK2-induced neuronal death, particularly in an in vivo setting. LRRK2 shares many common binding partners, such as the adaptor proteins, FADD and TRADD, and RIP1 itself; and certain features of death-inducing mutant forms of LRRK2 are common to RIP1 and RIP3 protein kinases, such as a propensity to form dimers and oligomers. Thus, LRRK2 may actively participate in RIP1/RIP3 death signaling pathways in general; and conversely, RIP1 and RIP3 necroptotic signaling may participate in the pathogenesis of LRRK2-PD.

The death-effector-like/skein-like filamentous structures formed by mutant LRRK2, and dephosphorylated wild-type LRRK2 for that matter, may be analogous to the amyloid filaments/oligomers formed by RIP1 and RIP3, which are required for RIP1-/RIP3-dependent necroptotic cell death under certain circumstances. Whether they are similarly required for LRRK2-dependent neuronal death remains to be determined. Likewise, whether there is any interaction with MLKL, as there is for RIP1 and RIP3, should also be investigated.

It also remains a distinct possibility that at the organism level, there is no cell-autonomous death in dopamine neurons expressing mutant LRRK2. In other words, does mutant LRRK2 have a *direct* consequence on the health of SNpc dopaminergic neurons? In rodents, LRRK2 expression in this region is low in comparison to striatum [67]; however, in primates higher expression levels have been reported in this region [68]. Another point to consider is that LRRK2 is also present in exosomes in human urine and CSF [69], and although the cell source of these exosomes is not yet known, they could certainly contribute to potential non-cell autonomous aspects of neuronal death. As more in vivo models where neurodegeneration is present become available, answers to these questions are now within reach and strengthen the foundation from which novel therapeutic targets are identified.

Conflict of Interest The author declares no conflicts of interest.

References

1. Smith WW, Pei Z, Jiang H, Dawson VL, Dawson TM, Ross CA (2006) Kinase activity of mutant LRRK2 mediates neuronal toxicity. Nat Neurosci 9(10):1231–1233. doi:10.1038/nn1776
2. Smith WW, Pei Z, Jiang H, Moore DJ, Liang Y, West AB, Dawson VL, Dawson TM, Ross CA (2005) Leucine-rich repeat kinase 2 (LRRK2) interacts with parkin, and mutant LRRK2 induces neuronal degeneration. Proc Natl Acad Sci U S A 102(51):18676–18681. doi:10.1073/pnas.0508052102

3. West AB, Moore DJ, Biskup S, Bugayenko A, Smith WW, Ross CA, Dawson VL, Dawson TM (2005) Parkinson's disease-associated mutations in leucine-rich repeat kinase 2 augment kinase activity. Proc Natl Acad Sci U S A 102(46):16842–16847. doi:10.1073/pnas.0507360102
4. Iaccarino C, Crosio C, Vitale C, Sanna G, Carri MT, Barone P (2007) Apoptotic mechanisms in mutant LRRK2-mediated cell death. Hum Mol Genet 16(11):1319–1326. doi:10.1093/hmg/ddm080
5. Ho CC, Rideout HJ, Ribe E, Troy CM, Dauer WT (2009) The Parkinson disease protein leucine-rich repeat kinase 2 transduces death signals via Fas-associated protein with death domain and caspase-8 in a cellular model of neurodegeneration. J Neurosci 29(4):1011–1016. doi:10.1523/JNEUROSCI.5175-08.2009
6. Kett LR, Boassa D, Ho CC, Rideout HJ, Hu J, Terada M, Ellisman M, Dauer WT (2012) LRRK2 Parkinson disease mutations enhance its microtubule association. Hum Mol Genet 21(4):890–899. doi:10.1093/hmg/ddr526
7. Skibinski G, Nakamura K, Cookson MR, Finkbeiner S (2014) Mutant LRRK2 toxicity in neurons depends on LRRK2 levels and synuclein but not kinase activity or inclusion bodies. J Neurosci 34(2):418–433. doi:10.1523/JNEUROSCI.2712-13.2014
8. Chen CY, Weng YH, Chien KY, Lin KJ, Yeh TH, Cheng YP, Lu CS, Wang HL (2012) (G2019S) LRRK2 activates MKK4-JNK pathway and causes degeneration of SN dopaminergic neurons in a transgenic mouse model of PD. Cell Death Differ 19(10):1623–1633. doi:10.1038/cdd.2012.42
9. Dusonchet J, Kochubey O, Stafa K, Young SM Jr, Zufferey R, Moore DJ, Schneider BL, Aebischer P (2011) A rat model of progressive nigral neurodegeneration induced by the Parkinson's disease-associated G2019S mutation in LRRK2. J Neurosci 31(3):907–912. doi:10.1523/JNEUROSCI.5092-10.2011
10. Lee BD, Shin JH, VanKampen J, Petrucelli L, West AB, Ko HS, Lee YI, Maguire-Zeiss KA, Bowers WJ, Federoff HJ, Dawson VL, Dawson TM (2010) Inhibitors of leucine-rich repeat kinase-2 protect against models of Parkinson's disease. Nat Med 16(9):998–1000
11. Ramonet D, Daher JP, Lin BM, Stafa K, Kim J, Banerjee R, Westerlund M, Pletnikova O, Glauser L, Yang L, Liu Y, Swing DA, Beal MF, Troncoso JC, McCaffery JM, Jenkins NA, Copeland NG, Galter D, Thomas B, Lee MK, Dawson TM, Dawson VL, Moore DJ (2011) Dopaminergic neuronal loss, reduced neurite complexity and autophagic abnormalities in transgenic mice expressing G2019S mutant LRRK2. PLoS One 6(4), e18568. doi:10.1371/journal.pone.0018568
12. Gillardon F, Schmid R, Draheim H (2012) Parkinson's disease-linked leucine-rich repeat kinase 2(R1441G) mutation increases proinflammatory cytokine release from activated primary microglial cells and resultant neurotoxicity. Neuroscience 208:41–48
13. Meylan E, Tschopp J (2005) The RIP kinases: crucial integrators of cellular stress. Trends Biochem Sci 30(3):151–159. doi:10.1016/j.tibs.2005.01.003
14. Degterev A, Huang Z, Boyce M, Li Y, Jagtap P, Mizushima N, Cuny GD, Mitchison TJ, Moskowitz MA, Yuan J (2005) Chemical inhibitor of nonapoptotic cell death with therapeutic potential for ischemic brain injury. Nat Chem Biol 1(2):112–119. doi:10.1038/nchembio711
15. Oberst A (2015) Death in the fast lane: what's next for necroptosis? FEBS J 283(14):2616–2625. doi:10.1111/febs.13520
16. Zong WX, Ditsworth D, Bauer DE, Wang ZQ, Thompson CB (2004) Alkylating DNA damage stimulates a regulated form of necrotic cell death. Genes Dev 18(11):1272–1282. doi:10.1101/gad.1199904
17. Wang Y, Huang ZH, Li YJ, He GW, Yu RY, Yang J, Liu WT, Li B, He QY (2016) Dynamic quantitative proteomics characterization of TNF-alpha-induced necroptosis. Apoptosis : an international journal on programmed cell death 21(12):1438–1446. doi:10.1007/s10495-016-1300-z
18. Boya P, Kroemer G (2008) Lysosomal membrane permeabilization in cell death. Oncogene 27(50):6434–6451. doi:10.1038/onc.2008.310

19. Schinzel AC, Takeuchi O, Huang Z, Fisher JK, Zhou Z, Rubens J, Hetz C, Danial NN, Moskowitz MA, Korsmeyer SJ (2005) Cyclophilin D is a component of mitochondrial permeability transition and mediates neuronal cell death after focal cerebral ischemia. Proc Natl Acad Sci U S A 102(34):12005–12010. doi:10.1073/pnas.0505294102
20. Laster SM, Wood JG, Gooding LR (1988) Tumor necrosis factor can induce both apoptic and necrotic forms of cell lysis. J Immunol 141(8):2629–2634
21. Vercammen D, Beyaert R, Denecker G, Goossens V, Van Loo G, Declercq W, Grooten J, Fiers W, Vandenabeele P (1998) Inhibition of caspases increases the sensitivity of L929 cells to necrosis mediated by tumor necrosis factor. J Exp Med 187(9):1477–1485
22. Moriwaki K, Bertin J, Gough PJ, Chan FK (2015) A RIPK3-caspase 8 complex mediates atypical pro-IL-1beta processing. J Immunol 194(4):1938–1944. doi:10.4049/jimmunol.1402167
23. Feng S, Yang Y, Mei Y, Ma L, Zhu DE, Hoti N, Castanares M, Wu M (2007) Cleavage of RIP3 inactivates its caspase-independent apoptosis pathway by removal of kinase domain. Cell Signal 19(10):2056–2067. doi:10.1016/j.cellsig.2007.05.016
24. Lin Y, Devin A, Rodriguez Y, Liu ZG (1999) Cleavage of the death domain kinase RIP by caspase-8 prompts TNF-induced apoptosis. Genes Dev 13(19):2514–2526
25. Orozco S, Yatim N, Werner MR, Tran H, Gunja SY, Tait SW, Albert ML, Green DR, Oberst A (2014) RIPK1 both positively and negatively regulates RIPK3 oligomerization and necroptosis. Cell Death Differ 21(10):1511–1521. doi:10.1038/cdd.2014.76
26. Alegre-Abarrategui J, Christian H, Lufino MM, Mutihac R, Venda LL, Ansorge O, Wade-Martins R (2009) LRRK2 regulates autophagic activity and localizes to specific membrane microdomains in a novel human genomic reporter cellular model. Hum Mol Genet 18(21):4022–4034
27. Reynolds A, Doggett EA, Riddle SM, Lebakken CS, Nichols RJ (2014) LRRK2 kinase activity and biology are not uniformly predicted by its autophosphorylation and cellular phosphorylation site status. Front Mol Neurosci 7:54. doi:10.3389/fnmol.2014.00054
28. Siegel RM, Martin DA, Zheng L, Ng SY, Bertin J, Cohen J, Lenardo MJ (1998) Death-effector filaments: novel cytoplasmic structures that recruit caspases and trigger apoptosis. J Cell Biol 141(5):1243–1253
29. Wu JR, Wang J, Zhou SK, Yang L, Yin JL, Cao JP, Cheng YB (2015) Necrostatin-1 protection of dopaminergic neurons. Neural Regen Res 10(7):1120–1124. doi:10.4103/1673-5374.160108
30. Houlden H, Singleton AB (2012) The genetics and neuropathology of Parkinson's disease. Acta Neuropathol 124(3):325–338. doi:10.1007/s00401-012-1013-5
31. Vitner EB, Salomon R, Farfel-Becker T, Meshcheriakova A, Ali M, Klein AD, Platt FM, Cox TM, Futerman AH (2014) RIPK3 as a potential therapeutic target for Gaucher's disease. Nat Med 20(2):204–208. doi:10.1038/nm.3449
32. Herzig MC, Kolly C, Persohn E, Theil D, Schweizer T, Hafner T, Stemmelen C, Troxler TJ, Schmid P, Danner S, Schnell CR, Mueller M, Kinzel B, Grevot A, Bolognani F, Stirn M, Kuhn RR, Kaupmann K, van der Putten PH, Rovelli G, Shimshek DR (2011) LRRK2 protein levels are determined by kinase function and are crucial for kidney and lung homeostasis in mice. Hum Mol Genet 20(21):4209–4223. doi:10.1093/hmg/ddr348
33. Liu Q, Qiu J, Liang M, Golinski J, van Leyen K, Jung JE, You Z, Lo EH, Degterev A, Whalen MJ (2014) Akt and mTOR mediate programmed necrosis in neurons. Cell Death Dis 5, e1084. doi:10.1038/cddis.2014.69
34. Matsuzawa Y, Oshima S, Nibe Y, Kobayashi M, Maeyashiki C, Nemoto Y, Nagaishi T, Okamoto R, Tsuchiya K, Nakamura T, Watanabe M (2015) RIPK3 regulates p62-LC3 complex formation via the caspase-8-dependent cleavage of p62. Biochem Biophys Res Commun 456(1):298–304. doi:10.1016/j.bbrc.2014.11.075
35. Hou W, Han J, Lu C, Goldstein LA, Rabinowich H (2010) Autophagic degradation of active caspase-8: a crosstalk mechanism between autophagy and apoptosis. Autophagy 6(7):891–900. doi:10.4161/auto.6.7.13038

36. Di Giorgio FP, Carrasco MA, Siao MC, Maniatis T, Eggan K (2007) Non-cell autonomous effect of glia on motor neurons in an embryonic stem cell-based ALS model. Nat Neurosci 10(5):608–614. doi:10.1038/nn1885
37. Nagai M, Re DB, Nagata T, Chalazonitis A, Jessell TM, Wichterle H, Przedborski S (2007) Astrocytes expressing ALS-linked mutated SOD1 release factors selectively toxic to motor neurons. Nat Neurosci 10(5):615–622. doi:10.1038/nn1876
38. Rojas F, Cortes N, Abarzua S, Dyrda A, van Zundert B (2014) Astrocytes expressing mutant SOD1 and TDP43 trigger motoneuron death that is mediated via sodium channels and nitroxidative stress. Front Cell Neurosci 8:24. doi:10.3389/fncel.2014.00024
39. Vargas MR, Pehar M, Cassina P, Beckman JS, Barbeito L (2006) Increased glutathione biosynthesis by Nrf2 activation in astrocytes prevents p75NTR-dependent motor neuron apoptosis. J Neurochem 97(3):687–696. doi:10.1111/j.1471-4159.2006.03742.x
40. Re DB, Le Verche V, Yu C, Amoroso MW, Politi KA, Phani S, Ikiz B, Hoffmann L, Koolen M, Nagata T, Papadimitriou D, Nagy P, Mitsumoto H, Kariya S, Wichterle H, Henderson CE, Przedborski S (2014) Necroptosis drives motor neuron death in models of both sporadic and familial ALS. Neuron 81(5):1001–1008. doi:10.1016/j.neuron.2014.01.011
41. Fricker M, Vilalta A, Tolkovsky AM, Brown GC (2013) Caspase inhibitors protect neurons by enabling selective necroptosis of inflamed microglia. J Biol Chem 288(13):9145–9152. doi:10.1074/jbc.M112.427880
42. Zhu S, Zhang Y, Bai G, Li H (2011) Necrostatin-1 ameliorates symptoms in R6/2 transgenic mouse model of Huntington's disease. Cell Death Dis 2, e115. doi:10.1038/cddis.2010.94
43. Dondelinger Y, Jouan-Lanhouet S, Divert T, Theatre E, Bertin J, Gough PJ, Giansanti P, Heck AJ, Dejardin E, Vandenabeele P, Bertrand MJ (2015) NF- κB-independent role of IKκ/IKKβ in Preventing RIPK1 kinase-dependent apoptotic and necroptotic cell death during TNF signaling. Mol Cell 60(1):63–76. doi:10.1016/j.molcel.2015.07.032
44. Dzamko N, Inesta-Vaquera F, Zhang J, Xie C, Cai H, Arthur S, Tan L, Choi H, Gray N, Cohen P, Pedrioli P, Clark K, Alessi DR (2012) The IκB kinase family phosphorylates the Parkinson's disease kinase LRRK2 at Ser935 and Ser910 during Toll-like receptor signaling. PLoS One 7(6), e39132. doi:10.1371/journal.pone.0039132
45. Chia R, Haddock S, Beilina A, Rudenko IN, Mamais A, Kaganovich A, Li Y, Kumaran R, Nalls MA, Cookson MR (2014) Phosphorylation of LRRK2 by casein kinase 1α regulates *trans*-Golgi clustering via differential interaction with ARHGEF7. Nat Commun 5:5827. doi:10.1038/ncomms6827
46. Dzamko N, Deak M, Hentati F, Reith AD, Prescott AR, Alessi DR, Nichols RJ (2010) Inhibition of LRRK2 kinase activity leads to dephosphorylation of Ser^{910}/Ser^{935}, disruption of 14-3-3 binding and altered cytoplasmic localization. Biochem J 430(3):405–413
47. Li X, Wang QJ, Pan N, Lee S, Zhao Y, Chait BT, Yue Z (2011) Phosphorylation-dependent 14-3-3 binding to LRRK2 is impaired by common mutations of familial Parkinson's disease. PLoS One 6(3), e17153. doi:10.1371/journal.pone.0017153
48. Nichols RJ, Dzamko N, Morrice NA, Campbell DG, Deak M, Ordureau A, Macartney T, Tong Y, Shen J, Prescott AR, Alessi DR (2010) 14-3-3 binding to LRRK2 is disrupted by multiple Parkinson's disease-associated mutations and regulates cytoplasmic localization. Biochem J 430(3):393–404
49. Yang JK (2015) Death effecter domain for the assembly of death-inducing signaling complex. Apoptosis 20(2):235–239. doi:10.1007/s10495-014-1060-6
50. Li J, McQuade T, Siemer AB, Napetschnig J, Moriwaki K, Hsiao YS, Damko E, Moquin D, Walz T, McDermott A, Chan FK, Wu H (2012) The RIP1/RIP3 necrosome forms a functional amyloid signaling complex required for programmed necrosis. Cell 150(2):339–350. doi:10.1016/j.cell.2012.06.019
51. Alappat EC, Feig C, Boyerinas B, Volkland J, Samuels M, Murmann AE, Thorburn A, Kidd VJ, Slaughter CA, Osborn SL, Winoto A, Tang WJ, Peter ME (2005) Phosphorylation of FADD at serine 194 by CKIα regulates its nonapoptotic activities. Mol Cell 19(3):321–332. doi:10.1016/j.molcel.2005.06.024

52. Jang MS, Lee SJ, Kim CJ, Lee CW, Kim E (2011) Phosphorylation by polo-like kinase 1 induces the tumor-suppressing activity of FADD. Oncogene 30(4):471–481. doi:10.1038/onc.2010.423
53. Jang MS, Lee SJ, Kang NS, Kim E (2011) Cooperative phosphorylation of FADD by Aur-A and Plk1 in response to taxol triggers both apoptotic and necrotic cell death. Cancer Res 71(23):7207–7215. doi:10.1158/0008-5472.CAN-11-0760
54. Su YC, Guo X, Qi X (2015) Threonine 56 phosphorylation of Bcl-2 is required for LRRK2 G2019S-induced mitochondrial depolarization and autophagy. Biochim Biophys Acta 1852(1):12–21. doi:10.1016/j.bbadis.2014.11.009
55. De Chiara G, Marcocci ME, Torcia M, Lucibello M, Rosini P, Bonini P, Higashimoto Y, Damonte G, Armirotti A, Amodei S, Palamara AT, Russo T, Garaci E, Cozzolino F (2006) Bcl-2 Phosphorylation by p38 MAPK: identification of target sites and biologic consequences. J Biol Chem 281(30):21353–21361. doi:10.1074/jbc.M511052200
56. Ho DH, Kim H, Kim J, Sim H, Ahn H, Kim J, Seo H, Chung KC, Park BJ, Son I, Seol W (2015) Leucine-Rich Repeat Kinase 2 (LRRK2) phosphorylates p53 and induces p21(WAF1/CIP1) expression. Mol Brain 8:54. doi:10.1186/s13041-015-0145-7
57. Dietrich P, Rideout HJ, Wang Q, Stefanis L (2003) Lack of p53 delays apoptosis, but increases ubiquitinated inclusions, in proteasomal inhibitor-treated cultured cortical neurons. Mol Cell Neurosci 24(2):430–441
58. Mogi M, Kondo T, Mizuno Y, Nagatsu T (2007) p53 protein, interferon-gamma, and NF-κB levels are elevated in the parkinsonian brain. Neurosci Lett 414(1):94–97. doi:10.1016/j.neulet.2006.12.003
59. Nair VD (2006) Activation of p53 signaling initiates apoptotic death in a cellular model of Parkinson's disease. Apoptosis 11(6):955–966. doi:10.1007/s10495-006-6316-3
60. Liang SH, Clarke MF (2001) Regulation of p53 localization. Eur J Biochem 268(10):2779–2783
61. Lin Y, Ma W, Benchimol S (2000) Pidd, a new death-domain-containing protein, is induced by p53 and promotes apoptosis. Nat Genet 26(1):122–127. doi:10.1038/79102
62. Cui J, Yu M, Niu J, Yue Z, Xu Z (2011) Expression of leucine-rich repeat kinase 2 (LRRK2) inhibits the processing of uMtCK to induce cell death in a cell culture model system. Biosci Rep 31(5):429–437. doi:10.1042/BSR20100127
63. Moehle MS, Daher JP, Hull TD, Boddu R, Abdelmotilib HA, Mobley J, Kannarkat GT, Tansey MG, West AB (2015) The G2019S LRRK2 mutation increases myeloid cell chemotactic responses and enhances LRRK2 binding to actin-regulatory proteins. Hum Mol Genet 24(15):4250–4267. doi:10.1093/hmg/ddv157
64. Moehle MS, Webber PJ, Tse T, Sukar N, Standaert DG, DeSilva TM, Cowell RM, West AB (2012) LRRK2 inhibition attenuates microglial inflammatory responses. J Neurosci 32(5):1602–1611. doi:10.1523/JNEUROSCI.5601-11.2012
65. Ofengeim D, Yuan J (2013) Regulation of RIP1 kinase signalling at the crossroads of inflammation and cell death. Nature reviews Molecular cell biology 14 (11):727–736. doi:10.1038/nrm3683
66. Kearney CJ, Cullen SP, Tynan GA, Henry CM, Clancy D, Lavelle EC, Martin SJ (2015) Necroptosis suppresses inflammation via termination of TNF- or LPS-induced cytokine and chemokine production. Cell Death Differ 22(8):1313–1327. doi:10.1038/cdd.2014.222
67. Davies P, Hinkle KM, Sukar NN, Sepulveda B, Mesias R, Serrano G, Alessi DR, Beach TG, Benson DL, White CL, Cowell RM, Das SS, West AB, Melrose HL (2013) Comprehensive characterization and optimization of anti-LRRK2 (leucine-rich repeat kinase 2) monoclonal antibodies. Biochem J 453(1):101–113. doi:10.1042/BJ20121742
68. Lee H, Melrose HL, Yue M, Pare JF, Farrer MJ, Smith Y (2010) Lrrk2 localization in the primate basal ganglia and thalamus: a light and electron microscopic analysis in monkeys. Exp Neurol 224(2):438–447. doi:10.1016/j.expneurol.2010.05.004
69. Fraser KB, Moehle MS, Daher JP, Webber PJ, Williams JY, Stewart CA, Yacoubian TA, Cowell RM, Dokland T, Ye T, Chen D, Siegal GP, Galemmo RA, Tsika E, Moore DJ, Standaert DG, Kojima K, Mobley JA, West AB (2013) LRRK2 secretion in exosomes is regulated by 14-3-3. Hum Mol Genet 22(24):4988–5000. doi:10.1093/hmg/ddt346

Chapter 11
Interaction of LRRK2 and α-Synuclein in Parkinson's Disease

João Paulo Lima Daher

Abstract Parkinson's disease (PD) is a progressively debilitating neurodegenerative syndrome. It is best described as a movement disorder characterized by motor dysfunctions, progressive degeneration of dopaminergic neurons of the substantia nigra pars compacta, and abnormal intraneuronal protein aggregates, named Lewy bodies and Lewy neurites. Nevertheless, knowledge of the molecular events leading to this pathophysiology is incomplete. To date, only mutations in the α-synuclein and LRRK2-encoding genes have been associated with typical findings of clinical and pathologic PD. LRRK2 appears to have a central role in the pathogenesis of PD as it is associated with α-synuclein pathology and other proteins implicated in neurodegeneration. Thus, LRRK2 dysfunction may influence the accumulation of α-synuclein and its pathology through diverse pathomechanisms altering cellular functions and signaling pathways, including immune system, autophagy, vesicle trafficking, and retromer complex modulation. Consequently, development of novel LRRK2 inhibitors can be justified to treat the neurodegeneration associated with abnormal α-synuclein accumulation.

Keywords Leucine-rich repeat kinase 2 • Alpha-synuclein (α-synuclein) • Neurodegeneration • Parkinson's disease

Introduction

Parkinson's disease (PD) is the most common neurodegenerative movement disorder, and resulting parkinsonism is characterized clinically by motor dysfunctions that manifest as resting tremor, bradykinesia, muscular rigidity, and often postural instability[1, 2]. Underlying motor symptoms are the progressive degeneration of dopaminergic neurons of the substantia nigra pars compacta (SNpc) and their

J.P.L. Daher (✉)
Faculty of Medicine, School of Medical Sciences, University of New South Wales, Sydney, NSW 2052, Australia

Neuroscience Research Australia, Barker St, Randwick, NSW 2031, Australia
e-mail: jpldaher@gmail.com

H.J. Rideout (ed.), *Leucine-Rich Repeat Kinase 2 (LRRK2)*,
Advances in Neurobiology 14, DOI 10.1007/978-3-319-49969-7_11

axonal projections to the caudate and putamen [1, 2]. Dopamine replacement therapy improves motor symptoms and is taken as supportive evidence for the clinical diagnosis. The hallmark neuropathology of PD is characterized by the formation of abnormal intraneuronal protein aggregates, named Lewy bodies (LBs) and Lewy neurites (LNs), of which α-synuclein is the major component [3]. The pathogenesis of PD requires further investigation, but it seems to constitute a multifactorial disorder driven by a combination of genes and environmental factors. The majority of PD cases are sporadic, but less than 10 % are due to inherited mutations found in a small number of families [4]. Interestingly, only mutations in the α-synuclein (PARK 1/4) and LRRK2 (PARK 8) genes have been associated with typical findings of clinical and pathologic PD [5]. Consequently, understanding the pathophysiological functions of α-synuclein and LRRK2 may reveal critical information and insight into the pathogenesis of PD and development of effective therapies that slow or halt the progression of this devastating disease (Fig. 11.1).

Genetic Role of α-Synuclein and LRRK2 to Parkinson's Disease

The *SNCA* gene encoding the presynaptic protein α-synuclein, located in the long arm of human chromosome 4, was the first gene associated to familial PD [3, 6]. The physiological function of α-synuclein remains undefined. Three missense autosomal-dominant point mutations (A53T, A30P, E46K) and multiplications (triplications and duplications) in *SNCA* have been linked to an early-onset familial parkinsonism that presents similar to the sporadic form of PD [6–10]. The discovery of familial *SNCA* mutations provided important insights into the pathobiology of the sporadic form of PD. Moreover, the increased severity of PD and earlier age of onset have been reported to correlate with increased α-synuclein dosage [11]. Autosomal-dominant mutations in the leucine-rich repeat kinase 2 (*LRRK2*) gene, located in the short and long arms of human chromosome 12, are the most common familial cause of PD and have been linked to a late-onset parkinsonism with clinical, neurochemical, and some neuropathological phenotypes that are largely indistinguishable from idiopathic PD [12–14].

Several missense mutations have been identified in *LRRK2*-linked families of various ethnic groups, including the R1441C/G/H, Y1699C, G2019S, and I2020T variants that are associated with increased PD pathogenesis [14, 15]. Of these, G2019S mutation is the most common variant that uniquely contributes to both familial and sporadic PD [14]. Interestingly, only familial cases with dominant genetic mutations in *SNCA* and *LRRK2* can potentially develop typical neuropathological features of PD, including SNpc doparminergic neuronal loss, and LB and LN formation [5].

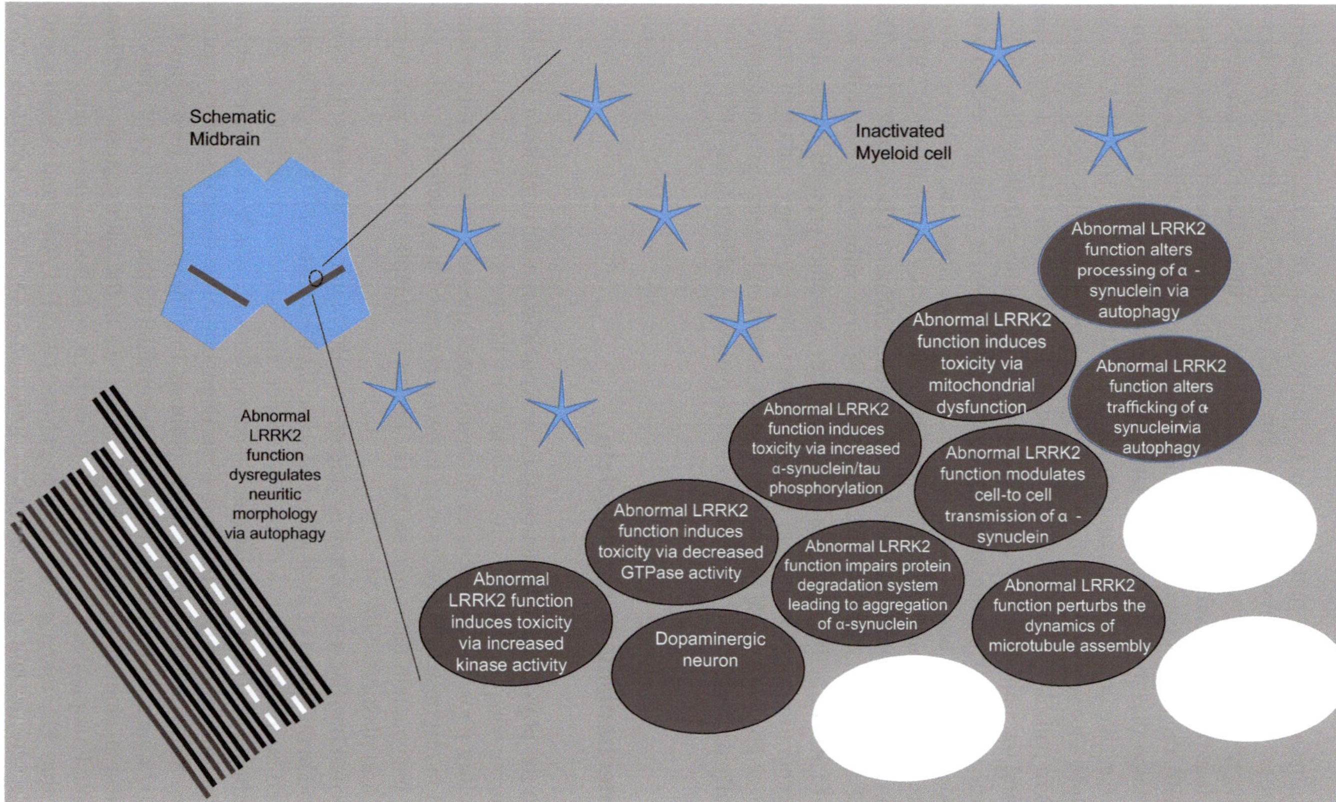

Fig. 11.1 Pathobiology of LRRK2 in neurodegeneration. Abnormal LRRK2 function triggers a series of pathomechanisms in dopaminergic neurons leading to an "intrinsic" nigral neurodegeneration

Pathological Link of α-Synuclein and LRRK2

LRRK2 appears to have a central role in the pathogenesis of several neurodegenerative disorders associated with parkinsonism, such as PD with LBs, diffuse Lewy body disease, nigral degeneration without distinctive histopathology, progressive supranuclear palsy-like pathology, and parkinsonism with dementia and/or amyotrophy, with their associated pathologies [13]. Several neuropathological studies have revealed that the LRRK2-associated neuropathology is fairly heterogeneous, showing variable degrees of α-synuclein and tau pathologies [13, 16–23]. Although it is mainly characterized by the loss of dopaminergic neurons and cumulative α-synuclein pathology in LBs and LNs, these are not present in all postmortem specimens, and the same mutation can cause quite diverse neuropathology [13, 16–23]. For instance, the most common LRRK2 variant, G2019S, is often associated with Lewy body pathology and neuronal loss restricted to the SNpc, tau pathology without LBs or LNs, and pure neuronal loss restricted to the SN, indicating that LRRK2 mutation does not always manifest as α-synucleinopathy or LB disease [18]. The fact that LRRK2 mutations are associated with diverse pathologies raises the possibility of LRRK2 acting upstream of α-synuclein and other proteins implicated in neurodegeneration. In conclusion, LRRK2-associated pathology, although mainly characterized by pure nigral neurodegeneration, is strikingly heterogeneous and can additionally present with α-synuclein-positive pathological inclusions [17].

Pathomechanisms of α-Synuclein

Although the physiological function(s) of α-synuclein remains unknown, its localization at presynaptic terminals, its association with the synaptic vesicles trafficking/recycling pool, and deficiencies in vesicle exocytosis/neurotransmitter release observed in response to knockdown or overexpression of α-synuclein together suggest that α-synuclein has a role in the regulation of synaptic function, neurotransmission, and plasticity [24–34].

Abnormal fibrillar α-synuclein aggregates are the major components of LBs and LNs found in the brains of idiopathic and some familial cases of PD, as well as other LB-related neurodegenerative disorders—hence, they are often referred to as synucleinopathies [3]. The molecular mechanism(s) through which α-synuclein abnormally accumulates and contributes to neurodegeneration in these disorders remains unknown. Many studies support the hypothesis that the processes of α-synuclein oligomerization and fibrillization—both fundamental to the formation of abnormal fibrillar α-synuclein aggregates—have central roles in the pathogenesis of PD and other synucleinopathies [35–39]. Of note, both in vitro and animal model studies indicate that PD-linked α-synuclein mutations (A53T, E46K, and A30P) greatly accelerate α-synuclein oligomerization in comparison to the wild-type (WT) protein, but only two of these (A53T and E46K) enhance fibrillization in vitro and in vivo in comparison to the wild-type (WT) protein [35, 40, 41]. Consequently, these studies suggest that fibrillar aggregation is necessary for the mechanism by which α-synuclein exerts its pathogenic actions.

According to the hypothesis of Braak et al., these pathological aggregates spread throughout the central nervous systems following a topographical sequence of six stages, which therefore may explain the symptomatic progression of the disease [42]. Stages in the evolution of PD-related pathology include:

Stage 1 in the medulla oblongata, with lesions in the dorsal IX/X motor nucleus and/or intermediate reticular zone.

Stage 2 in the medulla oblongata and pontine tegmentum, with pathology of stage 1 plus lesions in caudal raphe nuclei, gigantocellular reticular nucleus, and coeruleus-subcoeruleus complex.

Stage 3 in the midbrain, with pathology of stage 2 plus midbrain lesions, in particular, in the pars compacta of the substantia nigra.

Stage4 in the basal prosencephalon and mesocortex, with pathology of stage 3 plus prosencephalic lesions. Cortical involvement is confined to the temporal mesocortex (transentorhinal region) and allocortex (CA2-plexus), and the neocortex is unaffected.

Stage 5 in the neocortex, with pathology of stage 4 plus lesions in high-order sensory association areas of the neocortex and prefrontal neocortex.

Stage 6 in the neocortex, with pathology of stage 5 plus lesions in first-order sensory association areas of the neocortex and premotor areas and occasionally mild changes in primary sensory areas and the primary motor field [42].

In summary, fibrillary α-synuclein aggregation undergoes an ascending and highly predictable pattern of progression, spreading from the lower brainstem and olfactory bulb into the midbrain and, eventually, to the neocortex, suggesting a mechanism involving pathological propagation similar to the one observed in prion diseases.

The idea of pathological propagation has gained much attention after two studies reported the emergence of α-synuclein-positive Lewy-like pathology in long-term mesencephalic transplants in PD [43, 44]. Despite this, the underlying mechanism of the initiation and propagation of α-synuclein pathology remains unknown, but there is an increasing body of in vitro and in vivo studies that suggest a direct neuron-to-neuron transmission of α-synuclein aggregates as the underlying mechanism for spreading of Lewy pathology [27, 33, 34, 45, 46]. These studies suggested that α-synuclein fibrillar aggregates are released from neurons through exocytosis and transferred to neighboring neurons through endocytosis [27, 33, 34, 45–47]. Consequently, the mechanism underlying the propagation of α-synuclein fibrillar aggregates has become a critical question in understanding the mode of disease progression and may provide novel therapeutic approaches that may stop or halt this spreading process (Fig. 11.2).

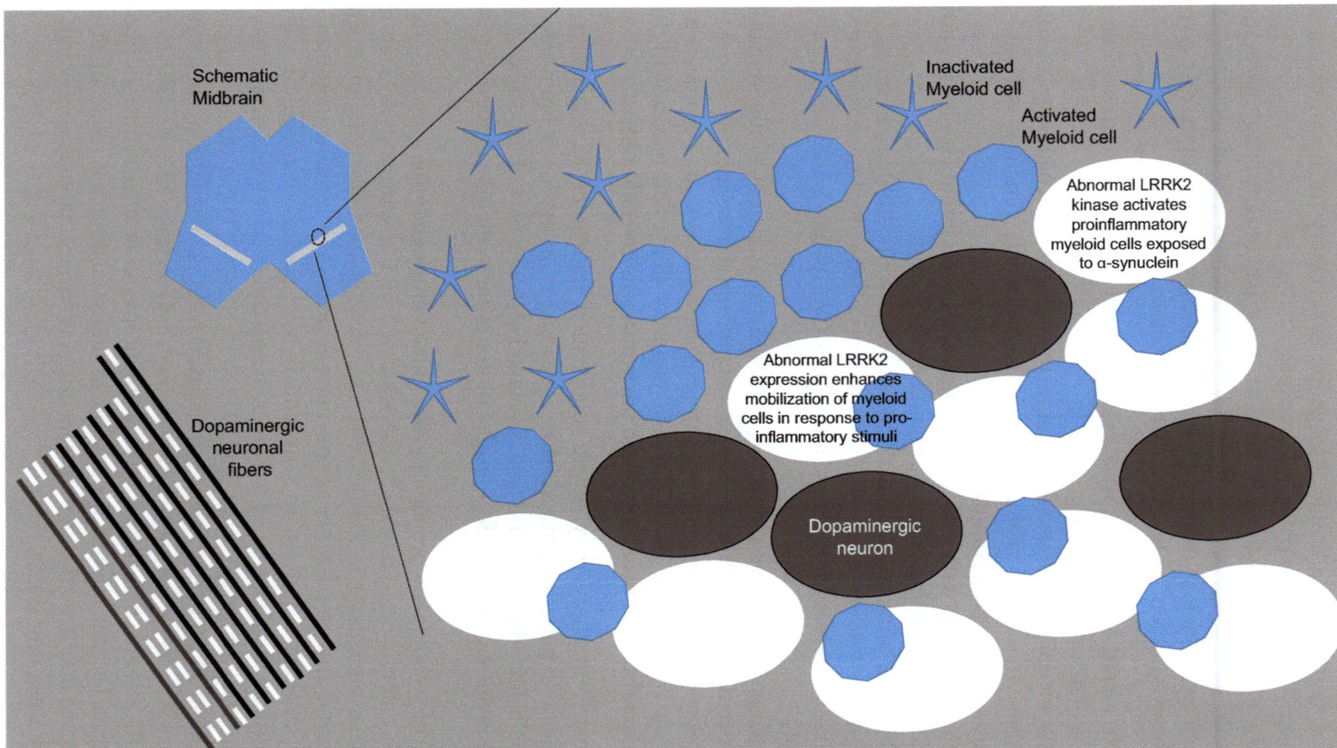

Fig. 11.2 Pathobiology of LRRK2 in neuroinflammation. "Intrinsic" nigral neurodegeneration triggers a series of pathomechanisms related to abnormal LRRK2 function in myeloid cells, leading to an "extrinsic" nigral neurodegeneration

Pathomechanisms of LRRK2

LRRK2 is a large, 2527 amino acid protein containing multiple functional domains, including a kinase and GTPase domains [48–50]. Most PD-linked mutations are located within these two enzymatic domains. The most prevalent mutations, G2019S (kinase domain) and R1441C (GTPase domain), enhanced kinase activity of the protein in vitro and in vivo [49–57]. These findings suggested that increased kinase activity may underlie the pathogenic properties of *LRRK2* mutations, most likely through a toxic gain-of-function mechanism.

Previous studies suggested that GTPase activity, that is GTP binding and GTP hydrolysis, is essential to regulate the protein kinase activity of LRRK2 [49, 50, 55, 58–62]. Importantly, two studies concluded that pathogenic kinase activity of LRRK2 and its associated cytotoxicity are dependent more upon GTP-binding capacity rather than GTP hydrolysis [61, 62]. Consequently, the development of strategies aimed at interfering with GTP-binding activity may provide an attractive therapeutic mechanism for inhibiting the pathogenic LRRK2 kinase activity and its associated neurodegeneration.

LRRK2 expression has been shown to regulate neuronal morphology in vitro, where familial LRRK2 mutants, G2019S and I2020T, induce progressive reduction of neuritic length and branching in primary neuronal cultures and intact rodent brains [63]. In contrast, LRRK2 deficiency leads to increased neuritic length and branching [63]. Autophagy may mediate neurite shortening induced by G2019S LRRK2 expression since inhibition of autophagy reversed, and its activation potentiated, the effects of G2019S LRRK2 on neurites [64]. These observations suggest a potential role for autophagy in mediating the pathogenic actions of LRRK2 mutations.

Recently, LRRK2 expression has been observed to be notably high in toll-like receptor 4 (TLR4)-activated microglia, macrophages, and monocytes, with age-dependent decreased expression in B lymphocytes and no expression in T cells, which suggests that LRRK2 might play a predominant modifying role to the innate immune system and inflammation in PD [65–67]. This also reflects an involvement of both peripheral and brain immune cells in the disease course of PD. The fact that LRRK2 is expressed in TLR4-activated microglia and that LRRK2 kinase activity modulates pro-inflammatory responses in these cells suggests that LRRK2 is an upstream-responsive kinase to TLR4 activation and that LRRK2 inhibition may prevent a full inflammatory response required for PD progression [65].

While previous studies have focused on how loss of LRRK2 expression or activity influences cells of innate immunity, only a few studies have evaluated the effects of pathogenic missense LRRK2 mutations. Through a combination of in vitro *and* in vivo models, Moehle and colleagues found that G2019S LRRK2 expression enhances mobilization of myeloid cells in response to a number of pro-inflammatory stimuli [68]. These findings revealed that in activated myeloid cells, the G2019S mutation robustly increases the association of LRRK2 with the actin regulatory network in a kinase-dependent manner [68]. As a result, the actin regulatory network that interacts with LRRK2 mediates chemotactic responses in myeloid cells that may lead to neurodegeneration [68]. Alternatively, Choi and colleagues found that LRRK2 G2019S mutation attenuates microglial motility by inhibiting focal adhesion kinase, thereby preventing these cells from an efficient response to brain damage and contributing to the development of PD [69].

A number of models in vivo have been developed to investigate the pathophysiology of LRRK2 and its familial mutations. Genetic disruption of LRRK2 or its paralogs in *Caenorhabditis elegans*, *Drosophila melanogaster*, mice, and rats suggests that LRRK2 is not essential for the survival of dopaminergic neurons [70–73]. However, transgenic expression of human LRRK2 bearing the G2019S mutation in *Drosophila* causes adult onset, selective degeneration of dopaminergic neurons, L-DOPA-responsive locomotor impairment, and early mortality [74, 75]. Several LRRK2 transgenic rodent models have been developed to model LRRK2-linked PD. BAC transgenic mice expressing R1441G mutant LRRK2 exhibit reduced striatal dopamine release, L-DOPA-sensitive motor deficits, dopaminergic neuritic atrophy/dystrophy, and increased tau phosphorylation [76]. Additional rodent models expressing G2019S or R1441C LRRK2 mutations display impairments or nigrostriatal dopaminergic neurotransmission and tau processing [77–79]. Moreover, CMVE-PDGFβ mice expressing G2019S LRRK2 display a progressive degeneration of nigrostriatal neurons, autophagic and mitochondrial abnormalities in the mouse brain, and reduced dopaminergic neurite complexity in primary cultures [80]. This study is in agreement with the results published by Dusonchet et al., in which adenoviral vectors were developed to induce expression of mutant G2019S human LRRK2 in the nigrostriatal system of adult rats. While wild-type LRRK2 did not induce any significant neuronal loss, the G2019S mutant LRRK2, under the same conditions and levels of expression, caused a progressive degeneration of nigral dopaminergic neurons [81]. Moreover, adenoviral-mediated expression of G2019S LRRK2 induced striatal pathology, with neuronal ubiquitin accumulation, neurite degeneration, and neurofilament reorganization, in a kinase-dependent manner in rats, supporting the development of kinase inhibitors as a potential therapeutical approach for treating LRRK2-associated PD [82]. Recently, transgenic rats expressing mutant LRRK2 have also been developed, with most models exhibiting locomotor dysfunction and/or changes in striatal dopamine release in the absence of progressive SNpc neurodegeneration or molecular neuropathology [83–85]. Consequently, these rodent models have provided important insight into the pathogenic effects of familial LRRK2 mutations in vivo related to the pathogenesis of PD and further supported a gain-of-function mechanism for these mutations.

Interaction Between LRRK2 and α-Synuclein

Mutations in *SCNA* and *LRRK2* have been linked to autosomal-dominant forms of PD. However, the pathological interplay between LRRK2 and α-synuclein at the protein level is poorly understood, and whether they function in a common pathway, in parallel pathways, or independently to mediate neurodegeneration in PD remains unclear. In favor of a common pathway is the observation that the majority of brains of PD cases with *LRRK2* mutations exhibit α-synuclein-positive Lewy body pathology, suggesting that LRRK2 might function upstream of α-synuclein to modulate its aggregation and toxicity [13].

In order to investigate whether α-synuclein and LRRK2 act synergistically in the pathogenesis of PD, Lin and colleagues characterized a series of double-mutant mice overexpressing PD-related A53T α-synuclein mutation in the forebrain with various forms of LRRK2. Although deletion or overexpression of LRRK2 alone failed to cause any overt gross neurodegenerative abnormalities in mutant mice, co-expression of WT or G2019S LRRK2 with PD-related A53T α-synuclein caused synergistic toxicity to neurons that accelerated the progression of α-synuclein-mediated neuropathology in the cortex and striatum [86]. The fact that mice lacking or overexpressing LRRK2 alone did not display neuropathological changes suggests that LRRK2 may function as a disease modifier rather than a direct cause of the disease progression [86]. Overexpression of WT LRRK2 in the forebrain of A53T transgenic mice resulted in neuropathological changes that include neurodegeneration, abnormal aggregation and somatic α-synuclein accumulation, astrogliosis, and microglial activation, compared to the A53T single transgenic mice [86]. Importantly, overexpression of PD-linked G2019S LRRK2 mutation produced a more severe neuropathology than that observed for WT LRRK2 in the A53T mice [86]. At the cellular level, overexpression of LRRK2 perturbed the dynamics of microtubule assembly, led to severe fragmentation of Golgi complex in neurons, impaired ubiquitin-proteasome system, and exacerbated mitochondrial structural and functional abnormalities in neurons of A53T transgenic mice [86]. It is likely that LRRK2 might have exacerbated A53T α-synuclein-induced cytotoxicity via promoting the abnormal somatic accumulation of α-synuclein [86]. In contrast, neuropathological changes produced by transgenic expression of A53T α-synuclein were significantly reduced in LRRK2 knockout mice. Genetic ablation of LRRK2 maintained the normal organization of Golgi complex, reduced the aggregation and somatic accumulation of A53T α-synuclein, and thereby significantly delayed the progression of A53T α-synuclein-induced neuropathology [86]. In conclusion, these findings clearly demonstrated that LRRK2 enhances α-synuclein-mediated cytotoxicity and suggest inhibition of LRRK2 expression as a potential therapeutic option for ameliorating α-synuclein-induced neurodegeneration. Moreover, these findings also suggest that LRKK2 may regulate A53T α-synuclein-mediated neuropathology though modulating the intracellular trafficking and accumulation of α-synuclein.

The role of LRRK2 in α-synuclein-induced neuropathology has been challenged by another study that modulated LRRK2 overexpression predominantly in the hindbrain of a well-established human A53T α-synuclein transgenic mouse model [87]. The overexpression of human G2019S LRRK2 or LRRK2 deletion failed to influence the premature lethality of A53T-α-synuclein transgenic mice, whereas LRRK2 deletion had no impact on presymptomatic behavioral deficits in these mice [87]. LRRK2 deletion modestly reduced human α-synuclein pathology in a number of brainstem nuclei, but not within the SNpc, of A53T-α-synuclein mice, whereas expression of G2019S-LRRK2 subtly enhances brainstem α-synuclein pathology in matched brainstem regions [87]. Furthermore, altering LRRK2 expression had no significant influence on glial pathology in the brainstem and SNpc of A53T α-synuclein mice [87]. Finally, human A53T α-synuclein and endogenous or human LRRK2 had no synergistic effect on the number of nigrostriatal dopaminergic neurons [87]. This study however failed to provide support for co-expression of LRRK2 and α-synuclein in similar neuronal populations. Herzig et al. approached this caveat by generating double transgenic mice co-expressing high levels of α-synuclein and LRRK2 variants in a large population of both forebrain and brainstem neurons [88]. High levels of these LRRK2 variants left endogenous α-synuclein and tau levels unaltered and did not exacerbate or otherwise modify α-synucleinopathy in mice that co-expressed high levels of LRRK2 and α-synuclein in brain regions [88]. On the contrary, in some lines, high LRRK2 levels improved motor skills in the presence and absence of α-synuclein transgene-induced disease [88]. Noteworthy, these α-synuclein and LRRK2 transgenic mice do not show loss of dopaminergic neurons in the SNpc.

The role of LRRK2 and α-synuclein in the SNpc is well described in two recent studies [73, 89]. The first study demonstrated that rats deficient in LRRK2 expression are protected from dopaminergic neurodegeneration caused by overexpression of human α-synuclein [73]. Additionally, G2019S-LRRK2 rats have exacerbated dopaminergic neurodegeneration in response to the overexpression of α-synuclein, in comparison to wild-type rats [89]. Taken together, these studies indicate that LRRK2-mediated exacerbation of α-synuclein neuropathology might be highly cell type and brain region dependent [73, 86–89].

Interaction of LRRK2 and α-Synuclein in the Immune System

Recently, Daher et al. reported LRRK2 knockout rats as resistant to dopaminergic neurodegeneration elicited by intracranial administration of lipopolysaccharide (LPS), a potent neuroinflammatory model. Such resistance to dopaminergic neurodegeneration correlated with reduced pro-inflammatory CD68-positive myeloid cells recruited in the brain [73]. Additionally, adeno-associated virus-mediated transduction of human α-synuclein also resulted in dopaminergic neurodegeneration in wild-type rats [73]. In contrast, LRRK2 knockout transduced animals had no significant loss of neurons and presented with reduced numbers of activated pro-inflammatory CD68-positive myeloid cells in the substantia nigra [73]. This study suggested that knocking down LRRK2 might protect from overt cell loss by inhibiting the recruitment of chronically activated pro-inflammatory myeloid cells. Although LRRK2 expression in the wild-type rat midbrain remained undetected under physiological conditions, LRRK2 became highly expressed in inducible nitric oxide synthase (iNOS)-positive myeloid cells in the substantia nigra in response to α-synuclein overexpression or LPS exposures [73]. This study extends observations made in mouse models in several ways:

Neuroprotective effects seen in the forebrain of LRRK2 KO mice are also observed in rat midbrain dopaminergic neurons.
LRRK2 expression in myeloid cells is correlated to pro-inflammatory responses induced by α-synuclein overexpression in the SNpc.
LRRK2 KO rats are protected from dopaminergic neurodegeneration induced by a potent myeloid cell agonist, LPS.
Neuroprotection in the LRRK2 KO was found in the rat midbrain where LRRK2 expression is normally low or nonexistent in both neurons and microglia.

Noteworthy, Russo and colleagues demonstrated that LRRK2 kinase activity modulated the induction of pro-inflammatory mediators (i.e., IL-1β cytokine, protein kinase A, NF-κB p50) in primary microglia cultures treated with LPS or α-synuclein fibrils [90]. In conclusion, these studies suggested that inhibition of LRRK2 might be an exciting therapeutic approach against neuroinflammation and a potential neuroprotective strategy in PD.

Selective LRRK2 Inhibition Blocks α-Synucleinopathy

Although therapeutic approaches to slow or block the progression of Parkinson's disease do not exist, LRRK2 kinase activity inhibition has been considered as a promising pharmacological approach in PD. To better understand the therapeutic potential of LRRK2 kinase inhibition in PD, Daher et al. evaluated the tolerability and efficacy of a LRRK2 kinase inhibitor, PF-06447475, in preventing α-synuclein-induced neurodegeneration in both wild-type and transgenic G2019S-LRRK2 rats [89]. This study demonstrated that G2019S-LRRK2 expression exacerbates neuroinflammation and dopaminergic neurodegeneration caused by α-synuclein overexpression and that these effects can be abated by the chronic administration of a potent LRRK2 kinase inhibitor [89]. The LRRK2 kinase inhibitor PF-06447475 was also effective at attenuating α-synuclein-induced dopaminergic neurodegeneration in wild-type rats [89]. Noteworthy, PF-06447475 treatment did not result in any of the expected adverse effects observed in LRRK2 KO rats or nonhuman primates treated with LRRK2 kinase inhibitors [89, 91, 92]. These results demonstrate that pharmacological inhibition of LRRK2 is well tolerated in rats and can counteract dopaminergic neurodegeneration caused by acute α-synuclein overexpression. One mechanism whereby LRRK2 kinase inhibitors could block α-synuclein-induced dopaminergic neurotoxicity could be reduction of inflammatory responses known to cause neurodegeneration of dopaminergic cells. The observation of reduced microgliosis and recruitment of CD68 cells to the midbrain in G2019S-LRRK2 rats treated with PF-06447475 are consistent with this mechanism of action [89]. Because LRRK2 is expressed in both neurons and activated myeloid cells in the rats utilized in this study, the interaction between G2019S-LRRK2 and α-synuclein-induced dopaminergic neurodegeneration awaits further clarification in models that conditionally restrict LRRK2 expression and/or inhibition. In summary, considering that α-synuclein-induced dopaminergic neurodegeneration is sensitive to microglia activation, LRRK2 inhibition may modify synucleinopathy by altering the inflammatory microenvironment surrounding neurons.

Interaction of LRRK2 and α-Synuclein in Autophagy

Several studies have reported the critical role of LRRK2 in regulating autophagy [64, 93–101]. However, the molecular mechanism through which LRRK2 orchestrates autophagy remains unclear. Perhaps the most informative report of the involvement of LRRK2 in autophagy was provided by the study of Orenstein et al. [98]. This study showed that LRRK2 is normally degraded in lysosomes by chaperone-mediated autophagy (CMA), whereas overexpression of G2019S or WT-LRRK2 reduced the functioning of this pathway [98]. In another study, deletion of LRRK2 gene resulted in reduced autophagy and age-dependent accumulation of α-synuclein and ubiquinated proteins in the kidneys of LRRK2 knockout mice [94]. In addition, evidences from an age-dependent analysis of these LRRK2 knockout kidneys indicated that LRRK2 mutations may cause PD and cell death by impairing protein degradation pathways which suggests that LRRK2 is required for normal regulation of the autophagy-lysosomal pathway [95]. However, evidence for such regulation by LRRK2 in the brain is lacking. Taken together, these studies suggest that the inhibitory effect of abnormal forms of LRRK2 mutations on CMA and/or lysosome function could underlie toxicity in Parkinson's disease by compromising the degradation of α-synuclein, which is a PD-related protein degraded by this pathway. Hence, LRRK2 mutations may cause defects in protein degradation system, leading to accumulation and aggregation of α-synuclein. Despite these findings, it should be noted that none of the currently reported cellular and animal models expressing either G2019S or WT LRRK2 show evidence of α-synuclein accumulation that is predicted in the study of Orenstein et al. [73, 84, 85, 89, 96, 97, 100, 101]. Therefore, the regulatory role of LRRK2 in autophagy requires further investigation, particularly in the critical pathways required for α-synuclein degradation.

Interaction of LRRK2 and α-Synuclein in Vesicle Dynamics and Retromer

Several studies suggested a role of LRRK2 in modifying pre- and postsynaptic phenotypes and vesicle trafficking [56, 57, 63, 102–106]. The interesting in vitro study of Kondo et al. reported that LRRK2 cotransfection with α-synuclein enhances the aggregate formation, phosphorylation, release, and cell-to-cell transmission of α-synuclein, resulting in the propagation of α-synuclein to neighboring cells and reduction of cell viability [104]. Furthermore, the cell-to-cell transmission of α-synuclein and the cell toxicity were more pronounced in G2019S mutant rather than in wild-type LRRK2, whereas the cell viability was not decreased in cells transfected with α-synuclein alone [104]. In human PD brains, Guerreiro et al. show co-localization of LRRK2 and α-synuclein as well as S129 phosphorylated α-synuclein in Lewy bodies localized in Lewy body-producing regions [106]. These results were replicated in established cell models for α-synuclein inclusion formation, suggesting that the interaction between LRRK2 and α-synuclein or S129 phosphorylated α-synuclein is enhanced prior to and during the formation of α-synuclein aggregation and fibrilization [106]. In addition, knocking down LRRK2 expression reduced α-synuclein aggregation in these cell models [106]. Collectively, these studies suggest a role of LRRK2 in the processing and/or cellular trafficking of α-synuclein.

We hypothesize that LRRK2 may control α-synuclein processing and/or trafficking by modulating the autophagy pathway. After the transfer to the recipient cells, α-synuclein is transported through the endolysosomal pathway, where LRRK2 may control the rate of degradation of transferred α-synuclein by modulating the trafficking pathways. By either promoting or inhibiting the encounter of the exogenous and endogenous α-synuclein proteins within the endosomal system, LRRK2 may also regulate seeded aggregation of α-synuclein. Subsequent secretion of the seeded aggregates may be another site of control by LRRK2. Moreover, two interesting studies showed a genetic interaction between LRRK2 and Rab7L1 (PARK16), a known genetic risk factor for sporadic PD [107, 108]. Rab7L deficiency resulted in neurodegeneration similar to the phenotype of LRRK2 mutant expression, whereas LRRK2-induced neurodegeneration was rescued by expression of Rab7L1 [107]. This study also showed defects of endolysosomal and Golgi-associated sorting and VPS35 defects in retromer complex by PD-associated defects in LRRK2 and Rab7L1. These defects were rescued by expression of WT VPS35. Therefore, LRRK2 collaborates with Rab7L1 and VPS35 in the endolysosomal pathway and the trafficking pathway that utilizes retromer complex, and defects in this system may increase accumulation of α-synuclein in neurons and, as a result, PD risk. At the moment we can only confidently state that all these studies suggest that LRRK2 may be associated with a complex array of cellular functions involving vesicle dynamics, *trans*-Golgi networks, and autophagy/lysosomal homeostasis. The intriguing hypothesis is that the synergism of all these membrane dynamics may be controlled by LRRK2 and may be at the molecular base of PD neurodegeneration.

Conclusion and Future Perspective

More than 190 years ago, James Parkinson first described the disorder that bears his name, and 40 years ago levodopa, still the most effective therapy, was introduced in the market. The etiology of late-onset PD is still unknown, but it seems to involve abnormal expression of both LRRK2 and α-synuclein genes. PD cases with LRRK2 mutations that show typical α-synuclein-positive Lewy body pathology may greatly benefit from LRRK2 inhibition. Should LRRK2 dysfunction be critical in the pathobiology of PD, LRRK2 inhibition may provide benefit to PD cases without LRRK2 mutations. Furthermore, inhibition of LRRK2 may also provide protection in other diseases in which neuroinflammation contributes to neurodegeneration. Some of the challenges for LRRK2 kinase inhibitors include the definition of efficacy and further refinement of the mechanism of action in the models described in this review as well as other rodent models of PD, such as the preformed fibril mode of α-synuclein neurotoxicity [109]. The tolerability and safety of LRRK2 kinase inhibitors need further evaluation, particularly in nonhuman primates, and the means and outcomes to assess on-target LRRK2 inhibition and desired clinical effect in humans need to be identified.

Acknowledgments I thank Dr. Andrew West for a critical review of this manuscript.

Conflict of Interest The author declares no conflicts of interest.

References

1. Lang AE, Lozano AM (1998) Parkinson's disease. First of two parts. N Engl J Med 339(15):1044–1053
2. Lang AE, Lozano AM (1998) Parkinson's disease. Second of two parts. N Engl J Med 339(16):1130–1143
3. Spillantini MG et al (1997) Alpha-synuclein in Lewy bodies. Nature 388(6645):839–840
4. Gasser T (2009) Mendelian forms of Parkinson's disease. Biochim Biophys Acta 1792(7): 587–596
5. Hardy J et al (2006) Genetics of Parkinson's disease and parkinsonism. Ann Neurol 60(4):389–398
6. Polymeropoulos MH (1997) Mutation in the -synuclein gene identified in families with Parkinson's disease. Science 276(5321):2045–2047
7. Kruger R et al (1998) Ala30Pro mutation in the gene encoding alpha-synuclein in Parkinson's disease. Nat Genet 18(2):106–108
8. Zarranz JJ et al (2004) The new mutation, E46K, of alpha-synuclein causes Parkinson and Lewy body dementia. Ann Neurol 55(2):164–173
9. Chartier-Harlin MC et al (2004) alpha-synuclein locus duplication as a cause of familial Parkinson's disease. Lancet 364(9440):1167–1169
10. Singleton AB et al (2003) α-Synuclein locus triplication causes Parkinson's disease. Science 302(5646):841
11. Ross OA et al (2008) Genomic investigation of alpha-synuclein multiplication and parkinsonism. Ann Neurol 63(6):743–750
12. Paisan-Ruiz C et al (2004) Cloning of the gene containing mutations that cause PARK8-linked Parkinson's disease. Neuron 44(4):595–600
13. Zimprich A et al (2004) Mutations in LRRK2 cause autosomal-dominant parkinsonism with pleomorphic pathology. Neuron 44(4):601–607
14. Healy DG et al (2008) Phenotype, genotype, and worldwide genetic penetrance of LRRK2-associated Parkinson's disease: a case-control study. Lancet Neurol 7(7):583–590

15. International Parkinson Disease Genomics Consortium, Nalls MA, Plagnol V, Hernandez DG, Sharma M, Sheerin UM, Saad M et al (2011) Imputation of sequence variants for identification of genetic risks for Parkinson's disease: a meta-analysis of genome-wide association studies. Lancet 377(9766):641–649
16. Santpere G, Ferrer I (2009) LRRK2 and neurodegeneration. Acta Neuropathol 117(3): 227–246
17. Wider C et al (2010) Leucine-rich repeat kinase 2 gene-associated disease: redefining genotype-phenotype correlation. Neurodegener Dis 7(1–3):175–179
18. Ruffmann C et al (2012) Atypical tauopathy in a patient with LRRK2-G2019S mutation and tremor-dominant Parkinsonism. Neuropathol Appl Neurobiol 38(4):382–386
19. Ujiie S et al (2012) LRRK2 I2020T mutation is associated with tau pathology. Parkinsonism Relat Disord 18(7):819–823
20. Khan NL et al (2005) Mutations in the gene LRRK2 encoding dardarin (PARK8) cause familial Parkinson's disease: clinical, pathological, olfactory and functional imaging and genetic data. Brain 128(Pt 12):2786–2796
21. Wszolek ZK et al (2004) Autosomal dominant parkinsonism associated with variable synuclein and tau pathology. Neurology 62(9):1619–1622
22. Marti-Masso JF et al (2009) Neuropathology of Parkinson's disease with the R1441G mutation in LRRK2. Mov Disord 24(13):1998–2001
23. Puschmann A et al (2012) First neuropathological description of a patient with Parkinson's disease and LRRK2 p.N1437H mutation. Parkinsonism Relat Disord 18(4):332–338
24. Iwai A et al (1995) The precursor protein of non-Aβ component of Alzheimer's disease amyloid is a presynaptic protein of the central nervous system. Neuron 14(2):467–475
25. Jakes R et al (1994) Identification of two distinct synucleins from human brain. FEBS Lett 345(1):27–32
26. Withers GS et al (1997) Delayed localization of synelfin (synuclein, NACP) to presynaptic terminals in cultured rat hippocampal neurons. Dev Brain Res 99(1):87–94
27. Lee SJ et al (2008) alpha-Synuclein is localized in a subpopulation of rat brain synaptic vesicles. Acta Neurobiol Exp 68(4):509–515
28. Cabin DE et al (2002) Synaptic vesicle depletion correlates with attenuated synaptic responses to prolonged repetitive stimulation in mice lacking alpha-synuclein. J Neurosci 22(20): 8797–8807
29. Abeliovich A et al (2000) Mice lacking α-synuclein display functional deficits in the nigrostriatal dopamine system. Neuron 25(1):239–252
30. Yavich L et al (2004) Role of alpha-synuclein in presynaptic dopamine recruitment. J Neurosci 24(49):11165–11170
31. Scott DA et al (2010) A pathologic cascade leading to synaptic dysfunction in alpha-synuclein-induced neurodegeneration. J Neurosci 30(24):8083–8095
32. Nemani VM et al (2010) Increased expression of alpha-synuclein reduces neurotransmitter release by inhibiting synaptic vesicle reclustering after endocytosis. Neuron 65(1):66–79
33. Lee HJ et al (2008) Assembly-dependent endocytosis and clearance of extracellular alpha-synuclein. Int J Biochem Cell Biol 40(9):1835–1849
34. Lee HJ et al (2008) Clearance and deposition of extracellular alpha-synuclein aggregates in microglia. Biochem Biophys Res Commun 372(3):423–428
35. Conway KA et al (1998) Accelerated in vitro fibril formation by a mutant alpha-synuclein linked to early-onset Parkinson disease. Nat Med 4(11):1318–1320
36. Tsigelny IF et al (2008) Mechanisms of hybrid oligomer formation in the pathogenesis of combined Alzheimer's and Parkinson's diseases. PLoS One 3(9), e3135
37. Oueslati A et al (2010) Role of post-translational modifications in modulating the structure, function and toxicity of α-synuclein. Prog Brain Res 183:115–145
38. Taschenberger G et al (2012) Aggregation of alphaSynuclein promotes progressive in vivo neurotoxicity in adult rat dopaminergic neurons. Acta Neuropathol 123(5):671–683
39. Galvin JE et al (2001) Synucleinopathies. Arch Neurol 58(2):186

40. Conway KA et al (2000) Accelerated oligomerization by Parkinson's disease linked α-synuclein mutants. Ann N Y Acad Sci 920(1):42–45
41. Lashuel HA et al (2013) The many faces of alpha-synuclein: from structure and toxicity to therapeutic target. Nat Rev Neurosci 14(1):38–48
42. Braak H et al (2003) Staging of brain pathology related to sporadic Parkinson's disease. Neurobiol Aging 24(2):197–211
43. Kordower JH et al (2008) Lewy body-like pathology in long-term embryonic nigral transplants in Parkinson's disease. Nat Med 14(5):504–506
44. Li JY et al (2008) Lewy bodies in grafted neurons in subjects with Parkinson's disease suggest host-to-graft disease propagation. Nat Med 14(5):501–503
45. Desplats P et al (2009) Inclusion formation and neuronal cell death through neuron-to-neuron transmission of alpha-synuclein. Proc Natl Acad Sci U S A 106(31):13010–13015
46. Lee HJ et al (2005) Intravesicular localization and exocytosis of alpha-synuclein and its aggregates. J Neurosci 25(25):6016–6024
47. Hansen C et al (2011) alpha-Synuclein propagates from mouse brain to grafted dopaminergic neurons and seeds aggregation in cultured human cells. J Clin Invest 121(2):715–725
48. Greggio E et al (2008) The Parkinson disease-associated leucine-rich repeat kinase 2 (LRRK2) is a dimer that undergoes intramolecular autophosphorylation. J Biol Chem 283(24):16906–16914
49. Stafa K et al (2012) GTPase activity and neuronal toxicity of Parkinson's disease-associated LRRK2 is regulated by ArfGAP1. PLoS Genet 8(2), e1002526
50. Sheng Z et al (2012) Ser1292 autophosphorylation is an indicator of LRRK2 kinase activity and contributes to the cellular effects of PD mutations. Sci Transl Med 4(164):164ra161
51. Greggio E et al (2006) Kinase activity is required for the toxic effects of mutant LRRK2/dardarin. Neurobiol Dis 23(2):329–341
52. Smith WW et al (2006) Kinase activity of mutant LRRK2 mediates neuronal toxicity. Nat Neurosci 9(10):1231–1233
53. Smith WW et al (2005) Leucine-rich repeat kinase 2 (LRRK2) interacts with parkin, and mutant LRRK2 induces neuronal degeneration. Proc Natl Acad Sci U S A 102(51): 18676–18681
54. West AB et al (2005) Parkinson's disease-associated mutations in leucine-rich repeat kinase 2 augment kinase activity. Proc Natl Acad Sci U S A 102(46):16842–16847
55. West AB et al (2007) Parkinson's disease-associated mutations in LRRK2 link enhanced GTP-binding and kinase activities to neuronal toxicity. Hum Mol Genet 16(2):223–232
56. Lee BD et al (2010) Inhibitors of leucine-rich repeat kinase-2 protect against models of Parkinson's disease. Nat Med 16(9):998–1000
57. Lee S et al (2010) LRRK2 kinase regulates synaptic morphology through distinct substrates at the presynaptic and postsynaptic compartments of the *Drosophila* neuromuscular junction. J Neurosci 30(50):16959–16969
58. Ito G et al (2007) GTP binding is essential to the protein kinase activity of LRRK2, a causative gene product for familial Parkinson's disease. Biochemistry 46(5):1380–1388
59. Guo L et al (2007) The Parkinson's disease-associated protein, leucine-rich repeat kinase 2 (LRRK2), is an authentic GTPase that stimulates kinase activity. Exp Cell Res 313(16): 3658–3670
60. Xiong Y et al (2010) GTPase activity plays a key role in the pathobiology of LRRK2. PLoS Genet 6(4), e1000902
61. Taymans JM et al (2011) LRRK2 kinase activity is dependent on LRRK2 GTP binding capacity but independent of LRRK2 GTP binding. PLoS One 6(8), e23207
62. Biosa A et al (2013) GTPase activity regulates kinase activity and cellular phenotypes of Parkinson's disease-associated LRRK2. Hum Mol Genet 22(6):1140–1156
63. MacLeod D et al (2006) The familial Parkinsonism gene LRRK2 regulates neurite process morphology. Neuron 52(4):587–593
64. Plowey ED et al (2008) Role of autophagy in G2019S-LRRK2-associated neurite shortening in differentiated SH-SY5Y cells. J Neurochem 105(3):1048–1056

65. Moehle MS et al (2012) LRRK2 inhibition attenuates microglial inflammatory responses. J Neurosci 32(5):1602–1611
66. Thevenet J et al (2011) Regulation of LRRK2 expression points to a functional role in human monocyte maturation. PLoS One 6(6), e21519
67. Maekawa T et al (2010) Age-dependent and cell-population-restricted LRRK2 expression in normal mouse spleen. Biochem Biophys Res Commun 392(3):431–435
68. Moehle MS et al (2015) The G2019S LRRK2 mutation increases myeloid cell chemotactic responses and enhances LRRK2 binding to actin-regulatory proteins. Hum Mol Genet 24(15):4250–4267
69. Choi I et al (2015) LRRK2 G2019S mutation attenuates microglial motility by inhibiting focal adhesion kinase. Nat Commun 6:8255
70. Saha S et al (2009) LRRK2 modulates vulnerability to mitochondrial dysfunction in Caenorhabditis elegans. J Neurosci 29(29):9210–9218
71. Wang D et al (2008) Dispensable role of *Drosophila* ortholog of LRRK2 kinase activity in survival of dopaminergic neurons. Mol Neurodegener 3:3
72. Andres-Mateos E et al (2009) Unexpected lack of hypersensitivity in LRRK2 knock-out mice to MPTP (1-methyl-4-phenyl-1,2,3,6-tetrahydropyridine). J Neurosci 29(50):15846–15850
73. Daher JP et al (2014) Abrogation of alpha-synuclein-mediated dopaminergic neurodegeneration in LRRK2-deficient rats. Proc Natl Acad Sci U S A 111(25):9289–9294
74. Liu Z et al (2008) A *Drosophila* model for LRRK2-linked parkinsonism. Proc Natl Acad Sci U S A 105(7):2693–2698
75. Ng CH et al (2009) Parkin protects against LRRK2 G2019S mutant-induced dopaminergic neurodegeneration in *Drosophila*. J Neurosci 29(36):11257–11262
76. Li Y et al (2009) Mutant LRRK2(R1441G) BAC transgenic mice recapitulate cardinal features of Parkinson's disease. Nat Neurosci 12(7):826–828
77. Li X et al (2010) Enhanced striatal dopamine transmission and motor performance with LRRK2 overexpression in mice is eliminated by familial Parkinson's disease mutation G2019S. J Neurosci 30(5):1788–1797
78. Melrose HL et al (2010) Impaired dopaminergic neurotransmission and microtubule-associated protein tau alterations in human LRRK2 transgenic mice. Neurobiol Dis 40(3): 503–517
79. Tong Y et al (2009) R1441C mutation in LRRK2 impairs dopaminergic neurotransmission in mice. Proc Natl Acad Sci U S A 106(34):14622–14627
80. Ramonet D et al (2011) Dopaminergic neuronal loss, reduced neurite complexity and autophagic abnormalities in transgenic mice expressing G2019S mutant LRRK2. PLoS One 6(4), e18568
81. Dusonchet J et al (2011) A rat model of progressive nigral neurodegeneration induced by the Parkinson's disease-associated G2019S mutation in LRRK2. J Neurosci 31(3):907–912
82. Tsika E et al (2015) Adenoviral-mediated expression of G2019S LRRK2 induces striatal pathology in a kinase-dependent manner in a rat model of Parkinson's disease. Neurobiol Dis 77:49–61
83. Walker MD et al (2014) Behavioral deficits and striatal DA signaling in LRRK2 p.G2019S transgenic rats: a multimodal investigation including PET neuroimaging. J Parkinsons Dis 4(3):483–498
84. Lee JW et al (2015) Behavioral, neurochemical, and pathologic alterations in bacterial artificial chromosome transgenic G2019S leucine-rich repeated kinase 2 rats. Neurobiol Aging 36(1):505–518
85. Sloan M et al (2016) LRRK2 BAC transgenic rats develop progressive, L-DOPA-responsive motor impairment, and deficits in dopamine circuit function. Hum Mol Genet 25(5): 951–963
86. Lin X et al (2009) Leucine-rich repeat kinase 2 regulates the progression of neuropathology induced by Parkinson's-disease-related mutant alpha-synuclein. Neuron 64(6):807–827
87. Daher JP et al (2012) Neurodegenerative phenotypes in an A53T alpha-synuclein transgenic mouse model are independent of LRRK2. Hum Mol Genet 21(11):2420–2431

88. Herzig MC et al (2012) High LRRK2 levels fail to induce or exacerbate neuronal alpha-synucleinopathy in mouse brain. PLoS One 7(5), e36581
89. Daher JP et al (2015) Leucine-rich Repeat Kinase 2 (LRRK2) pharmacological inhibition abates alpha-synuclein gene-induced neurodegeneration. J Biol Chem 290(32):19433–19444
90. Russo I et al (2015) Leucine-rich repeat kinase 2 positively regulates inflammation and down-regulates NF-kappaB p50 signaling in cultured microglia cells. J Neuroinflammation 12(1):230
91. Boddu R et al (2015) Leucine-rich repeat kinase 2 deficiency is protective in rhabdomyolysis-induced kidney injury. Hum Mol Genet 24(14):4078–4093
92. Fuji RN et al (2015) Effect of selective LRRK2 kinase inhibition on nonhuman primate lung. Sci Transl Med 7(273):273ra215
93. Alegre-Abarrategui J et al (2009) LRRK2 regulates autophagic activity and localizes to specific membrane microdomains in a novel human genomic reporter cellular model. Hum Mol Genet 18(21):4022–4034
94. Tong Y et al (2010) Loss of leucine-rich repeat kinase 2 causes impairment of protein degradation pathways, accumulation of alpha-synuclein, and apoptotic cell death in aged mice. Proc Natl Acad Sci U S A 107(21):9879–9884
95. Tong Y et al (2012) Loss of leucine-rich repeat kinase 2 causes age-dependent bi-phasic alterations of the autophagy pathway. Mol Neurodegener 7:2
96. Bravo-San Pedro JM et al (2013) The LRRK2 G2019S mutant exacerbates basal autophagy through activation of the MEK/ERK pathway. Cell Mol Life Sci 70(1):121–136
97. Manzoni C et al (2013) Inhibition of LRRK2 kinase activity stimulates macroautophagy. Biochim Biophys Acta 1833(12):2900–2910
98. Orenstein SJ et al (2013) Interplay of LRRK2 with chaperone-mediated autophagy. Nat Neurosci 16(4):394–406
99. Dodson MW et al (2014) Novel ethyl methanesulfonate (EMS)-induced null alleles of the *Drosophila* homolog of LRRK2 reveal a crucial role in endolysosomal functions and autophagy in vivo. Dis Model Mech 7(12):1351–1363
100. Schapansky J et al (2014) Membrane recruitment of endogenous LRRK2 precedes its potent regulation of autophagy. Hum Mol Genet 23(16):4201–4214
101. Saez-Atienzar S et al (2014) The LRRK2 inhibitor GSK2578215A induces protective autophagy in SH-SY5Y cells: involvement of Drp-1-mediated mitochondrial fission and mitochondrial-derived ROS signaling. Cell Death Dis 5, e1368
102. Pfeffer S, Aivazian D (2004) Targeting Rab GTPases to distinct membrane compartments. Nat Rev Mol Cell Biol 5(11):886–896
103. Shin N et al (2008) LRRK2 regulates synaptic vesicle endocytosis. Exp Cell Res 314(10): 2055–2065
104. Kondo K et al (2011) alpha-Synuclein aggregation and transmission are enhanced by leucine-rich repeat kinase 2 in human neuroblastoma SH-SY5Y cells. Biol Pharm Bull 34(7): 1078–1083
105. Matta S et al (2012) LRRK2 controls an EndoA phosphorylation cycle in synaptic endocytosis. Neuron 75(6):1008–1021
106. Guerreiro PS et al (2013) LRRK2 interactions with alpha-synuclein in Parkinson's disease brains and in cell models. J Mol Med (Berl) 91(4):513–522
107. MacLeod DA et al (2013) RAB7L1 interacts with LRRK2 to modify intraneuronal protein sorting and Parkinson's disease risk. Neuron 77(3):425–439
108. Beilina A et al (2014) Unbiased screen for interactors of leucine-rich repeat kinase 2 supports a common pathway for sporadic and familial Parkinson disease. Proc Natl Acad Sci U S A 111(7):2626–2631
109. Luk KC et al (2012) Pathological alpha-synuclein transmission initiates Parkinson-like neurodegeneration in nontransgenic mice. Science 338(6109):949–953

Chapter 12
Mechanisms of Mutant LRRK2 Neurodegeneration

Mark R. Cookson

Abstract LRRK2 mutations are associated with the loss of neurons, that is to say toxicity, in patients and in experimental model systems. However, the mechanisms by which mutations can be linked to neurodegeneration are not fully defined. Here I will argue that mechanism in this context encompasses a variety of levels of information. Mutations or alterations in gene expression at a genetic level are one set of mechanisms that are reflected at the biochemical level likely in enhanced or persistent function of LRRK2. By impacting cellular pathways, prominently including changes in autophagy but also microtubule function, mitochondria and protein synthesis, in neurons and immune cells, the LRRK2 brain is primed for neurodegeneration in an age-dependent manner. These concepts have implications for not only modeling LRRK2 disease but also perhaps for Parkinson's disease more generally, including the development of therapeutic modalities.

Keywords α-Synuclein • Gene expression • GTPase • Kinase • Mutations • Pleomorphic risk locus • Rab7L1

Abbreviations

BAC	Bacterial artificial chromosome
COR	C-terminal of ROC
GWAS	Genome-wide association study
LRRK2	Leucine-rich repeat kinase 2
MAPT	Microtubule-associated protein tau
PD	Parkinson's disease
ROC	Ras of complex proteins
SNP	Single-nucleotide polymorphism

M.R. Cookson (✉)
Laboratory of Neurogenetics, National Institute on Aging, National Institutes of Health, Room 1A116, 35 Convent Drive, Bethesda, MD 20892-3707, USA
e-mail: cookson@mail.nih.gov

H.J. Rideout (ed.), *Leucine-Rich Repeat Kinase 2 (LRRK2)*,
Advances in Neurobiology 14, DOI 10.1007/978-3-319-49969-7_12

Introduction

The initial reports of mutations in *leucine-rich repeat kinase 2* (*LRRK2*) associated with Parkinson's disease (PD) [1, 2] were rapidly followed by the identification of the G2019S variant that accounts for an appreciable percentage of all PD cases, depending on population [3, 4]. Additionally, association of amino acid-changing variants in *LRRK2* with PD was noted [5–7]. Finally, genome-wide association studies (GWAS) have reliably nominated the locus that contains *LRRK2* on chromosome 12 as contributing to risk of PD [8]. Therefore, the genomic region encompassing *LRRK2* is an example of a pleomorphic risk locus, containing several different types of genetic risk factors [9].

Importantly for the data to be discussed here, the clinical phenotype related to *LRRK2* variants in all these situations is similar to sporadic PD. Pathologically, there is a consistent loss of pigmented dopamine neurons that arise in the *substantia nigra* and project to the striatum. Other aspects of pathology, particularly the presence of α-synuclein-positive Lewy bodies or tau-positive structures, are variable [10]. It is therefore reasonable to infer that mutations and other variation in *LRRK2* are associated with toxic effects on neurons.

Here, I will discuss some of the proposed mechanisms by which LRRK2 can have toxic effects on neurons. As the first part of this discussion, I will discuss potential mechanisms that arise from the central dogma of molecular biology, namely, that while risk of PD is partially encoding in DNA, the mechanisms of neurodegeneration are largely evoked at the level of the proteins, at least for an enzyme like LRRK2.

Genetic Variants and Their Effects on Biochemical Function

As briefly outlined above, the *LRRK2* locus contains at least three types of variation that can be associated with disease. First, amino acid variants that segregate with PD in a Mendelian manner (albeit with reduced age-dependent penetrance) are likely to have strongest effect on phenotypes. Second, variants that do not clearly segregate with familial disease but show association at the population level are, by definition, likely to have smaller effects on disease risk. Third, the variants nominated by GWAS are associated with very modest effects on risk of disease but to date are not shown to affect amino acid composition of the encoded protein. Because the mechanisms by which an amino acid-changing variant and noncoding variants that are likely to act are different, they will be discussed separately.

Amino Acid-Changing Mutations

As has been discussed elsewhere, there are two clearly agreed-upon effects of mutations in the enzymatic domains of LRRK2. Mutations in the Ras of complex proteins (ROC) and C-terminal of ROC (COR) domains, particularly R1441C/G and Y1699C, diminish the normally modest GTPase activity of LRRK2 [11–15]. In the kinase domain, G2019S clearly increases kinase activity when measured against any number of substrates [16], while I2020T has a more modest effect but likely works in a similar manner [17].

I have argued previously [18] that these observations are consistent with the concept that all mutations *increase* signaling from LRRK2, predicated on two assumptions that are important to clearly label. Lower GTPase activity is predicted to extend the period of time that LRRK2 exists in its GTP-bound state. If GTP-bound LRRK2 is the predominantly active signaling molecule, then lower GTPase activity would therefore be associated with the persistence of signaling. Equally, the kinase hyperactive mutations would be predicted to also have higher net signaling, making the assumption that kinase and GTPase activities are not antagonistic in some way. These arguments incidentally suggest that distinctions such as "loss" or "gain" of function are unhelpful for a complex molecule.

A test of the prediction that mutations enhance a normal function of LRRK2, then blocking that function, would limit any detrimental effects. We tested that idea several years ago by making kinase-dead versions of LRRK2 that contained mutations in the ROC or COR domains and transfecting these into primary neurons in culture [19]. We found that inactivating the kinase domain limited the amount of cell death in these relatively simple experiments. Similar results were reported by other groups [20], confirmed with more sophisticated longitudinal approaches in culture [21], and extended to animal models in various species [22–24]. These results strongly suggest that kinase activity of LRRK2 is critical for the neurotoxic effects of mutations in the gene and have lead to the concept that kinase inhibition might be therapeutically useful [25].

Despite the consistent results regarding LRRK2 kinase inhibition and neuronal toxicity, there are some important caveats that may limit interpretation. In some of the early experiments, cells were counted as not viable if their neurites were retracted. Subsequently, it was shown that LRRK2 has an effect on neurite length that is largely independent of cell viability [26]. The consequent concern that the toxic effects might be confounded by loss of neurites has, however, being partially mitigated by following cells over time and monitoring loss of membrane integrity in live cells [21]. However, even with these improvements to the measurement of "viability," these are still relatively acute events compared to the decade-long development of disease in PD patients. Therefore, to what extent normal function relates to toxicity in the disease itself is difficult to evaluate, and this question may remain unresolved until inhibitors are in clinical usage.

One difficulty with the available data on normal function is that it does not explain the association of disease with all variants. For example, it was noted that modifications at the C-terminus of LRRK2 including the risk variant G2385R have lower kinase activity than the wild-type protein [27, 28]. We were able to confirm these results and show that the effect of G2385R is sufficient to overcome the activating effects of G2019S when the two mutations are present in the same molecule [29]. Reconciling these results with the more general observation that enhanced function underpins pathogenesis is difficult. One possibility is that both higher and lower LRRK2 activity can result in PD, which has implications for potential treatments [30]. Alternatively, G2385R may alter some pathogenic function of LRRK2 to a greater extent than the diminishment of kinase activity. In this case, the observation that G2385R is a risk factor and not a penetrant variant might relate to having more than one effect on protein function.

A final problem is that some of the manipulations used in these experiments can alter LRRK2 protein levels, but this will be discussed separately below as it is helpful in the interpretation of expression results.

Expression

In the work discussed above where hypothesis-testing mutations were used to probe protein function, it was generally assumed that those mutations did not change protein properties other than to be tested. However, subsequent studies have shown that there are effects on LRRK2 other than simply changing enzyme activity and that a major effect occurs at the level of protein stability.

As an example of this problem, some mutations that remove the ability of LRRK2 to bind guanosine nucleotides strongly destabilize the protein. Two very commonly used hypothesis-testing mutations, K1357A and T1358N, are highly unstable when expressed in cells [31]. Therefore, a lack of toxicity from constructs containing these mutations [20] could be due to lower expression levels rather than enzyme activity per se. Along the same lines, careful examination of kinase-dead mutations suggested that lower steady-state levels explain the diminished toxicity associated with inactivation of kinase [21]. Conversely, manipulations that lower LRRK2 protein levels can limit toxicity in cells [32]. These events are not driven purely by artificial situations such as transfection of tagged constructs, as they may also occur in vivo [33].

Although this discussion might appear to relate to artifacts in the experimental systems used for assessing mutations, it illustrates a potentially important point about LRRK2. Toxicity, at least in these situations, depends on expression levels, and lower expression levels result in lower toxicity. If the extrapolation from cells is valid, then the implication is that expression levels of mutations may contribute to risk of disease in humans. This might explain why some variants that tend to be unstable are risk factors rather than penetrant mutations.

More generally, this thinking provides a testable hypothesis as to why *LRRK2* variants that do not change amino acids are identified in GWAS for PD. If the normal function of LRRK2 is important in toxicity, and if expression levels influence that toxicity, then altered expression of wild-type LRRK2 without mutations might act as a subtle risk factor for PD. Thus, variants around the gene such as promoter elements might be associated with altered expression, and hence disease risk, without changing the amino acid sequence. This is a testable hypothesis, as LRRK2 expression levels could be measured in people carrying the risk allele identified by GWAS. It might even be possible to perform such experiments under conditions where LRRK2 expression is induced rather than at basal levels.

These considerations suggest that LRRK2 toxicity relates to normal function and that persistent signaling and/or altered protein levels are causative for disease. In the next two sections, I will discuss how LRRK2 might act at a cellular level.

Cellular Mechanisms: Neuronal Events

LRRK2 is widely expressed in most brain regions with a pattern that is consistent with broad expression in neurons [34]. Expression is particularly high in the striatum, likely in medium spiny neurons in the striosome [35], but not particularly high in the substantia nigra [36, 37]. There are some interesting differences between species, with a potential primate-specific promoter sequence that may enhance expression in nigral neurons [38]. Given that LRRK2 is therefore a neuronal protein, it is reasonable to look at how expression of mutations in neurons affects viability.

As discussed above, transfection of LRRK2 directly into neurons in culture produces acute toxicity that is enhanced by mutations [19, 20, 39]. Although these models are limited in that they measure acute events in a context that lacks the normally extensive cellular interactions of the adult brain, they can be useful in discerning mechanistic aspects of cell-autonomous toxicity. For example, early results suggested that the mode of cell death induced by LRRK2 mutations was via apoptosis [40].

Several transgenic models have been reported using different promoters that direct expression in various neuronal types. Some of these can induce neuronal damage and associated behavioral phenotypes. A transgenic mouse line expressing the R1441G mutation from a bacterial artificial chromosome (BAC) was reported to show akinesia [41] but not cell death [42]. The behavioral phenotype is not fully penetrant and appears to be related to protein expression levels [42]. While the lack of consistency for an important phenotype is frustrating, these reports further support the idea that higher levels of LRRK2 expression are more toxic.

Other mouse models have shown some neuronal toxicity, including extensive loss of striatal neurons, but not nigral cells, using a calcium/calmodulin protein kinase II promoter [43], but other mouse models generally have only modest levels of neuronal cell death [44–46]. Similarly, BAC-transgenic rats do not show overt degeneration [47]. Stronger toxicity is seen with acute viral expression of LRRK2 in dopamine neurons [48, 49].

Despite the relatively modest effects on cell death, there are some effects in these animals that might be considered neurotoxic. For example, there are axonal abnormalities in projection neurons from the substantia nigra in one R1441G BAC mouse [41, 42]. LRRK2 mutations can also induce alterations in neurotransmitter systems, particularly in the turnover of dopamine in the striatum [50, 51] although these results have not been confirmed in all studies [52]. These results suggest that mutations in LRRK2 can cause damage to neuronal-specific systems.

There are several candidates for subcellular mechanisms that could mediate LRRK2 toxicity in neurons. Particular attention has been focused on alterations in vesicular trafficking systems such as autophagy [44, 53–55], mitochondrial damage [51, 56–59], protein translation [60, 61], or abnormalities of the nuclear envelope [62, 63]. The causal relationships between these different observations are difficult to disentangle; these cellular processes tend to be interrelated. For example, a generalized defect in autophagy could easily result in changed turnover of organelles such as the mitochondria with attendant effects on protein synthesis. In some cases, there have been claims of direct substrates of LRRK2 that might lead directly to cell death [61], although these need to be confirmed independently. Nonetheless, collectively, these results suggest a model whereby kinase activity of LRRK2 leads to altered regulation of one or more substrates, then to alterations in key cellular pathways, and finally to neuronal cell death.

Non-cell-Autonomous Mechanisms

The focus on neurons in the above discussion is reasonable given that dopamine cell loss is a prominent feature of LRRK2-associated PD and that some of the systems that are affected by LRRK2 mutations, particularly synapses, are neuronal specific. However, there are reasons to think that cells other than neurons might also contribute to neurodegeneration.

LRRK2 is highly expressed in immune cells, particularly in macrophages and microglia [64]. Microglia expressing mutant LRRK2 have enhanced cytokine expression [65] and altered motility [66]. Conversely, inhibition or knockout of LRRK2 limits microglial responses both in vitro and in vivo [64, 67, 68]. Mechanistically, these observations are probably related to some of the cellular processes discussed above. Specifically, LRRK2 regulates autophagy in microglia as it does in neurons [69].

The importance of altered microglia activation by LRRK2 is that enhanced neuroinflammation is likely to provoke excessive neurodegeneration. Therefore, one model is that LRRK2 mutations alter fundamental cellular pathways, for example, trafficking of lipid vesicles, that change the way in which neurons and non-neuronal cells communicate. As neuroinflammatory processes change during aging [70], this may explain why mutations in LRRK2 show age-dependent penetrance. Again, a likely mechanism involved is via some of the cellular processes that LRRK2 regulates such as autophagy [71].

A practical implication of this idea is that models where LRRK2 is restricted to neurons, such as neuron-specific promoters or highly purified cell cultures, are likely not to have access to non-neuronal mechanisms of toxicity. Furthermore, it might be appropriate to stimulate neuroimmune cells using stimulants such as the bacterial lipopolysaccharide or extracellular proteins such as α-synuclein [72]. This argument leads to a final important consideration about LRRK2 toxicity, which is whether LRRK2 acts alone or has important relationships to other proteins implicated in PD.

Relationships to Other Proteins and Genes Associated with PD

There are several reasons to expect the mechanisms by which LRRK2 evokes neurodegeneration will be related to other genes and proteins. For example, the acute toxic effects of LRRK2 in culture appear to require full-length protein that includes the nonenzymatically active C-terminus [73]. Several binding partners of LRRK2 are also reported to be important in mediating toxicity, including MKK6 [74], ArfGAP1 [75, 76], PAK6 [77], or 14-3-3 proteins [78]. The long list of candidate modifiers of LRRK2 toxicity is impressive, but it might be important to focus on a smaller number that relate more specifically to PD. Of these, those that have proven genetic links to PD might be especially important.

The most prominently examined relationship for toxicity is between LRRK2 and α-synuclein. Like LRRK2, mutations in α-synuclein or altered expression of the wild-type protein are associated with neuronal toxicity in various systems [79]. Therefore, several laboratories have examined whether there are relationships between LRRK2 and α-synuclein. The neurodegenerative actions of α-synuclein have been shown to be attenuated by LRRK2 knockout in mice [43] or rats [80], and LRRK2 kinase inhibitors may have a similar effect [81]. This is not seen in all models [82, 83], suggesting that the mechanism(s) involved is restricted to some forms of neurodegeneration that remain to be defined. Why LRRK2 and a-synuclein are interrelated is also unclear, although changes in vesicular trafficking are currently being examined in many contexts [84].

Another important gene for neurodegeneration in general is *MAPT*, mutated in some forms of frontotemporal dementia, which encodes the protein tau that is deposited in Alzheimer's disease and other tauopathies [85]. Perhaps surprisingly, variation around the *MAPT* locus has been nominated as contributing to PD risk in several GWAS [8]. The main function of tau is to bind and stabilize microtubules, which is thought to be important in the maintenance of long axons. LRRK2 also has roles in the cytoskeletal system in general and in the regulation of microtubules in particular [86]. Therefore, several labs have examined the relationship between LRRK2 and tau, with particularly compelling suggestions that LRRK2 might regulate tau via scaffolding of glycogen synthase kinase 3β [87, 88]. What is particularly important about tau is that, at least in the CNS, expression is restricted to neurons, reinforcing the point that a least some aspects of LRRK2 neurodegeneration are neuron autonomous.

If LRRK2 has relationships with a risk factor gene like MAPT, it might be reasonable to look more globally at other genes nominated by GWAS. In two independent approaches, we and others have examined physical and genetic interactions of LRRK2 [89, 90]. One candidate that was identified by both approaches was Rab7L1, a small GTPase that is resident at the endoplasmic reticulum. Expression levels of Rab7L1 were proposed to explain the GWAS-linked risk [8, 89] and thereby modulate LRRK2-mediated effects in cells.

Conclusions

From the above discussion, we can see that the mechanisms by which LRRK2 provokes neurodegeneration act a number of levels. These vary from the most proximal effects of mutations or expression on biochemical activities, which converge on cellular activities in both neurons and non-neuronal cells. In particular, changes in microtubule-dependent cytoskeletal functions and on secretory vesicles in neurons are combined with changes in neuroinflammatory cells that interact with aging to produce loss of neurons in specific brain areas.

Although this might appear to be a conceptually complete explanation for PD, in practice, there are several aspects that remain to be clarified. For example, because LRRK2 kinase inhibition has been proposed to be helpful in limiting neurodegeneration, there must be an immediately downstream target of kinase activity that is an important key mediator of neurotoxic events. Related to this, it remains difficult to explain why if the processes in which LRRK2 is involved are general to many cells and certainly most neurons then why do only some neurons degenerate in this disease. Finally, while there is an argument as to why aging acts as a strong modifier of LRRK2 phenotypes, it is a difficult concept to test experimentally, and how this impacts incomplete penetrance is also unclear.

Overall, there have been great strides in the understanding of the mechanisms related to LRRK2 neurodegeneration, and these are leading to testable hypotheses that might support clinical development of new approaches to limit cell death in PD. Future studies will be very important in filling those gaps that we have at the moment and hopefully further improving our models.

Acknowledgments This research was supported entirely by the Intramural Research Program of the NIH and the National Institute on Aging.

Conflict of Interest The author declares no conflicts of interest.

References

1. Zimprich A, Biskup S, Leitner P et al (2004) Mutations in LRRK2 cause autosomal-dominant parkinsonism with pleomorphic pathology. Neuron 44:601–607
2. Paisán-Ruíz C, Jain S, Evans EW et al (2004) Cloning of the gene containing mutations that cause PARK8-linked Parkinson's disease. Neuron 44:595–600
3. Benamer HTS, de Silva R (2010) LRRK2 G2019S in the North African population: a review. Eur Neurol 63:321–325
4. Inzelberg R, Hassin-Baer S, Jankovic J (2014) Genetic movement disorders in patients of Jewish ancestry. JAMA Neurol 71:1567–1572
5. Peeraully T, Tan EK (2012) Genetic variants in sporadic Parkinson's disease: East vs West. Parkinsonism Relat Disord 18(Suppl 1):S63–S65
6. Tan E-K (2007) The role of common genetic risk variants in Parkinson disease. Clin Genet 72:387–393
7. Bardien S, Lesage S, Brice A, Carr J (2011) Genetic characteristics of leucine-rich repeat kinase 2 (LRRK2) associated Parkinson's disease. Parkinsonism Relat Disord 17:501–508
8. Nalls MA, Pankratz N, Lill CM et al (2014) Large-scale meta-analysis of genome-wide association data identifies six new risk loci for Parkinson's disease. Nat Genet 46:989–993
9. Singleton A, Hardy J (2011) A generalizable hypothesis for the genetic architecture of disease: pleomorphic risk loci. Hum Mol Genet 20:R158–R162
10. Cookson MR, Hardy J, Lewis PA (2008) Genetic neuropathology of Parkinson's disease. Int J Clin Exp Pathol 1:217–231
11. Lewis PA, Greggio E, Beilina A et al (2007) The R1441C mutation of LRRK2 disrupts GTP hydrolysis. Biochem Biophys Res Commun 357:668–671
12. Li X, Tan Y-C, Poulose S et al (2007) Leucine-rich repeat kinase 2 (LRRK2)/PARK8 possesses GTPase activity that is altered in familial Parkinson's disease R1441C/G mutants. J Neurochem 103:238–247
13. Daniëls V, Vancraenenbroeck R, Law BMH et al (2011) Insight into the mode of action of the LRRK2 Y1699C pathogenic mutant. J Neurochem 116:304–315
14. Guo L, Gandhi PN, Wang W et al (2007) The Parkinson's disease-associated protein, leucine-rich repeat kinase 2 (LRRK2), is an authentic GTPase that stimulates kinase activity. Exp Cell Res 313:3658–3670
15. Liao J, Wu C-X, Burlak C et al (2014) Parkinson disease-associated mutation R1441H in LRRK2 prolongs the "active state" of its GTPase domain. Proc Natl Acad Sci U S A 111:4055–4060
16. Greggio E, Cookson MR (2009) Leucine-rich repeat kinase 2 mutations and Parkinson's disease: three questions. ASN Neuro. doi:10.1042/AN20090007
17. Ray S, Bender S, Kang S et al (2014) The Parkinson disease-linked LRRK2 protein mutation I2020T stabilizes an active state conformation leading to increased kinase activity. J Biol Chem 289:13042–13053
18. Cookson MR (2010) The role of leucine-rich repeat kinase 2 (LRRK2) in Parkinson's disease. Nat Rev Neurosci 11:791–797
19. Greggio E, Jain S, Kingsbury A et al (2006) Kinase activity is required for the toxic effects of mutant LRRK2/dardarin. Neurobiol Dis 23:329–341
20. Smith WW, Pei Z, Jiang H et al (2006) Kinase activity of mutant LRRK2 mediates neuronal toxicity. Nat Neurosci 9:1231–1233
21. Skibinski G, Nakamura K, Cookson MR, Finkbeiner S (2014) Mutant LRRK2 toxicity in neurons depends on LRRK2 levels and synuclein but not kinase activity or inclusion bodies. J Neurosci Off J Soc Neurosci 34:418–433
22. Yao C, Johnson WM, Gao Y et al (2012) Kinase inhibitors arrest neurodegeneration in cell and *C. elegans* models of LRRK2 toxicity. Hum Mol Genet 22(2):328–344. doi:10.1093/hmg/dds431

23. Liu Z, Hamamichi S, Lee BD et al (2011) Inhibitors of LRRK2 kinase attenuate neurodegeneration and Parkinson-like phenotypes in *Caenorhabditis elegans* and *Drosophila* Parkinson's disease models. Hum Mol Genet 20:3933–3942
24. Lee BD, Shin J-H, VanKampen J et al (2010) Inhibitors of leucine-rich repeat kinase-2 protect against models of Parkinson's disease. Nat Med 16:998–1000
25. West AB (2014) Ten years and counting: moving leucine-rich repeat kinase 2 inhibitors to the clinic. Mov Disord 30(2):180–189. doi:10.1002/mds.26075
26. MacLeod D, Dowman J, Hammond R et al (2006) The familial Parkinsonism gene LRRK2 regulates neurite process morphology. Neuron 52:587–593
27. Nichols RJ, Dzamko N, Morrice NA et al (2010) 14-3-3 binding to LRRK2 is disrupted by multiple Parkinson's disease-associated mutations and regulates cytoplasmic localization. Biochem J 430:393–404
28. Jaleel M, Nichols RJ, Deak M et al (2007) LRRK2 phosphorylates moesin at threonine-558: characterization of how Parkinson's disease mutants affect kinase activity. Biochem J 405:307–317
29. Rudenko IN, Kaganovich A, Hauser DN et al (2012) The G2385R variant of leucine-rich repeat kinase 2 associated with Parkinson's disease is a partial loss-of-function mutation. Biochem J 446:99–111
30. Rudenko IN, Chia R, Cookson MR (2012) Is inhibition of kinase activity the only therapeutic strategy for LRRK2-associated Parkinson's disease? BMC Med 10:20
31. Biosa A, Trancikova A, Civiero L et al (2013) GTPase activity regulates kinase activity and cellular phenotypes of Parkinson's disease-associated LRRK2. Hum Mol Genet 22:1140–1156
32. Ko HS, Bailey R, Smith WW et al (2009) CHIP regulates leucine-rich repeat kinase-2 ubiquitination, degradation, and toxicity. Proc Natl Acad Sci U S A 106:2897–2902
33. Herzig MC, Kolly C, Persohn E et al (2011) LRRK2 protein levels are determined by kinase function and are crucial for kidney and lung homeostasis in mice. Hum Mol Genet 20:4209–4223
34. Westerlund M, Belin AC, Anvret A et al (2008) Developmental regulation of leucine-rich repeat kinase 1 and 2 expression in the brain and other rodent and human organs: Implications for Parkinson's disease. Neuroscience 152:429–436
35. Mandemakers W, Snellinx A, O'Neill MJ, de Strooper B (2012) LRRK2 expression is enriched in the striosomal compartment of mouse striatum. Neurobiol Dis 48:582–593
36. Melrose H, Lincoln S, Tyndall G et al (2006) Anatomical localization of leucine-rich repeat kinase 2 in mouse brain. Neuroscience 139:791–794
37. Simón-Sánchez J, Herranz-Pérez V, Olucha-Bordonau F, Pérez-Tur J (2006) LRRK2 is expressed in areas affected by Parkinson's disease in the adult mouse brain. Eur J Neurosci 23:659–666
38. West AB, Cowell RM, Daher JPL et al (2014) Differential LRRK2 expression in the cortex, striatum, and substantia nigra in transgenic and nontransgenic rodents. J Comp Neurol 522:2465–2480
39. West AB, Moore DJ, Choi C et al (2007) Parkinson's disease-associated mutations in LRRK2 link enhanced GTP-binding and kinase activities to neuronal toxicity. Hum Mol Genet 16:223–232
40. Iaccarino C, Crosio C, Vitale C et al (2007) Apoptotic mechanisms in mutant LRRK2-mediated cell death. Hum Mol Genet 16:1319–1326
41. Li Y, Liu W, Oo TF et al (2009) Mutant LRRK2(R1441G) BAC transgenic mice recapitulate cardinal features of Parkinson's disease. Nat Neurosci 12:826–828
42. Tagliaferro P, Kareva T, Oo TF et al (2015) An early axonopathy in a hLRRK2(R1441G) transgenic model of Parkinson disease. Neurobiol Dis 82:359–371
43. Lin X, Parisiadou L, Gu X-L et al (2009) Leucine-rich repeat kinase 2 regulates the progression of neuropathology induced by Parkinson's-disease-related mutant alpha-synuclein. Neuron 64:807–827

44. Ramonet D, Daher JPL, Lin BM et al (2011) Dopaminergic neuronal loss, reduced neurite complexity and autophagic abnormalities in transgenic mice expressing G2019S mutant LRRK2. PLoS One 6(4), e18568. doi:10.1371/journal.pone.0018568
45. Garcia-Miralles M, Coomaraswamy J, Häbig K et al (2015) No dopamine cell loss or changes in cytoskeleton function in transgenic mice expressing physiological levels of wild type or G2019S mutant LRRK2 and in human fibroblasts. PLoS One 10, e0118947. doi:10.1371/journal.pone.0118947
46. Maekawa T, Mori S, Sasaki Y et al (2012) The I2020T Leucine-rich repeat kinase 2 transgenic mouse exhibits impaired locomotive ability accompanied by dopaminergic neuron abnormalities. Mol Neurodegener 7:15
47. Lee J-W, Tapias V, Di Maio R et al (2015) Behavioral, neurochemical, and pathologic alterations in bacterial artificial chromosome transgenic G2019S leucine-rich repeated kinase 2 rats. Neurobiol Aging 36:505–518
48. Dusonchet J, Kochubey O, Stafa K et al (2011) A rat model of progressive nigral neurodegeneration induced by the Parkinson's disease-associated G2019S mutation in LRRK2. J Neurosci Off J Soc Neurosci 31:907–912
49. Tsika E, Nguyen APT, Dusonchet J et al (2015) Adenoviral-mediated expression of G2019S LRRK2 induces striatal pathology in a kinase-dependent manner in a rat model of Parkinson's disease. Neurobiol Dis 77:49–61
50. Beccano-Kelly DA, Volta M, Munsie LN et al (2015) LRRK2 overexpression alters glutamatergic presynaptic plasticity, striatal dopamine tone, postsynaptic signal transduction, motor activity and memory. Hum Mol Genet 24:1336–1349
51. Yue M, Hinkle KM, Davies P et al (2015) Progressive dopaminergic alterations and mitochondrial abnormalities in LRRK2 G2019S knock-in mice. Neurobiol Dis 78:172–195
52. Sanchez G, Varaschin RK, Büeler H et al (2014) Unaltered striatal dopamine release levels in young Parkin knockout, Pink1 knockout, DJ-1 knockout and LRRK2 R1441G transgenic mice. PLoS One 9, e94826
53. Plowey ED, Cherra SJ, Liu Y-J, Chu CT (2008) Role of autophagy in G2019S-LRRK2-associated neurite shortening in differentiated SH-SY5Y cells. J Neurochem 105:1048–1056
54. Manzoni C, Mamais A, Dihanich S et al (2013) Pathogenic Parkinson's disease mutations across the functional domains of LRRK2 alter the autophagic/lysosomal response to starvation. Biochem Biophys Res Commun 441:862–866
55. Sánchez-Danés A, Richaud-Patin Y, Carballo-Carbajal I et al (2012) Disease-specific phenotypes in dopamine neurons from human iPS-based models of genetic and sporadic Parkinson's disease. EMBO Mol Med 4:380–395
56. Cherra SJ, Steer E, Gusdon AM et al (2013) Mutant LRRK2 elicits calcium imbalance and depletion of dendritic mitochondria in neurons. Am J Pathol 182:474–484
57. Mortiboys H, Furmston R, Bronstad G et al (2015) UDCA exerts beneficial effect on mitochondrial dysfunction in LRRK2(G2019S) carriers and in vivo. Neurology 85:846–852
58. Su Y-C, Qi X (2013) Inhibition of excessive mitochondrial fission reduced aberrant autophagy and neuronal damage caused by LRRK2 G2019S mutation. Hum Mol Genet 22:4545–4561
59. Wang X, Yan MH, Fujioka H et al (2012) LRRK2 regulates mitochondrial dynamics and function through direct interaction with DLP1. Hum Mol Genet 21:1931–1944
60. Gehrke S, Imai Y, Sokol N, Lu B (2010) Pathogenic LRRK2 negatively regulates microRNA-mediated translational repression. Nature 466:637–641
61. Martin I, Kim JW, Lee BD et al (2014) Ribosomal protein s15 phosphorylation mediates LRRK2 neurodegeneration in Parkinson's disease. Cell 157:472–485
62. Liu G-H, Qu J, Suzuki K et al (2012) Progressive degeneration of human neural stem cells caused by pathogenic LRRK2. Nature 491:603–607
63. Tsika E, Kannan M, Foo CS-Y et al (2014) Conditional expression of Parkinson's disease-related R1441C LRRK2 in midbrain dopaminergic neurons of mice causes nuclear abnormalities without neurodegeneration. Neurobiol Dis 71:345–358

64. Moehle MS, Webber PJ, Tse T et al (2012) LRRK2 inhibition attenuates microglial inflammatory responses. J Neurosci Off J Soc Neurosci 32:1602–1611
65. Gillardon F, Schmid R, Draheim H (2012) Parkinson's disease-linked leucine-rich repeat kinase 2(R1441G) mutation increases proinflammatory cytokine release from activated primary microglial cells and resultant neurotoxicity. Neuroscience 208:41–48
66. Choi I, Kim B, Byun J-W et al (2015) LRRK2 G2019S mutation attenuates microglial motility by inhibiting focal adhesion kinase. Nat Commun 6:8255
67. Kim B, Yang M-S, Choi D et al (2012) Impaired inflammatory responses in murine lrrk2-knockdown brain microglia. PLoS One 7, e34693
68. Russo I, Berti G, Plotegher N et al (2015) Leucine-rich repeat kinase 2 positively regulates inflammation and down-regulates NF-κB p50 signaling in cultured microglia cells. J Neuroinflammation 12:230
69. Schapansky J, Nardozzi JD, Felizia F, Lavoie MJ (2014) Membrane recruitment of endogenous LRRK2 precedes its potent regulation of autophagy. Hum Mol Genet 23(16):4201–4214. doi:10.1093/hmg/ddu138
70. Matt SM, Johnson RW (2015) Neuro-immune dysfunction during brain aging: new insights in microglial cell regulation. Curr Opin Pharmacol 26:96–101
71. Saha S, Ash PEA, Gowda V et al (2015) Mutations in LRRK2 potentiate age-related impairment of autophagic flux. Mol Neurodegener 10:26
72. Schapansky J, Nardozzi JD, LaVoie MJ (2015) The complex relationships between microglia, alpha-synuclein, and LRRK2 in Parkinson's disease. Neuroscience 302:74–88
73. Jorgensen ND, Peng Y, Ho CC-Y et al (2009) The WD40 domain is required for LRRK2 neurotoxicity. PLoS One 4, e8463.1
74. Hsu CH, Chan D, Greggio E et al (2010) MKK6 binds and regulates expression of Parkinson's disease-related protein LRRK2. J Neurochem 112:1593–1604
75. Xiong Y, Yuan C, Chen R et al (2012) ArfGAP1 is a GTPase activating protein for LRRK2: reciprocal regulation of ArfGAP1 by LRRK2. J Neurosci Off J Soc Neurosci 32:3877–3886
76. Stafa K, Trancikova A, Webber PJ et al (2012) GTPase activity and neuronal toxicity of Parkinson's disease-associated LRRK2 is regulated by ArfGAP1. PLoS Genet 8, e1002526.1
77. Civiero L, Cirnaru MD, Beilina A et al (2015) Leucine-rich repeat kinase 2 interacts with p21-activated kinase 6 to control neurite complexity in mammalian brain. J Neurochem 135:1242–1256. doi:10.1111/jnc.13369
78. Lavalley NJ, Slone SR, Ding H et al (2015) 14-3-3 Proteins regulate mutant LRRK2 kinase activity and neurite shortening. Hum Mol Genet 25(1):109–122. doi:10.1093/hmg/ddv453
79. Cookson MR (2009) alpha-Synuclein and neuronal cell death. Mol Neurodegener 4:9
80. Daher JPL, Volpicelli-Daley LA, Blackburn JP et al (2014) Abrogation of α-synuclein-mediated dopaminergic neurodegeneration in LRRK2-deficient rats. Proc Natl Acad Sci U S A 111:9289–9294
81. Daher JPL, Abdelmotilib HA, Hu X et al (2015) Leucine-rich Repeat Kinase 2 (LRRK2) Pharmacological Inhibition Abates α-Synuclein Gene-induced Neurodegeneration. J Biol Chem 290:19433–19444
82. Daher JPL, Pletnikova O, Biskup S et al (2012) Neurodegenerative phenotypes in an A53T α-synuclein transgenic mouse model are independent of LRRK2. Hum Mol Genet 21:2420–2431
83. Herzig MC, Bidinosti M, Schweizer T et al (2012) High LRRK2 levels fail to induce or exacerbate neuronal alpha-synucleinopathy in mouse brain. PLoS One 7, e36581
84. Beilina A, Cookson MR (2015) Genes associated with Parkinson's disease: regulation of autophagy and beyond. J Neurochem 139(Suppl 1):91–107. doi:10.1111/jnc.13266
85. Goedert M (2015) NEURODEGENERATION. Alzheimer's and Parkinson's diseases: the prion concept in relation to assembled Aβ, tau, and α-synuclein. Science 349:1255555
86. Wallings R, Manzoni C, Bandopadhyay R (2015) Cellular processes associated with LRRK2 function and dysfunction. FEBS J 282:2806–2826

87. Kawakami F, Yabata T, Ohta E et al (2012) LRRK2 phosphorylates tubulin-associated tau but not the free molecule: LRRK2-mediated regulation of the tau-tubulin association and neurite outgrowth. PLoS One 7, e30834
88. Lin C-H, Tsai P-I, Wu R-M, Chien C-T (2010) LRRK2 G2019S mutation induces dendrite degeneration through mislocalization and phosphorylation of tau by recruiting autoactivated GSK3β. J Neurosci Off J Soc Neurosci 30:13138–13149
89. Beilina A, Rudenko IN, Kaganovich A et al (2014) Unbiased screen for interactors of leucine-rich repeat kinase 2 supports a common pathway for sporadic and familial Parkinson disease. Proc Natl Acad Sci U S A 111(7):2626–2631. doi:10.1073/pnas.1318306111
90. MacLeod DA, Rhinn H, Kuwahara T et al (2013) RAB7L1 interacts with LRRK2 to modify intraneuronal protein sorting and Parkinson's disease risk. Neuron 77:425–439

Chapter 13
Small-Molecule Inhibitors of LRRK2

John M. Hatcher, Hwan Geun Choi, Dario R. Alessi, and Nathanael S. Gray

Abstract Mutations in the leucine-rich repeat kinase 2 (LRRK2) protein have been genetically and functionally linked to Parkinson's disease (PD). The kinase activity of LRRK2 is increased by pathogenic mutations; therefore, modulation of LRRK2 kinase activity by a selective small-molecule inhibitor has been proposed as a potentially viable treatment for Parkinson's disease. This chapter presents a historical overview of the development and bioactivity of several small-molecule LRRK2 inhibitors that have been used to inhibit LRRK2 kinase activity in vitro or in vivo. These compounds are important tools for understanding the cellular biology of LRRK2 and for evaluating the potential of LRRK2 inhibitors as disease-modifying PD therapies.

Keywords Parkinson's disease • Leucine-rich repeat kinase 2 (LRRK2) • LRRK2 inhibitors • Mutations

J.M. Hatcher • N.S. Gray (✉)
Department of Cancer Biology, Dana Farber Cancer Institute,
450 Brookline Ave, Boston, MA 02215, USA
e-mail: Johnm_Hatcher@dfci.harvard.edu; nathanael_gray@dfci.harvard.edu

H.G. Choi
New Drug Development Center, Daegu Gyeongbuk Medical Innovation Foundation,
80 Cheombok-ro, Dong-gu, Daegu 41061, South Korea
e-mail: hgchoi@dgmif.re.kr; hwangeun.choi@gmail.com

D.R. Alessi
MRC Protein Phosphorylation and Ubiquitylation Unit, University of Dundee,
Sir James Black Centre, Dow Street, Dundee, DD1 5EH, UK
e-mail: d.r.alessi@dundee.ac.uk

H.J. Rideout (ed.), *Leucine-Rich Repeat Kinase 2 (LRRK2)*,
Advances in Neurobiology 14, DOI 10.1007/978-3-319-49969-7_13

Introduction

Parkinson's disease (PD) is the second most common neurodegenerative disease in the world. It affects over one million Americans and more than 60,000 patients are newly diagnosed each year [1, 2]. Recent genetic studies have revealed an underlying genetic cause in at least 10% of all PD cases [3], which provides new opportunities for the discovery of molecularly targeted therapeutics that may ameliorate neurodegeneration. Among the genes associated with PD, leucine-rich repeat kinase 2 (LRRK2) is unique because a missense mutation, G2019S, is frequently found in both familial and sporadic Parkinson's disease cases [4, 5]. The G2019S mutation increases kinase activity which may result in activation of the neuronal death signal pathway, suggesting that small-molecule LRRK2 kinase inhibitors may be able to serve as a new class of therapeutics for the treatment of Parkinson's disease [6, 7]. Transgenic G2019S LRRK2 mice aged 12–16 months display progressive degeneration of the substantia nigra pars compacta (SNpc) dopaminergic neurons and Parkinson's phenotypes of motor dysfunction suggesting that this mutation may be functionally relevant to the disease [8]. Recent work indicates that LRRK2 directly phosphorylates a subset of the Rab GTPases on an evolutionary conserved residue within the effector interacting-switch II domain [9]. Pathogenic LRRK2 including G2019S mutation increases phosphorylation of endogenous Rabs, and this strongly decreases their affinity to regulatory proteins that bind to the switch II domain including Rab GDP dissociation inhibitors [9]. Given the dearth of potential targets for the treatment of PD and the popularity of kinases as therapeutic targets, it is perhaps not surprising that there has been a very substantial effort to develop potent and selective LRRK2 inhibitors. This chapter presents a historical overview of the development of small-molecule LRRK2 inhibitors from a chemist's perspective.

The first reported LRRK2 inhibitors in 2009 were several ROCK (Rho kinase) inhibitors such as Y-27632 and H-1152, which suppressed LRRK2 with similar potency to which they inhibited ROCK2 as well as sunitinib, a structurally unrelated multikinase inhibitor that suppresses LRRK2, but not ROCK [10]. This study also described for the first time the mutant LRRK2[A2016T] that was normally active but resistant to H-1152, Y-27632, and sunitinib [10]. Prior to the recent identification of Rab GTPase substrates, the effectiveness of inhibitors was assessed by monitoring the dephosphorylation of two phosphorylation sites (Ser910 and Ser935) that mediate binding to 14-3-3 as the phosphorylation of these residues is indirectly controlled by LRRK2 kinase activity through a mechanism that is still not understood [11] (Table 13.1).

Subsequent screening of libraries resulted in compounds that were quite promiscuous kinase inhibitors [12]. A significant advance in the field of small-molecule LRRK2 inhibitors came in 2011 when more selective inhibitors LRRK2-IN-1 (**1**) [13] and CZC-25146 (**2**) [14] were developed through screening of historical kinase inhibitor libraries. LRRK2-IN-1 inhibited both truncated wild-type LRRK2 and LRRK2-G2019S with IC50 values of 13 and 6 nM, respectively, but LRRK2-A2016T and LRRK2-A2016T + G2019S mutants were found to be ~400-fold more

Table 13.1 Early selective LRRK2 inhibitors

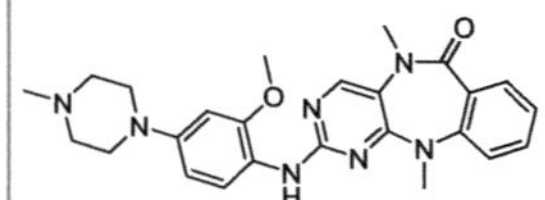

1 (LRRK2-IN-1)

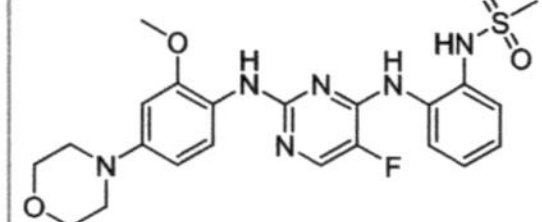

2 (CZC-25146)

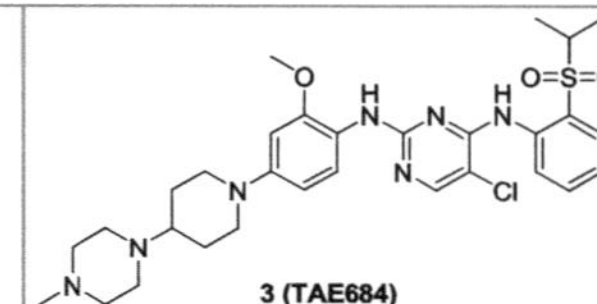

3 (TAE684)

Compd	MPO score	MW	Clog*P*	LRRK2 IC50 (nM)[a]	Kinase selectivity	Mouse Cl (mL min^{-1} kg^{-1})[b]	Total B/P[c]
1	4.4	571	2.9	6	12 of 440 < 10% control[d]	5.6	Negligible
2	2.4	489	3.5	7	4 of 184 < 300 nM[e]	2.3	Negligible
3	1.8	614	5.8	6	27 of 440 < 10% control[d]	17	1.7

[a]Biochemical assay; LRRKtide used for compound **1** and Nictide used for compounds **2** and **3**; G2019S LRRK2 mutant IC_{50} for compounds **1**–**3**
[b]Compound **1** was dosed i.v. (1 mg kg^{-1}). Compound **2** was dosed i.v. (1 mg kg^{-1}) for both systemic and brain PK. Compound **3** was dosed i.v. (1 mg kg^{-1}) for both systemic and brain PK
[c]Total brain/plasma AUC ratio.
[d]KINOMEscan profiling (1 μM)
[e]Kinobeads profiling (2 μM)

resistant to LRRK2-IN-1. The reason for this resistance was explained using a molecular docking study of LRRK2-IN-1 bound to a homology model of LRRK2-A2016T, which revealed an unfavorable interaction between the anthranilic acid moiety and the A2016T residue. The large degree of resistance conferred by the A2016T mutation provides a convenient method for assessing the degree to which LRRK2-IN-1 observed pharmacology is "on target" to LRRK2 through rescue experiments. LRRK2-IN-1 was moderately selective inhibiting 12 of 440 kinases with <10% activity of control using KINOMEscan profiling. LRRK2-IN-1 induced dose-dependent Ser910 and Ser935 dephosphorylation and loss of 14-3-3 binding to endogenous LRRK2 in human-derived neuroblastoma SHSY5Y cells and Swiss3T3 mouse cells. PK studies of LRRK2-IN-1 showed a half-life of 4.5 h and a bioavailability of 49.4% in mice. However, insufficient blood–brain barrier penetration was found based on evaluation of Ser910 and S935 phosphorylation status in the kidney versus the brain. Subsequent characterization of LRRK2-IN-1 has revealed significant off-target activity on ERK5 [15], DCLK1 [16], and Brd4 (unpublished observations). The activity of LRRK2-IN-1 on Brd4, a general transcriptional coactivator protein [17], confounds the use of LRRK2-IN-1 as a selective LRRK2 inhibitor, and newer compounds described below represent superior pharmacological probes. CZC-25146 inhibited the activity of recombinant human wild-type LRRK2 with an IC50 ranging from 1 to 5 nM. The LRRK2-G2019S mutant was inhibited with an IC50 ranging from 2 to 7 nM in a TF-FRET assay. CZC-21546 was screened against a kinase panel of 184 kinases and only inhibited 4 other kinases with high potency. Additionally, it protects against mutant LRRK2-induced injury of cultured human and rodent neurons with nanomolar potency [14]. Unfortunately, this compound also suffered from poor brain penetration of only 4% [14]. TAE684 (3) [18], which was initially reported as an ALK inhibitor [19], was also found to inhibit LRRK2. It possesses similar structural features as LRRK2-IN-1 and CZC-25146 and displayed similar biochemical potency against LRRK2. TAE684 also inhibits in vivo phosphorylation of mouse Ser910 and Ser935 in the kidney and spleen in a dose-dependent manner following oral administration. Unfortunately, this compound is highly promiscuous inhibitor which binds tightly to 27 of 440 kinases with <10% activity of control.

Diaminopyrimidines

The major drawback of compounds **1–3** is their inability to probe the effects of LRRK2 inhibition in the CNS. As a result, simplified hybrid structures of **1**, **2**, and **3** were created leading to a variety of lower molecular weight diaminopyrimidines. Compound **4** was reported independently by Gray et al. [20] and Genentech [21] and was found to maintain the ability to potently inhibit the biochemical activity of both wild-type and G2019S mutant LRRK2. Compound **4** exhibited biochemical IC50 values of 20.3 and 3.2 nM against LRRK2-wt and LRRK2-G2019S, respectively. Compound **4** also maintains inhibition of the A2016T mutant due to the removal of the 4-anilino moiety. Compound **4** was profiled against a panel of 451 kinases using KINOMEscan technology at a concentration of 0.1 μM, which revealed no other interactions except for a mutant form of c-Kit (L576P). The mouse pharmacokinetic profile showed good oral bioavailability (%*F* = 67), a short half-life (0.13 h), and a low plasma exposure [502 h*ng/mL, area under the concentration–time curve $(AUC)_{last}$]. The short half-life was attributed to rapid first-pass metabolism because incubation with mouse liver microsomes revealed a half-life of 13 min. In vivo testing of compound **4** showed LRRK2 inhibition of ~40% at 30 mg/kg and ~70% at 50 and 100 mg/kg in the mouse brain following intraperitoneal administration. An independent medicinal chemical effort by researchers at Genentech resulted in the identification of compound **5** [22]. Compound **5** also featured a high degree of kinase selectivity with no other interactions found in a panel of 178 kinases with >50% inhibition. Compound **5** showed a threefold improvement in the pS1292 cellular assay and an improved in vivo rat PK profile. Compound **5** was used as an in vivo tool to demonstrate the inhibition of in vivo kinase activity in G2019S transgenic mouse brains (pS1292) and the reversal of cellular effects of LRRK2 PD mutations in cultured primary hippocampal neurons. The in vivo unbound brain concentration required to effectively reduce pS1292 autophosphorylation (IC50 = 12 nM) was similar to compound **4**. Additionally, a statistically significant reversal of the neurite outgrowth defects associated with the LRRK2-G2019S mutant [23] was observed upon treatment of LRRK2-G2019S mouse embryonic hippocampal neurons with compound **5** at 100 nM. Compound **5** was screened against a targeted subset of kinases that share a similar ATP-binding site sequence to LRRK2 and showed strong inhibition of TTK. Genentech used the JAK2-based homology model and the TTK cocrystal structures to determine that substitution on the phenyl ring para to the methoxy group would produce steric clash with Asp608 of TTK to increase the selectivity for LRRK2 over TTK. Thus, a few minor structural modifications, including fluorine substitution para to the methoxy group on the phenyl ring position, resulted in the discovery of compound **6** [22]. Compound **6** possessed a nearly identical activity and DMPK profile as compound **5** but with an improved TTK selectivity index (53-fold) and good kinome selectivity at a concentration (0.1 μM) [2/449 kinases with <30% activity of control]. The excellent selectivity profile of optimized inhibitor **6** at 0.1 μM was further established in an Invitrogen kinase selectivity panel [1/197 kinases with >50% inhibition (TTK)].

In an effort to increase the structural diversity and selectivity of the brain-penetrant diaminopyrimidine compounds, Genentech explored aniline bioisosteres, which led to the identification of aminopyrazoles GNE-0877 (**7**) [24] and GNE-9605 (**8**) [24] (Table 13.2) as highly potent and specific LRRK2 inhibitors. Compound **7** was tested in a panel of 188 kinases and showed >50% inhibition of 4 other kinases (Aurora B, RSK2, RSK3, RSK4). Compound **8** inhibited no other kinases to greater than >70% inhibition when tested at a concentration of 1 μM. Compound **8** was tested in a panel of 178 kinases and inhibited one other kinase with >50% inhibition (TAK1-TAB1). Compound **8** inhibited no other kinases with >50% inhibition. Compounds **7** and **8** also showed excellent DMPK profiles (Table 13.3).

Table 13.2 Highly selective aminopyrimidine LRRK2 inhibitors

	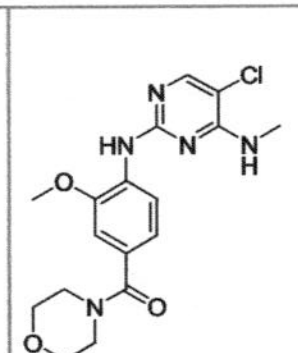 4 (HG-10-102-01)		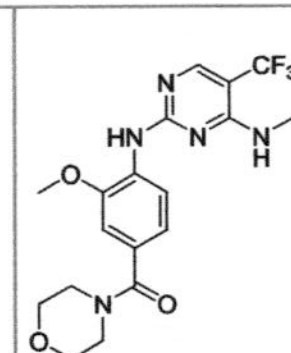 5 (G1023)	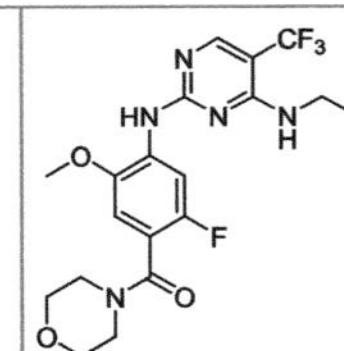 6 (GNE-7915)	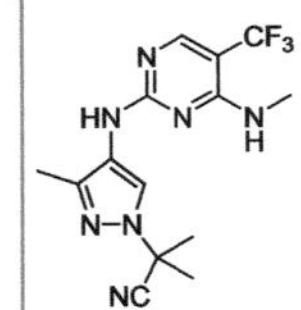 7 GNE-0877	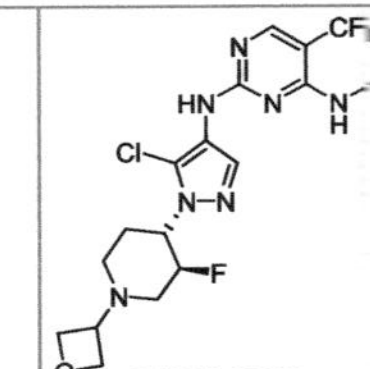 8 (GNE-9605)	
Compd	MPO score	MW	Clog*P*	LRRK2 IC50 (nM)[a]	Kinase selectivity	pLRRK2 IC_{50}[b]	Total B/P[c]
4	5.4	378	1.9	3	1 of 451 < 30% of control[d]	29	0.23
5	5.1	411	2.1	2	0 of 178 >50%[e]	9	0.9
6	4.5	443	2.8	1	2 of 449 < 30% of control[d]	9	0.9
7	5.0	339	3.0	0.7	4 of 188 > 50%[e]	3	1.2
8	4.5	450	2.8	2	1 of 178 > 50%[e]	19	0.7

[a]Biochemical assay; LRRKtide and G2019S mutant used for compounds **4–8**
[b]pSer1292 triple mutant (G2019S/R1441G/Y1699C) cellular assay used for compounds **4–8**
[c]Total brain/plasma AUC ratio
[d]KINOMEscan profiling (0.1 μM)
[e]Invitrogen profiling (0.1 μM)

Table 13.3 In vivo and in vitro rat DMPK profiles of diaminopyrimidine LRRK2 inhibitors

Compd	hep Cl_{hep}(mL min^{-1} kg^{-1})[b] h/r[c]	% rat PPB	Cl (Cl_u) (mL min^{-1} kg^{-1})[d]	iv $t_{1/2}$ (h)	%*F*	MDR1[e] P-gp ER[f] (B:A/A:B)[g]
4	4.9/24	92	51 (616)	0.5	67	2.8
5	1.8/7.6	86	24 (156)	1.2	80	1.2
6	3.2/14	97	8.3 (244)	3.1	40	0.9
7	3/25	79	44 (210)	1.3	90	0.8
8	1/21	79	26 (261)	1.5	90	0.8

[a]Compounds **4–8** were dosed p.o. (1 mg kg^{-1}) as an aqueous suspension with 1% methylcellulose, i.v. (0.5 mg kg^{-1}) as a 60% PEG solution or 20–60% NMP solution for systemic PK, and i.v. (0.5 mg kg^{-1}) as a 60% NMP solution for brain PK [21–24]
[b]In vitro stability in cryopreserved hepatocytes
[c]h/r = human/rat
[d]Cl_u = unbound clearance = total clearance/f_{up}, where f_{up} is the unbound plasma free fraction
[e]MDCK-MDR1 human P-gp transfected cell line
[f]Efflux ratio
[g]Basolateral to apical/apical to basolateral

Table 13.4 In vivo NHP PK profiles of Genentech aminopyrimidine LRRK2 inhibitors

Compd	hep Cl_{hep} (mL min^{-1} kg^{-1})[b] NHP	% NHP PPB	Cl (Cl_u) (mL min^{-1} kg^{-1})[c]	iv $t_{1/2}$ (h)	%*F*	MDR1[d] P-gp ER[e] (BA/A:B)[f]	B_u/P_u[g]	CSF/P_u[h]
6	14	95	11 (200)	7.7	17	0.9	0.6	1.1
7	19	80	20 (100)	2.2	35	0.8	0.7	1.2
8	13	82	8 (43)	4.0	74	0.8	n/a	1.1

[a]Compounds **6–8** were dosed p.o. (1 mg kg^{-1}) as an aqueous suspension with 1% methylcellulose and i.v. (0.5 mg kg^{-1}) as a 20–60% NMP solution [22–24]
[b]In vitro stability in cryopreserved hepatocytes
[c]Cl_u = unbound clearance = total clearance/f_{up}, where f_{up} is the unbound plasma free fraction
[d]MDCK-MDR1 human P-gp transfected cell line
[e]Efflux ratio
[f]Basolateral to apical/apical to basolateral
[g]Unbound brain/unbound plasma AUC ratio
[h]CSF/unbound plasma AUC ratio

In vivo nonhuman primate (NHP) PK profiles of compounds **6**, **7**, and **8** are summarized in Table 13.4. All three compounds demonstrated good in vitro–in vivo correlation, moderate plasma clearance rates, and good i.v. half-lives. Additionally, CSF/P ratios extracted from low-dose PK studies suggested that all three compounds possessed approximately equal free brain to free plasma distribution [25]. Desirable B ratios were later confirmed for compounds **7** and **8** during NHP safety assessments (vide infra).

In the absence of a LRRK2-dependent PD efficacy model, in vivo PD knockdown for compounds **6–8** was assessed through the use of LRRK2 BAC transgenic mice expressing human LRRK2 protein with the G2019S mutation [23]. Inhibitors were evaluated for their ability to inhibit LRRK2 S1292 autophosphorylation in vivo. Tissue samples were harvested and examined from the hippocampus and spleen [22]. Dose-dependent inhibition of S1292 phosphorylation was observed for all inhibitors tested in both the brain and peripheral tissues. In vivo unbound brain IC50 values of 7, 3, and 20 nM were calculated for **6**, **7**, and **8**, respectively.

Toxicities Observed with Aminopyrimidine LRRK2 Inhibitors

In 2011, Novartis published a report indicating that lung and kidney abnormalities existed in LRRK2 genetic knockout mice [26]. In order to assess these potential liabilities, Genentech advanced their aminopyrimidine compounds into safety studies. Due to low anti-pS1292 antibody sensitivity, in vivo PD target engagement in all of the studies was assessed using pS935. Male C57BL/6 mice were dosed p.o. with compound **6** (200 and 300 mg kg^{-1} b.i.d.) and compound **7** (30 and 65 mg kg^{-1} b.i.d.) for 15 days. Toxicokinetic analysis showed dose-dependent increases in plasma and brain levels with average free drug exposures of 5- and 36-fold above pS1292 cellular IC50 values for the higher doses. While evidence of lung, kidney, and brain PD knockdown was observed with both inhibitors, no microscopic effects were observed in the lung or kidney, and both compounds were well tolerated. Similar 7-day repeat-dosing studies were conducted in male and female Sprague–Dawley rats. Once daily p.o. administration of compound **6** (10, 50, and 100 mg kg^{-1}) and compound **7** (30, 75, and 200 mg kg^{-1}) showed dose-dependent exposure increases. The highest tolerated doses for compounds **6** (100 mg kg^{-1}) and 7 (30 mg kg^{-1}) translated to maximum free drug exposures of 22- and 195-fold over the cellular IC50 of compound **6** and **7**, respectively. No macroscopic or microscopic effects were seen in the lung or kidney with either compound. Finally, NHPs were dosed orally with compound **6** (10, 25, and 65 mg kg^{-1} q.d.) for 7 days. Toxicokinetic analysis showed a dose-dependent increase in plasma exposures with average free drug levels correlating to 4-, 14-, and 35-fold above the cellular IC50s. Terminal B_u/P_u and CSF/P_uAUC ratios of 0.6 and 0.8 were achieved, respectively, showing a high degree of brain penetration. Statistically significant PD inhibition of pS935 was observed at all doses tested. Transient and reversible clinical observations included tremors (=10 mg kg^{-1}) and hypoactivity and decreased reactivity to stimulus (25 and 65 mg kg^{-1}). The only anatomic pathology observation linked to compound **6** was limited to the lung and characterized by abnormal cytoplasmic accumulation of secretory lysosome-related organelles known as lamellar bodies in type II pneumocytes of all animals administered 25 and 65 mg kg^{-1} in both sexes. Comparison of these findings with the published LRRK2 knockout mouse data showed that the lung phenotypes were essentially identical. With the goal of examining potential on-target- versus off-target-related effects, structurally distinct aminopyrazole 7 (6 and 20 mg kg^{-1} b.i.d.) and anilinoaminopyrimidine 6 (30 mg kg^{-1} b.i.d.) were administered to NHPs in a 29-day repeat-dose study. Significant free drug coverage above cellular IC50 values and excellent brain penetration were achieved in both tests. PK/PD knockdown of pS935 in the brain, kidney, and lung was confirmed at all doses. Upon microscopic evaluation of the lung, abnormalities identical to those observed in LRRK2 knockout mice were observed at all doses with both inhibitors.

These findings are consistent with an on-target effect of reduction of LRRK2 kinase activity that leads to lamellar body accumulation in type II pneumocytes in the lung of certain species. Interestingly, the morphologic abnormality described for the LRRK2 knockout mouse kidney was absent in both NHP studies.

Arylbenzamides

Researchers at GSK developed a series of arylbenzamides exemplified by GSK2578215A which were reported on in 2012 [27] (**9**, Table 13.5). Compound **9** is a structurally distinct, potent, and highly selective LRRK2 kinase inhibitor that shows inhibition of only 2 other kinases with <10% activity of control (ALK and FLT3) in a panel of 449 kinases. Compound **9** demonstrates moderate mouse plasma clearance (30 mL min^{-1} kg^{-1}), low oral bioavailability (12% at 10 mg/kg), and a total B/P ratio of 1.4 (Table 13.5). In addition, Gray and coworkers were able to demonstrate dose-dependent inhibition of pS910 and pS935 in stably transfected HEK293 cells (wild-type and G2019S LRRK2), with endogenous LRRK2 from lymphoblastoid cells (G2019S PD patient sample) and in mouse Swiss 3T3 cells at approximately 0.3–1.0 μM. In vivo PK/PD analysis of normal mice revealed strong inhibition of pS910 and pS935 in peripheral tissues (spleen and kidney lysates); however, no significant inhibition was detected in the mouse brain despite a total brain to plasma ratio of 1.4. One possible explanation for the lack of brain target engagement could be insufficient free drug levels in the brain relative to the cellular potency. Compound **9** has recently been used to probe the relationship between kinase activity and synaptic vesicle release [28]. Additionally, a patented N-methylpiperazine inhibitor BMPBB-32 [29, 30] (**10**, Table 13.5) closely related to 2-fluoropyridine **9** was profiled. It was reported that piperazine **10** could improve contrast sensitivity in *Drosophila* that overexpress the human LRRK2[G2019S] transgene. Additionally, it was found that N-methylpiperazine benzamide **10** lacks the off-target effects seen with the less selective inhibitor **1** in *Drosophila* in vivo models.

Quinolines and Cinnolines

Elan Pharmaceuticals published their findings on a series of cinnoline [31] and quinoline [32] LRRK2 small-molecule inhibitors, which were identified from a kinase-focused HTS of an in-house library. The screen used a homogeneous time-resolved fluorescence (HTRF) assay measuring the inhibition of phosphorylation of LRRKtide. Hits from this scaffold were known P-gp substrates and inhibitors of PDE4 and CSF1R. Using a MLK1-based homology model, Garofalo and coworkers employed structure-based drug design to develop potent and modestly LRRK2-selective quinolines represented by compound **11** and cinnolines represented by compound **12** (Table 13.6). Quinoline **11** had a biochemical IC50 = 3 nM against

Table 13.5 Arylbenzamide LRRK2 inhibitors

				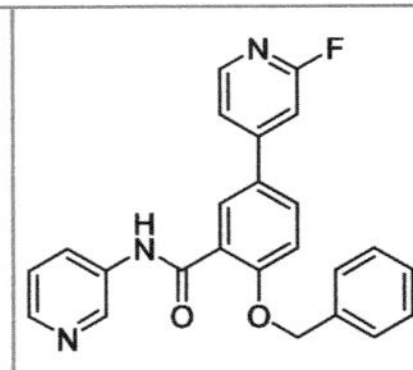 9 (GSK2578215A)		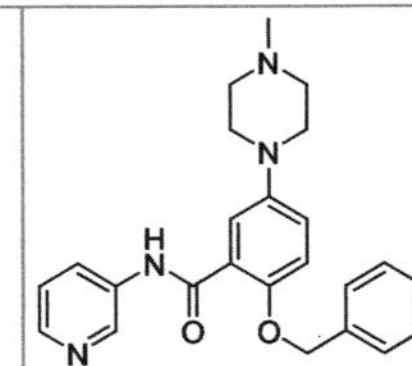 10 (BMPPB-32)				
Compd	MPO score	MW	Clog*P*	LRRK2 IC50 (nM)[a]	Kinase selectivity	pLRRK2 IC_{50}[b]	Cl (mL min^{-1} kg^{-1})	iv $t_{1/2}$ (h)	*F* (%)	Total B/P[c]
9	3.9	399	4.3	9	2 of 449 < 10% of control[e]	n/a	30	1.1	12	1.4
10	4.8	402	3.5	6[d]	3 of 449 < 10% of control[f]	94	n/a	n/a	n/a	0.9

[a]Biochemical assay; Nictide used for compound **9** and LRRKtide used for compound **10**; LRRK2 (G2019S) IC_{50} at 100 μM ATP for compounds **9** and **10**
[b]pSer935 G2019S mutant cellular assay (HEK293 cells)
[c]Total brain/plasma AUC ratio
[d]LanthaScreen LRRK2 K_i
[e]KINOMEscan profiling (10 μM)
[f]KINOMEscan profiling (1 μM). Compound **9** was dosed p.o. (10 mg kg^{-1}) and i.v. (1 mg kg^{-1}) for both systemic and brain PK.

Table 13.6 Quinoline and cinnoline LRRK2 inhibitors

			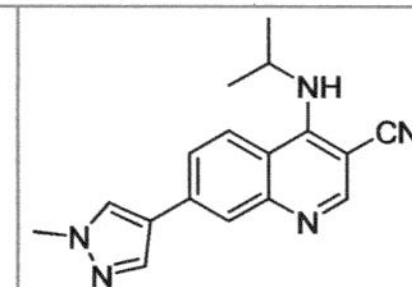 **11 (Elan)**		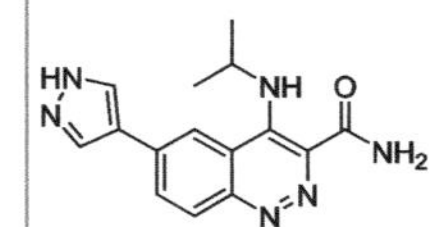 **12 (Elan)**		
Compd	MPO score	MW	Clog*P*	LRRK2 IC50 (nM)[a]	Kinase selectivity	pLRRK2 IC_{50}[b]	Total B/P[c]
11	5.4	291	2.8	3	11 of 39 > 50%[d]	140	1.2
12	4.7	295	2.9	5	n/a	62	1.5

[a]Biochemical assay; LRRKtide used for compound **11** and **12**; LRRK2 (G2019S) IC_{50} at 100 μM ATP
[b]pSer935 G2019S mutant cellular assay (HEK293 cells)
[c]Total brain/plasma AUC ratio
[d]Invitrogen profiling (1 μM)

LRRK2-G2019S and a cellular pS935 EC50 = 140 nM using HEK293 cells stably transfected with LRRK2-G2019S. Quinoline **11** was screened against a panel of 39 kinases and showed inhibition of 11 kinases with >50% inhibition. Guided by the hypothesis that an amide functionality in the original HTS hit was linked to the P-gp-mediated efflux, suitable replacements were discovered, leading to the 3-cyano substitution in **11**, which removed this liability. This also led to a total mouse B/P ratio of 1.2 for quinoline **11** and statistically significant reduction of pS935 in the brains of G2019S LRRK2 transgenic mice at 3 h following oral doses of 30 and 100 mg/kg [31]. Cinnoline **12** had a biochemical IC50 = 5 nM against LRRK2-G2019S and a cellular pS935 EC50 = 62 nM using HEK293 cells stably transfected with LRRK2-G2019S. Cinnoline **12** was found to have a total mouse B/P ratio of 1.5; however, when cinnoline **12** was screened against a panel of 40 kinases, it was found to be highly promiscuous [32].

Triazolopyridazine and Indazole

Elan Pharmaceuticals discovered a series of triazolopyridazines based on a kinase-focused HTS of an in-house library [33]. The screen used a homogeneous time-resolved fluorescence (HTRF) assay measuring the inhibition of phosphorylation of LRRKtide. Based on a homology model built using MLK1, they found that the triazolopyridazines made a single hydrogen bond contact between the triazolo ring and A1950. These compounds displayed a moderate level of selectivity for LRRK2-G2019S over LRRK2-wt with the lead compound **13** having a biochemical IC50 = 76 nM for LRRK2-wt and a biochemical IC50 = 10 nM for LRRK2-G2019S. However, these compounds suffered from oxidative metabolism due to the sulfur linker as well as poor permeability. Pfizer later reported a series of related triazolopyridazines exemplified by compound **14** [34]. This compound was also developed by optimizing a hit from an HTS library screen. Using TYK2 cocrystal structures, they found a single point contact between the kinase hinge (N-H of Val981) and a nitrogen atom of the triazole ring. The absence of hydrogen bond donors in compound **14** is notable, since hydrogen bond donors increase the probability of P-gp recognition and reduce CNS penetration [35]. Through the application of CNS physicochemical property constraints and by use of TYK2 cocrystal structures to facilitate structure-based drug design, compound **14** (Table 13.7) was developed. Triazolopyridazine **14** has a LRRK2-wt IC50 of 64 nM. Compound **14** was screened against a panel of 391 kinases and was found to be highly selective inhibiting only 2 other kinases with <30% of control (TYK2 and JAK3). The in vivo tolerability profile of **14** was assessed in a repeat-dose 14-day PK/PD study in mice at 30 and 300 mg kg^{-1} (n = 5 males/dose). No test article-related findings were reported from the examined tissues, which included microscopic examinations of the lung and kidney. This compound did not show inhibition of pS935 or pS1292 in mouse brains at any dose levels [34].

Table 13.7 Triazolopyridazine and indazole LRRK2 inhibitors

		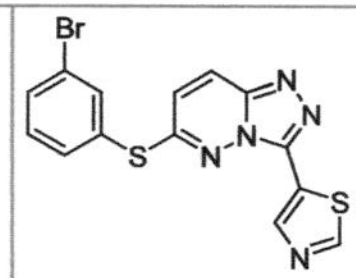 **13 (Elan)**		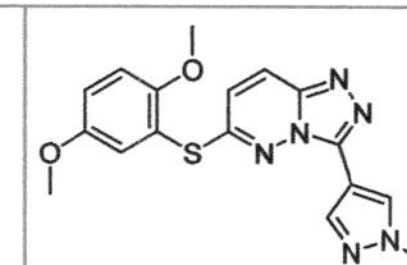 **14 (Pfizer)**	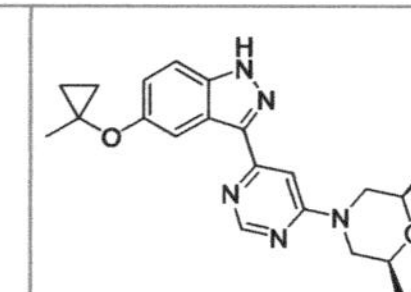 **15 MLi-2**		
Compd	MPO score	MW	Clog*P*	LRRK2 IC50 (nM)[a]	Kinase selectivity	pLRRK2 IC_{50}	Total B/P
13	5.4	390	3.5	10	n/a	683[b]	n/a
14	5.6	368	2.7	64	2 of 391 < 30% of control[f]	n/a	0.1[d]
15	5.6	379	4.0	0.7	1 of 144 < 20% of control[g]	1.4[c]	0.008[e]

[a]Biochemical assay; LRRKtide used for compounds **13**, **14**, and **15**; LRRK2 (G2019S) IC_{50} at 100 µM ATP for compound **13**, LRRK2 (G2019S) IC_{50} at 134 µM ATP for compounds **15,** and LRRK2 (wt) IC50 at 100 µM ATP for compound **14**

[b]pSer935 G2019S mutant cellular assay (HEK293 cells)

[c]pSer935 G2019S mutant cellular assay (SH-SY5Y cells)

[d]Total brain/plasma AUC ratio

[e]Unbound fraction in plasma and brain

[f]KINOMEscan profiling (1 µM)

[g]Dundee profiling (10 µM)

Researchers at Merck recently disclosed the indazole MLi-2 (**15**) [44], a structurally novel, highly potent, and selective LRRK2 kinase inhibitor with central nervous system activity. MLi-2 exhibits exceptional potency in a purified LRRK2 kinase assay in vitro (IC50 = 0.76 nM), a cellular assay monitoring dephosphorylation of LRRK2 pSer935 LRRK2 (IC50 = 1.4 nM), and a radioligand competition binding assay (IC50 53.4 nM). MLi-2 showed inhibition of only 1 other kinase with $<20\%$ activity of control in a panel of 144 kinases at a concentration of 10 μM. MLi-2 suppresses LRRK2 Ser935 phosphorylation as well as Rab10 Thr73 phosphorylation at 1–10 nM in mouse embryonic fibroblasts (MEFs) [9]. In mice, a dose of 3 mg/kg of MLi-2 reduces Ser395 phosphorylation as well as Rab8 (Thr72) and Rab12 (Ser105) phosphorylation in the brain to undetectable levels. Treatment of mice with MLi-2 was found to be well tolerated, with no adverse effects of MLi-2 on body weight, food intake, or behavioral activity observed at brain and plasma exposures 100 x the in vivo IC50 for CNS LRRK2 kinase inhibition. Morphologic changes in the lung, consistent with enlarged type II pneumocytes, were observed in MLi-2-treated MitoPark mice. Moreover, the A2016T mutation renders LRRK2 nearly tenfold resistant to MLI-2. Consistent with this, 10 nM MLi-2 induces dephosphorylation of Ser935 and Rab10 (Thr73) in mouse embryonic fibroblasts of wild type but not in LRRK2[A2016T] knockin cells [44]. Furthermore, MLi-2 at 3 mg/kg injected into mice induces complete dephosphorylation of Ser935 in the brain, kidney, lung, and spleen of wild-type mice but not of littermate LRRK2[A2016T] knockin animals (unpublished data).

Indolinones

Sunitinib (**16**) was originally identified as a potent inhibitor of LRRK2; however, high kinase promiscuity precluded the use of this compound for reliable model studies of LRRK2 function. Recently, scientists at Novartis reported the optimization of sunitinib to indolinones **17** [36] and **18** [36] (Table 13.8) with single-digit nanomolar LRRK2 biochemical activity and modest in vivo pharmacokinetic properties. By using an IRAK4-based homology model, the 5-position of the indolinone core was targeted for improving kinase selectivity. This strategy led to the introduction of 5-alkoxy substituents in **16** and **17** that demonstrated improved selectivity profiles over other indolinone-based inhibitors against a small panel of off-target kinases (ALK, KDR, LCK, PDGFRa, and RET). When administered to mice, compound **17** reduced LRRK2 protein levels in the kidney and in the brain. Inhibitor **17**, along with the less selective LRRK2 kinase inhibitor H-1152, was subsequently used by Longo et al. to ameliorate the observed age-dependent hyperkinetic phenotype of LRRK2-G2019S knockin mice [37]. These results suggest that the enhanced kinase activity of the LRRK2[G2019S] protein is responsible for the observed lack of age-related decline in stepping activity and immobility time that was demonstrated by wild type.

Table 13.8 Indolinone LRRK2 inhibitors

	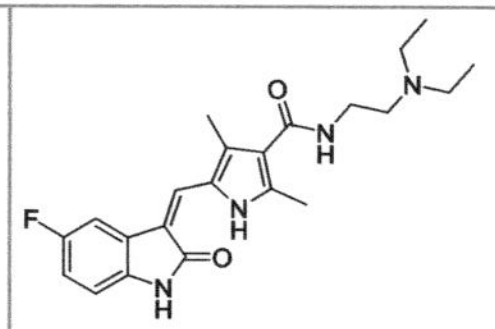 16 (Sunitinib)			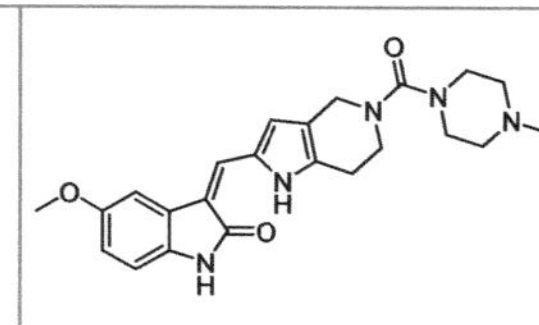 17 (Novartis)	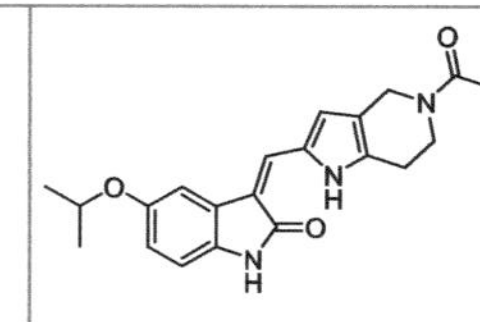 18 (Nov-LRRK2-11)	
Compd	MPO score	MW	Clog*P*	LRRK2 IC50 (nM)[a]	Kinase selectivity	pLRRK2 IC_{50}[b]
16	4.2	398	2.9	19	30 of 140 < 20% of control[c]	370
17	5.1	421	1.6	9	n/a	n/a
18	5.5	365	2.4	4	n/a	380

[a]Biochemical assay IC_{50} at 200 μM ATP for compounds **15–16**
[b]pSer935 cellular assay (NIH3T3 Cells)
[c]Dundee profiling (1 μM)

Pyrrolopyrimidines

The Gray lab reported the discovery of a highly potent and selective brain-penetrant pyrrolopyrimidine LRRK2 inhibitor JH-II-127 (**19**) [38]. This compound was designed based on an intramolecular hydrogen bond between a fluorine of the –CF3 group and the –NHMe group on the pyrimidine of GNE-7915 (**6**) forming a pseudobicycle. Researchers in the Gray lab constructed the ring-closed version to give the pyrrolopyrimidine **19**. They proposed that a fused bicyclic analogue would increase the binding affinity due to the additional third hydrogen bond donor at the 7 position, which is predicted to hydrogen bond with M1949. In addition, they reasoned that a fused bicyclic compound should be able to better fill the hydrophobic area around the hinge region, thus leading to an increase in binding affinity. Their molecular docking study based on the crystal structure of Roco kinase also predicted a sulfur–halogen interaction between the chlorine of the pyrrole ring and the Met1947 residue. Compound **19** had a biochemical IC50 of 6.6 nM against LRRK2-wt and 2.2 nM against LRRK2-G2019S. Compound **19** induced a dose-dependent inhibition of Ser910 and Ser935 phosphorylation in both wild-type LRRK2 and LRRK2-G2019S stably transfected into HEK293 cells. Substantial dephosphorylation of Ser910 and Ser935 was observed at approximately 0.3 μM concentrations of **19** for wild-type LRRK2 and LRRK2-G2019S, which is a similar potency to that observed for LRRK2-IN-1 (**1**). Consistent with the biochemical results, **19** also induced dephosphorylation of Ser910 and Ser935 at a concentration of 0.3—1 μM in the drug-resistant LRRK2[A2016T + G2019S] and LRRK2[A2016T] mutants, revealing that the A2016T mutation does not induce resistance to **19**. Compound **19** was tested on endogenously expressed LRRK2 in human lymphoblastoid cells derived from a control and Parkinson's patient homozygous for the LRRK2-G2019S mutation. Increasing doses of **19** led to similar dephosphorylation of endogenous LRRK2 at Ser910 and Ser935, as was observed in HEK293 cells stably expressing wild-type LRRK2 or LRRK2-G2019S. Moreover, endogenous LRRK2 was also more sensitive to **19** than LRRK2-IN-1 (**1**), which is consistent with the trend observed in HEK293 cells. It was also found that **19** induced similar dose-dependent Ser935 dephosphorylation of endogenous LRRK2 in mouse Swiss 3T3 cells. The mouse pharmacokinetic profile of **19** demonstrated good oral bioavailability (116 %*F*), a half-life of 0.66 h, and a plasma exposure of 3094 (hr * ng/mL, AUC_{last}) following 10 mg/kg p.o. dosing (Table 13.9). Additionally, following 2 mg/kg i.v. dosing, **19** showed a plasma exposure of 533 (hr * ng/mL, AUC_{last}) and a brain exposure of 239 (hr * ng/mL, AUC_{last}), which equates to a brain/plasma concentration ratio of 0.45. They compared the pharmacodynamic properties of **19** with GNE-7915 (**6**) by monitoring inhibition of LRRK2 Ser910/Ser935 phosphorylation in the kidney, spleen, and brain following intraperitoneal delivery of each compound at 100 mg/kg. They observed near-complete dephosphorylation of Ser935 of LRRK2 in all tissues including the brain at this dose for both compounds. They repeated the study at lower doses of 50, 30, and 10 mg/kg of **19** and **6**. With **19**, they observed near-complete inhibition in all tissues at 30 mg/kg but only partial

Table 13.9 Pyrrolopyrimidine LRRK2 inhibitors

				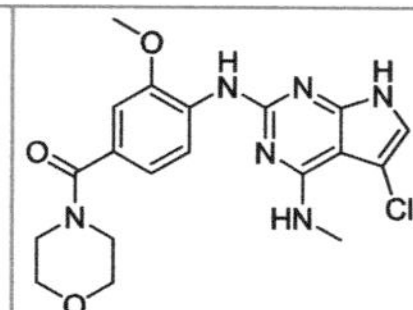 19 (JH-II-127)		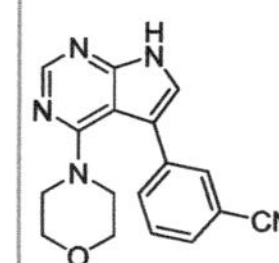 20 (PF-06447475)				
Compd	MPO score	MW	Clog*P*	LRRK2 IC50 (nM)[a]	Kinase selectivity	pLRRK2 IC_{50}[b]	Cl (mL min^{-1} kg^{-1})	iv $t_{1/2}$ (h)	*F* (%)	Total B/P[c]
19	4.8	416	3.3	2	2 of 451 < 10% of control[d]	7	62	1.1	116	0.45
20	5.7	305	2.2	11	3 of 449 < 30% of control[d]	25	n/a	n/a	n/a	0.9

[a]Biochemical assay; Nictide used for compound **19**. LRRKtide used for compound **20**; LRRK2 (G2019S) IC_{50} at 100 μM ATP for compound **19** and LRRK2 (G2019S) IC_{50} at 50 μM ATP for compound **20**

[b]pSer935 G2019S mutant cellular assay (HEK293 cells)

[c]Total brain/plasma AUC ratio

[d]KINOMEscan profiling (1 μM)

inhibition in the brain at the 10 mg/kg dose. However, with **6**, complete inhibition in the brain was only observed at the 100 mg/kg. These results indicate that **19** is a promising chemo-type for achieving dephosphorylation of Ser935 in the brain. KINOMEscan analysis against a near comprehensive panel of 451 kinases at a concentration of 1 μM resulted in no interactions with kinases other than LRRK2[G2019S] with the exception of TTK and RPS6KA4 [38]

Pfizer reported the identification of a pyrrolopyrimidine scaffold that provided a highly efficient starting point with favorable CNS properties for lead optimization. Using MST3 as a crystallographic surrogate for LRRK2 (reported MST3-LRRK2 ATP-binding site residue similarity of 73%), they improved the off-target liabilities of early HTS leads. This led to the discovery of PF-06447475 (**20**, Table 13.9) [39] with in vitro LRRK2-wt and LRRK2-G2019S biochemical IC50 values of 3 and 11 nM, respectively, and a pS935 cellular IC50 of 25 nM. KINOMEscan profiling of 449 kinases at 1 μM showed inhibition of 3 kinases with <30% activity of control. Selectivity profiling of pyrrolopyrimidine **20** in a cellular context was performed using the ActivX KiNativ technology demonstrating good selectivity in human peripheral blood mononuclear cells at 1 μM and an ActivX LRRK2 cellular IC of 15 nM. Compound **20** is not a P-gp substrate and has moderate and high turnover in human and rat liver microsomes, respectively. The oral in vivo PK profile for **20** in rat, dog, and NHP was also determined. While low oral exposure in higher species precluded further clinical advancement, sufficient oral bioavailability in rodents was achieved, which enabled in vivo mouse PK/PD studies. Additionally, inhibitor **20** exhibited approximate equal distribution of free drug between the rat brain, plasma, and kidney tissues. Dose-dependent reduction in phosphorylation in the brain and kidney was observed in wild-type and G2019S BAC transgenic mice [39]. From these experiments, in vivo unbound brain EC values of 8 nM LRRK2-wt and 103 nM LRRK2-G2019S were calculated using the pS935 biomarker, and 21 nM (G2019S) was calculated for pS1292. On the basis of a clean in vitro safety profile (Ames, MNT, THLE cell viability), pyrrolopyrimidine **20** was evaluated in a 14-day repeat-dose rat toxicity study (3, 10, and 30 mg kg^{-1} b.i.d.). No major test article-related findings were observed with day 11 maximum free drug exposures 6-, 25-, and 70-fold over the cellular EC (15 nM), respectively. Similar to the rat toxicity studies carried out with Genentech aminopyrimidines, no changes were noted upon close examination of the kidney and lung tissue.

Docking Studies

Molecular docking studies were recently conducted by our lab on representative classes of LRRK2 inhibitors, including HG-10-102-01 (**4**), JH-II-127 (**19**), GSK2578215A (**9**), and compound **13** (Elan), based on a crystal structure of Roco kinase (PDB accession code: 4F1T). These studies predicted two hydrogen bonds between the hinge region A1950 and the aminopyrimidine motif of HG-2-102-01 (**4**) as well as a sulfur–halogen interaction between the 5-chloro substituent on the pyrimidine ring and M1947 (Fig. 13.1a). The docking model of JH-II-127 (**19**) predicted the same two hydrogen bonds between the hinge region A1950 and the

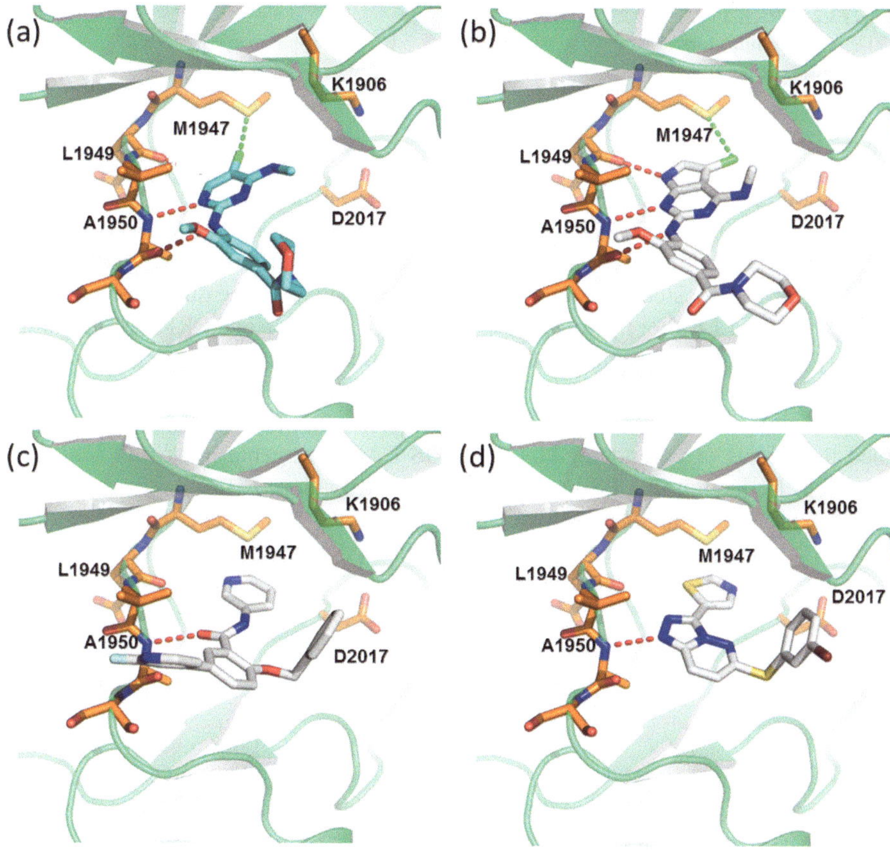

Fig. 13.1 Molecular model of HG-10-102-01 (**a**), JH-II-127 (**b**), GSK2578215A (**c**), and compound **13** from Elan (**d**)

aminopyrrolopyrimidine motif and the sulfur–halogen interaction between the 5-chloro substituent on the pyrrolopyrimidine ring and M1947; however, the docking study also predicted an additional hydrogen bond between the –NH of the pyrrolopyrimidine ring and the carbonyl group of L1949 (Fig. 13.1b). Surprisingly, the docking study for both GSK2578215A (**9**) and compound **13** (Elan) predicted only a single hydrogen bond between the hinge region A1950 and the carbonyl group of GSK2578215A (Fig. 13.1c) and the –N- of the triazolopyridazine of compound **13** (Elan) (Fig. 13.1d).

LRRK2 GTP-Binding Inhibitors

Li et al. recently discovered a series of novel LRRK2 GTP-binding inhibitors **20** and **21** (Fig. 13.2) [40] through virtual screening using an LRRK2 GTPase ROC domain crystal structure (PDB code 2zej) [41]. Both compounds demonstrate in vitro LRRK2

Fig. 13.2 LRRK2 GTP-binding inhibitors

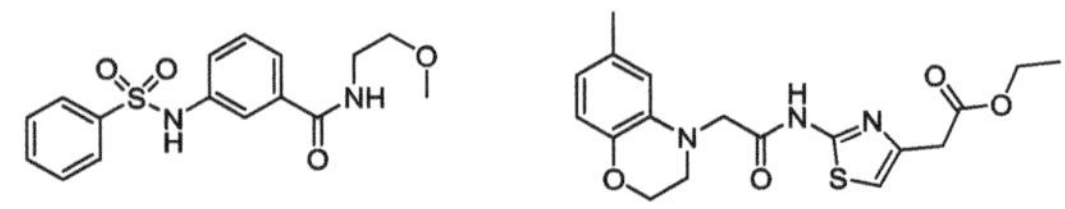

21

22

GTP-binding and kinase inhibitory activities in the nanomolar range. Inhibitors **21** and **22** also inhibited LRRK2-induced neuronal degeneration in SH-SY5Y cells and mouse primary cortical neurons at nanomolar concentrations. Compound **21** was selected for in vivo studies using G2019S BAC transgenic mice due to a better solubility profile. At 1 h after i.p. injection of 20 mg/kg, reduction of LRRK2 GTP-binding activity and LRRK2 phosphorylation in mouse brains was observed, indicating CNS penetration. Based on a recent study that found increases in LRRK2 expression and kinase activity following lipopolysaccharide (LPS) stimulation [42, 43], they examined the effects of compound **21** in G2019S BAC transgenic mice following LPS injection. Reduction in LPS-induced microglia activation, LRRK2 expression, and LRRK2 phosphorylation in activated microglia cells was observed. These studies support a potentially promising and orthogonal approach, compared to ATP kinase inhibitors for the use of LRRK2 GTP inhibitors in LRRK2-associated PD.

Conclusions

LRRK2 was found to be linked to PD in several studies and by the examination of families with high incidence of disease. Over the past decade, researchers have been working to advance our understanding of LRRK2 protein and its function. Several structurally diverse inhibitors of LRRK2 function have been discovered, which can be used to monitor LRRK2 in vivo. Given the significant expense and complexity of clinical studies assessing inhibitors of neurodegeneration, a more complete understanding of the cellular function of LRRK2 will most likely be required before studies can be performed to assess the clinical utility of LRRK2 inhibitors.

Although initially it seemed that LRRK2-focused genetic manipulation of rodents would cause accumulation of deficits that resemble PD with age, these animals have exhibited fairly subtle and variable phenotypes. Only recently has a selective LRRK2 kinase inhibitor been reported to modulate some of these deficits. More importantly, LRRK2 knockout rodents and the LRRK2 kinase-dead knockin mice display unusual lung and/or kidney pathology. While several LRRK2 kinase inhibitors have been well tolerated in rats, two molecules have been found to be associated with lung toxicities in NHPs that are histopathologically identical to the LRRK2 knockout rodent phenotype, suggesting an LRRK2-related effect. There is some uncertainty surrounding the clinical consequences of this type II pneumocyte lamellar body accumulation in the lungs of patients, and pulmonary toxicities are challenging to monitor in the clinic. These findings indicate the need for mutant-selective LRRK2 inhibitors, which may not cause the accumulation of lamellar bodies.

Further examination of the toxicities associated with structurally diverse LRRK2 inhibitors in higher species will permit a more complete assessment of the risks of inhibiting LRRK2 kinase activity in man; however, a large degree of uncertainty regarding the therapeutic index of LRRK2 inhibitors is likely to remain due to the absence of direct biomarkers of LRRK2 function and the lack of suitable efficacy models.

To study the impact of inhibiting LRRK2, we would strongly recommend the use of MLi-2 inhibitor which is the most selective or potent in conjunction with one of the other highly characterized structurally diverse LRRK2 inhibitors such as **19** and **20**. In addition we would strongly advocate where possible to combine the power or pharmacological and genetic approaches by exploiting inhibitor-resistant LRRK2[A2016T] knockin cells or mice model. As the A2016T mutation renders LRRK2 ~10-fold resistant to MLi-2 [44], genuine effects that are mediated through inhibition of LRRK2 would be suppressed at a ~10-fold lower dose in wild-type compared to LRRK2[A2016T] mice or cells.

From a kinase inhibitor pharmacology perspective, LRRK2 is amenable to inhibition by a surprisingly diverse array of structurally distinct ATP-competitive inhibitors. In addition, these LRRK2 inhibitors are among the most selective and "drug-like" inhibitors reported for any kinase. To date no LRRK2 kinase structures have been solved, but eventually acquisition of such structures will provide a means to rationalize the remarkable kinase selectivity achieved by some of these inhibitors.

Conflict of Interest The author declares no conflicts of interest.

References

1. Gandhi PN, Chen SG, Wilson-Delfosse AL (2009) Leucine-rich repeat kinase 2 (LRRK2): a key player in the pathogenesis of Parkinson's disease. J Neurosci Res 87(6):1283–1295
2. Dorsey ER, Constantinescu R, Thompson JP, Biglan KM, Holloway RG, Kieburtz K, Marshall FJ, Ravina BM, Schifitto G, Siderowf A, Tanner CM (2007) Projected number of people with Parkinson disease in the most populous nations, 2005 through 2030. Neurology 68(5):384–386
3. Daniëls V, Baekelandt V, Taymans JM (2011) On the road to Leucine-Rich repeat Kinase 2 signalling: evidence from cellular and in vivo studies. Neurosignals 19(1):1–15
4. Healy DG, Falchi M, O'Sullivan SS, Bonifati V, Durr A, Bressman S, Brice A, Aasly J, Zabetian CP, Goldwurm S, Ferreira JJ, Tolosa E, Kay DM, Klein C, Williams DR, Marras C, Lang AE, Wszolek ZK, Berciano J, Schapira AHV, Lynch T, Bhatia KP, Gasser T, Lees AJ, Wood NW (2008) Phenotype, genotype, and worldwide genetic penetrance of LRRK2-associated Parkinson's disease: a case–control study. Lancet Neurol 7(7):583–590
5. Dachsel JC, Farrer MJ (2010) LRRK2 and Parkinson disease. Arch Neurol 67(5):542–547
6. Greggio E, Cookson MR (2009) Leucine-rich repeat kinase 2 mutations and Parkinson's disease: three questions. ASN Neuro 1(1), e00002
7. Kumar A, Cookson MR (2011) Role of LRRK2 kinase dysfunction in Parkinson disease. Expert Rev Mol Med 13:e20. doi: 10.1017/S146239941100192X

8. Chen CY, Weng YH, Chien KY, Lin KJ, Yeh TH, Cheng YP, Lu CS, Wang HL (2012) (G2019S) LRRK2 activates MKK4-JNK pathway and causes degeneration of SN dopaminergic neurons in a transgenic mouse model of PD. Cell Death Differ 19(10):1623–1633
9. Steger M, Tonelli F, Ito G, Davies P, Trost M, Vetter M, Wachter S, Lorentzen E, Duddy G, Wilson S, Baptista M, Fiske B, Fell M, Morrow J, Reith A, Alessi D, Mann M (2016) Phosphoproteomics reveals that Parkinson's disease kinase LRRK2 regulates a subset of Rab GTPases. Elife 10:12813
10. Nichols RJ, Dzamko N, Hutti JE, Cantley LC, Deak M, Moran J, Bamborough P, Reith AD, Alessi DR (2009) Substrate specificity and inhibitors of LRRK2, a protein kinase mutated in Parkinson's disease. Biochem J 424:47–60
11. Dzamko N, Deak M, Hentati F, Reith AD, Prescott AR, Alessi DR, Nichols RJ (2010) Inhibition of LRRK2 kinase activity leads to dephosphorylation of Ser(910)/Ser(935), disruption of 14-3-3 binding and altered cytoplasmic localization. Biochem J 430:405–413
12. Kramer T, Lo Monte F, Goring S, Amombo GMO, Schmidt B (2012) Small molecule kinase inhibitors for LRRK2 and their application to Parkinson's disease models. ACS Chem Nerosci 3(3):151–160
13. Deng XM, Dzamko N, Prescott A, Davies P, Liu QS, Yang QK, Lee JD, Patricelli MP, Nomanbhoy TK, Alessi DR, Gray NS (2011) Characterization of a selective inhibitor of the Parkinson's disease kinase LRRK2. Nat Chem Biol 7(4):203–205
14. Ramsden N, Perrin J, Ren Z, Lee BD, Zinn N, Dawson VL, Tam D, Bova M, Lang M, Drewes G, Bantscheff M, Bard F, Dawson TM, Hopf C (2011) Chemoproteomics-based design of potent LRRK2-selective lead compounds that attenuate Parkinson's disease-related toxicity in human neurons. ACS Chem Biol 6(10):1021–1028
15. Deng XM, Elkins JM, Zhang JW, Yang QK, Erazo T, Gomez N, Choi HG, Wang JH, Dzamko N, Lee JD, Sim T, Kim N, Alessi DR, Lizcano JM, Knapp S, Gray NS (2013) Structural determinants for ERK5 (MAPK7) and leucine rich repeat kinase 2 activities of benzo e pyrimido-5,4-b diazepine-6(11H)-ones. Eur J Med Chem 70:758–767
16. Weygant N, Qu DF, Berry WL, May R, Chandrakesan P, Owen DB, Sureban SM, Ali N, Janknecht R, Houchen CW (2014) Small molecule kinase inhibitor LRRK2-IN-1 demonstrates potent activity against colorectal and pancreatic cancer through inhibition of doublecortin-like kinase 1. Mol Cancer 13:14
17. Fu LL, Tian M, Li X, Li JJ, Huang J, Ouyang L, Zhang YH, Liu B (2015) Inhibition of BET bromodomains as a therapeutic strategy for cancer drug discovery. Oncotarget 6(8):5501–5516
18. Zhang JW, Deng XM, Choi HG, Alessi DR, Gray NS (2012) Characterization of TAE684 as a potent LRRK2 kinase inhibitor. Bioorg Med Chem Lett 22(5):1864–1869
19. Galkin AV, Melnick JS, Kim S, Hood TL, Li NX, Li LT, Xia G, Steensma R, Chopiuk G, Jiang JQ, Wan YQ, Ding P, Liu Y, Sun FX, Schultz PG, Gray NS, Warmuth M (2007) Identification of NVP-TAE684, a potent, selective, and efficacious inhibitor of NPM-ALK. Proc Natl Acad Sci U S A 104(1):270–275
20. Choi HG, Zhang JW, Deng XM, Hatcher JM, Patricelli MP, Zhao Z, Alessi DR, Gray NS (2012) Brain penetrant LRRK2 inhibitor. ACS Med Chem Lett 3(8):658–662
21. Chen HF, Chan BK, Drummond J, Estrada AA, Gunzner-Toste J, Liu XR, Liu YC, Moffat J, Shore D, Sweeney ZK, Tran T, Wang SM, Zhao GL, Zhu HT, Burdick DJ (2012) Discovery of selective LRRK2 inhibitors guided by computational analysis and molecular modeling. J Med Chem 55(11):5536–5545
22. Estrada AA, Liu XR, Baker-Glenn C, Beresford A, Burdick DJ, Chambers M, Chan BK, Chen HF, Ding X, Di Pasquale AG, Dominguez SL, Dotson J, Drummond J, Flagella M, Flynn S, Fuji R, Gill A, Gunzner-Toste J, Harris SF, Heffron TP, Kleinheinz T, Lee DW, Le Pichon CE, Lyssikatos JP, Medhurst AD, Moffat JG, Mukund S, Nash K, Scearce-Levie K, Sheng ZJ, Shore DG, Tran T, Trivedi N, Wang SM, Zhang S, Zhang XL, Zhao GL, Zhu HT, Sweeney ZK (2012) Discovery of highly potent, selective, and brain-penetrable leucine-rich repeat kinase 2 (LRRK2) small molecule inhibitors. J Med Chem 55(22):9416–9433

23. Sheng ZJ, Zhang SO, Bustos D, Kleinheinz T, Le Pichon CE, Dominguez SL, Solanoy HO, Drummond J, Zhang XL, Ding X, Cai F, Song QH, Li XT, Yue ZY, van der Brug MP, Burdick DJ, Gunzner-Toste J, Chen HF, Liu XR, Estrada AA, Sweeney ZK, Scearce-Levie K, Moffat JG, Kirkpatrick DS, Zhu HT (2012) Ser(1292) autophosphorylation is an indicator of LRRK2 kinase activity and contributes to the cellular effects of PD mutations. Sci Transl Med 4(164):12
24. Estrada AA, Chan BK, Baker-Glenn C, Beresford A, Burdick DJ, Chambers M, Chen HF, Dominguez SL, Dotson J, Drummond J, Flagella M, Fuji R, Gill A, Halladay J, Harris SF, Heffron TP, Kleinheinz T, Lee DW, Le Pichon CE, Liu XR, Lyssikatos JP, Medhurst AD, Moffat JG, Nash K, Scearce-Levie K, Sheng ZJ, Shore DG, Wong S, Zhang S, Zhang XL, Zhu HT, Sweeney ZK (2014) Discovery of highly potent, selective, and brain-penetrant aminopyrazole leucine-rich repeat kinase 2 (LRRK2) small molecule inhibitors. J Med Chem 57(3):921–936
25. (a) Liu XR, Smith BJ, Chen C, Callegari E, Becker SL, Chen X, Cianfrogna J, Doran AC, Doran SD, Gibbs JP, Hosea N, Liu JH, Nelson FR, Szewc MA, Van Deusen J (2006) Evaluation of cerebrospinal fluid concentration and plasma free concentration as a surrogate measurement for brain free concentration. Drug Metab Dispos 34(9):1443–1447; (b) Lin JH (2008) CSF as a surrogate for assessing CNS exposure: an industrial perspective. Curr Drug Metab 9(1):46–59
26. Herzig MC, Kolly C, Persohn E, Theil D, Schweizer T, Hafner T, Stemmelen C, Troxler TJ, Schmid P, Danner S, Schnell CR, Mueller M, Kinzel B, Grevot A, Bolognani F, Stirn M, Kuhn RR, Kaupmann K, van der Putten PH, Rovelli G, Shimshek DR (2011) LRRK2 protein levels are determined by kinase function and are crucial for kidney and lung homeostasis in mice. Hum Mol Genet 20(21):4209–4223
27. Reith AD, Bamborough P, Jandu K, Andreotti D, Mensah L, Dossang P, Choi HG, Deng XM, Zhang JW, Alessi DR, Gray NS (2012) GSK2578215A; a potent and highly selective 2-arylmethyloxy-5-substitutent-N-arylbenzamide LRRK2 kinase inhibitor. Bioorg Med Chem Lett 22(17):5625–5629
28. Cirnaru MD, Marte A, Belluzzi E, Russo I, Gabrielli M, Longo F, Arcuri L, Murru L, Bubacco L, Matteoli M, Fedele E, Sala C, Passafaro M, Morari M, Greggio E, Onofri F, Piccoli G (2014) LRRK2 kinase activity regulates synaptic vesicle trafficking and neurotransmitter release through modulation of LRRK2 macro-molecular complex. Front Mol Neurosci 7:12
29. Nichols PL, Eatherton AJ, Bamborough P, Jandu KS, Philps OJ, Andreotti D (2011) Novel compounds. Google Patents WO 2011038572 A1, 27 Sept 2010
30. Afsari F, Christensen KV, Smith GP, Hentzer M, Nippe OM, Elliott CJH, Wade AR (2014) Abnormal visual gain control in a Parkinson's disease model. Hum Mol Genet 23(17):4465–4478
31. Garofalo AW, Adler M, Aubele DL, Bowers S, Franzini M, Goldbach E, Lorentzen C, Neitz RJ, Probst GD, Quinn KP, Santiago P, Sham HL, Tam D, Truong AP, Ye XCM, Ren Z (2013) Novel cinnoline-based inhibitors of LRRK2 kinase activity. Bioorg Med Chem Lett 23(1):71–74
32. Garofalo AW, Adler M, Aubele DL, Brigham EF, Chian D, Franzini M, Goldbach E, Kwong GT, Motter R, Probst GD, Quinn KP, Ruslim L, Sham HL, Tam D, Tanaka P, Truong AP, Ye XCM, Ren Z (2013) Discovery of 4-alkylamino-7-aryl-3-cyanoquinoline LRRK2 kinase inhibitors. Bioorg Med Chem Lett 23(7):1974–1977
33. Franzini M, Ye XCM, Adler M, Aubele DL, Garofalo AW, Gauby S, Goldbach E, Probst GD, Quinn KP, Santiago P, Sham HL, Tam D, Truong A, Ren Z (2013) Triazolopyridazine LRRK2 kinase inhibitors. Bioorg Med Chem Lett 23(7):1967–1973
34. Galatsis P, Henderson JL, Kormos BL, Han S, Kurumbail RG, Wager TT, Verhoest PR, Noell GS, Chen Y, Needle E, Berger Z, Steyn SJ, Houle C, Hirst WD (2014) Kinase domain inhibition of leucine rich repeat kinase 2 (LRRK2) using a 1,2,4 triazolo 4,3-b pyridazine scaffold. Bioorg Med Chem Lett 24(17):4132–4140
35. Rankovic Z (2015) CNS drug design: balancing physicochemical properties for optimal brain exposure. J Med Chem 58(6):2584–2608

36. Troxler T, Greenidge P, Zimmermann K, Desrayaud S, Druckes P, Schweizer T, Stauffer D, Rovelli G, Shimshek DR (2013) Discovery of novel indolinone-based, potent, selective and brain penetrant inhibitors of LRRK2. Bioorg Med Chem Lett 23(14):4085–4090
37. Longo F, Russo I, Shimshek DR, Greggio E, Morari M (2014) Genetic and pharmacological evidence that G2019S LRRK2 confers a hyperkinetic phenotype, resistant to motor decline associated with aging. Neurobiol Dis 71:62–73
38. Hatcher JM, Zhang JW, Choi HG, Ito G, Alessi DR, Gray NS (2015) Discovery of a pyrrolopyrimidine (JH-II-127), a highly potent, selective, and brain penetrant LRRK2 inhibitor. ACS Med Chem Lett 6(5):584–589
39. Henderson JL, Kormos BL, Hayward MM, Coffman KJ, Jasti J, Kurumbail RG, Wager TT, Verhoest PR, Noell GS, Chen Y, Needle E, Berger Z, Steyn SJ, Houle C, Hirst WD, Galatsis P (2015) Discovery and preclinical profiling of 3–4-(Morpholin-4-yl)-7H-pyrrolo 2,3-d pyrimidin-5-yl benzonitrile (PF-06447475), a highly potent, selective, brain penetrant, and in vivo active LRRK2 kinase inhibitor. J Med Chem 58(1):419–432
40. Li TX, Yang DJ, Zhong SJ, Thomas JM, Xue FT, Liu JN, Kong LB, Voulalas P, Hassan HE, Park JS, MacKerell AD, Smith WW (2014) Novel LRRK2 GTP-binding inhibitors reduced degeneration in Parkinson's disease cell and mouse models. Hum Mol Genet 23(23): 6212–6222
41. Deng JP, Lewis PA, Greggio E, Sluch E, Beilina A, Cookson MR (2008) Structure of the ROC domain from the Parkinson's disease-associated leucine-rich repeat kinase 2 reveals a dimeric GTPase. Proc Natl Acad Sci U S A 105(5):1499–1504
42. Moehle MS, Webber PJ, Tse T, Sukar N, Standaert DG, DeSilva TM, Cowell RM, West AB (2012) LRRK2 inhibition attenuates microglial inflammatory responses. J Neurosci 32(5):1602–1611
43. Gardet A, Benita Y, Li C, Sands BE, Ballester I, Stevens C, Korzenik JR, Rioux JD, Daly MJ, Xavier RJ, Podolsky DK (2010) LRRK2 is involved in the IFN-gamma response and host response to pathogens. J Immunol 185(9):5577–5585
44. Fell MJ, Mirescu C, Basu K, Cheewatrakoolpong B, DeMong DE, Ellis JM, Hyde LA, Lin YH, Markgraf CG, Mei H, Miller M, Poulet FM, Scott JD, Smith MD, Yin ZZ, Zhou XP, Parker EM, Kennedy ME, Morrow JA (2015) MLi-2, a potent, selective, and centrally active compound for exploring the therapeutic potential and safety of LRRK2 kinase inhibitions. J Pharmacol Exp Ther 355(3):397–409

Index

H.J. Rideout (ed.), *Leucine-Rich Repeat Kinase 2 (LRRK2)*,
Advances in Neurobiology 14, DOI 10.1007/978-3-319-49969-7

MIX
Papier aus verantwortungsvollen Quellen
Paper from responsible sources
FSC® C105338

If you have any concerns about our products,
you can contact us on
ProductSafety@springernature.com

In case Publisher is established outside the EU,
the EU authorized representative is:
Springer Nature Customer Service Center GmbH
Europaplatz 3, 69115 Heidelberg, Germany

Printed by Libri Plureos GmbH
in Hamburg, Germany